Österreichische Akademie der Wissenschaften
Schriftenreihe der Erdwissenschaftlichen Kommissionen
Band 9

New Aspects on Tethyan Cretaceous Fossil Assemblages

Edited by

H. A. Kollmann and H. Zapfe

In Kommission bei

Springer-Verlag Wien New York

1992

ISSN 0171-2225

ISBN-13: 978-3-211-86555-2 e-ISBN-13: 978-3-7091-5644-5
DOI: 10.1007/978-3-7091-5644-5

Contents

Preface

The geologic term "Tethys" introduced by Eduard Suess (1893) into science was originally a paleogeographic one. Since its original description it has been used in various ways. One of them is Tethys as a paleo-biogeographic concept. This concept was adopted for IGCP Project 262, Tethyan Cretaceous Correlation. It was proposed by G. Csaszar and Heinz A. Kollmann and has been approved by the IGCP Board in February 1987.

The project has its aims primarily in the improvement of the stratigraphic correlation between the heterogenous facies of the Tethyan realm. The requests to paleontology in this programme are of various kinds: The delimitation of the Tethyan realm in space and time needs a clear statements on the composition of Tethyan faunal or floral assemblages. Biozones based on various fossil groups have to be established for biostratigraphic correlation. Finally, Tethyan bioprovinces have to be established.

These problems have been discussed in a special meeting of the palaeontological group of the project which was held on January 25–27, 1988, in Vienna. Papers presented at this meeting are published in this volume.

Heinz A. Kollmann

1. Cretaceous Tethys versus Mésogée

New Aspects on Tethyan Cretaceous Fossil Assemblages.
Band 9 Schriftenreihe der Erdwissenschaftlichen Kommissionen der Österreichischen Akademie der Wissenschaften, 15–33.

1.1. Tethys – the evolution of an idea

By Heinz A. KOLLMANN*)

Abstract

Tethys in its original meaning was understood by Eduard SUESS as the ancient sea separating Angaraland from Gondwanaland. Contrasting to this paleogeographic conception, "Tethys" and "Tethyan" are currently used with different meanings in tectonics and paleo-biogeography. In paleobiogeography, Tethys is understood as a realm with varying extension. This dynamic conception is in contrast to the conception of the stable Cretaceous Mesogée by DOUVILLÉ (1900).

Introduction

The terms Tethys and Tethyan were introduced to science by Eduard SUESS (1893). Since this time not only the meaning of these terms has changed considerably. To make the confusion perfect they are also used in various ways in Earth Sciences. It is therefore of interest to compare the different meanings of Tethys and Tethyan in modern literature. Completeness has not been attempted in this account. The aim is a general outline of how these terms have evolved.

In an address to the Geological Society of London, E. SUESS, (1893) stated: "Modern geology permits us to follow the first outlines of the history of a great ocean which once stretched across part of Eurasia. The folded and crumpled deposits of this ocean stand forth to heaven in Tibet, Himalaya, and the Alps. This ocean we designate by the name Tethys, after the sister and consort of Oceanus." This name was accepted in science immediately. As BITTNER (1896) remarked sarcastically it was also mis-spelled as Thetis by authors from the beginning and therefore confused with the mother of Achilles. The tran-

*) Naturhistorisches Museum A-1014 Vienna, Austria.

scriptions into the latin alphabet are very similar, indee, but there is no way of confusion in the original Greek.

In the third volume of his synthesis on the geology of the earth, "The Face of the Earth", Eduard SUESS, (1901), gave a more detailed description of this ocean: "Gondwanaland is bound to the north by a broad zone of marine sediments of Mesozoic age: From Sumatra and Timor over Tonking, Yunan, the Himalaya and Pamir, the Hindukush into Smaller Asia. As a whole they have to be considered as the remains of a sea which extended through Asia." Later in the same book, SUESS decribes the prolongation of this hypothetical sea into Mexico and the Caribbean. The land north of Tethys was named by him Angaraland.

Actually, Eduard SUESS was not the first to draw attention to the existence of this former sea. It had already been deduced by NEUMAYR (1887) from the distribution of Mesozoic marine sediments. The later Tethys had been named by NEUMAYR Central Mediterranean (Zentrales Mittelmeer). He had already come to the conclusion that this sea was not very broad and extended from West to East between Central America and India.

This original conception of the Tethys by NEUMAYR and SUESS was therefore an exclusively paleogeographic one. It was understood in this sense by UHLIG (1911), DIENER (1925), and DAQUÉ (1926). This was also pointed out by NAIDIN (1986) in his account on the term Tethys and by YENKINS (1980). RAKUS, DERCOURT & NAIRN (1990) have discussed the northern margin of the Tethys in a paleogeographic conception but in the light of plate tectonics.

Tethys as a tectonic concept

This is based primarily on HAUG's (1900) interpretation of the concept of geosynclines of HALL (1859) and DANA (1875). In the sense of Hall a geosyncline (the original term geosynclinal was created by DANA) is an extraordinary accumulation of sediments of shallow water origin. HAUG (1900), restricted the geosynclines to depressional zones of great depth between continental masses where thick series of deep-water sediments were deposited.

In HAUG's figure a geosynclinal area is situated between an Afro-Brasilean continent and a North Atlantic continent. It was considered by STAUB (1924) as the central zone of the Tethys ocean of E. SUESS. According to STAUB this ocean covered broad areas of the adjoining continental masses. Opposed to the designation in the text the name Tethys was applied exclusively to the oceanic area in the table on the evolution of the alpine system. In 1928, STAUB states that two types of mountain chains may be distinguished in the Alpine orogenetic zone. One of them stems from a broad marine basin. STAUB says: "This is the so-called Tethys by SUESS." He further points out that ophiolites generally occur together with deep-water sediments in the sections.

Although this interpretation of the Tethys does not agree with the original concept of Eduard SUESS it has found entrance in virtually all synthetic work on the plate tectonics of the Mediterranean area. Based on STAUB's work, Tethys is considered as an oceanic plate by SMITH (1971), DEWEY, PITMAN, RYAN & BONNIN (1973), LAUBSCHER & BERNOULLI (1973), BIJU-DUVAL, DERCOURT & PICHON (1977), and others.

Temperature-controlled realms

In 1883 NEUMAYR had established a latitudinal zonation of the Jurassic and the lower Cretaceous which was based on marine faunas, mainly ammonites. He related the paleobiologically defined zones to temperature-controlled realms. After NEUMAYR, the following realms can be distinguished:

1. The tropical equatorial realm with *Phylloceras, Lytoceras,* and *Simoceras;*

2. the (north) temperate realm with *Oppelia* and *Peltoceras;*

3. the boreal realm with Aucellids and the "group of *Belemnites excentricus*". *Oppelia* and *Aspidoceras* are rare in this realm; *Phylloceras, Lytoceras* and *Simoceras* are missing as are reef corals;

4. the south temperate realm.

Only five years later GUEMBEL (1888) published a lateral zonation for the Upper Cretaceous. He distinguished the following provinces in Europe:

1. The North province characterized by *Belemnitella;*

2. the Hercynic province with *Exogyra columba;*

3. the Moscow province with "*Aucella*";

4. the province characterized by abundant rudists. GUEMBEL gives the following distribution of this province (translated from the German): Alps, Italy, Greece, Crimea, Caucasus, Asia Minor, Palestine, through Persia to the Persian Gulf. He further includes the Cretaceous of Africa beginning with Egypt through the Libyan desert and the whole of North Africa.

This zonation of GUEMBEL (1888) was already quoted by NEUMAYR (1887) who distinguished for the Upper Cretaceous two temperature-controlled realms: A temperate realm with Belemnitella and the equatorial realm with rudists, Actaeonella, Nerinea and Lytoceras. He was followed by UHLIG (1911), who pointed out that like modern coral reefs the Cretaceous coral and rudist reefs were developed in the Tropical belt which was much broader then. The same was emphasized by DIENER (1925), and DACQUÉ (1926).

The Mésogée

All authors mentioned above have strictly kept apart the paleogeographic concept of Tethys in the sense of Suess from the temperature-controlled realms deduced from the distribution of fossils. This is different in the concept of the Mésogée by H. DOUVILLÉ (1900), who defined it as follows: «La Mésogée correspond à une phase particulière de la Méditerranée centrale de NEUMAYR ou de la Tethys de SUESS: c'est uniquement la mer dans laquelle les Rudists ont vecu et se sont developpés.» In addition to rudists, DOUVILLÉ listed a number of other fossil groups resticted in his opinion to the Mésogée. These include especially orbitoids and orbitolinids among the larger foraminifera, a number of ammonite families and genera as well as the echinoid *Entallaster*.

The Mésogée although a linguistic monstrosity (GIGNOUX, 1950, has pointed out that Mésogée means continent in the middle which is just the opposite of what DOUVILLÉ wanted to say) is therefore basically a paleo-biogeographic concept. DOUVILLÉ demonstrated that in the Cretaceous a number of fossil groups is restricted to the Tethys in the sense of NEUMAYR and SUESS. This concept has been referred to outside of France by DIENER (1925).

Generally, Cretaceous faunas of low latitudes were called Mediterranean by SCHUCHERT (1910), KOSSMAT (1936) and SCHUCHERT (1935). It was SCHUCHERT who obviously first used the term Tethyan realm in the same work. Tethyan is herewith first employed for the low latitude belt defined with fossil assemblages by NEUMAYR, GUEMBEL, DOUVILLÉ and others. This application does definitely not agree with the original meaning of this term. Nevertheless, it is widely used in this sense, as a biologically defined circumequatoreal belt by paleontologists among them AGER (1967), DONOVAN (1967), SOHL (1971, 1987), KAUFFMAN (1973), BERGGREN & HOLLISTER (1974), KENNEDY & COBBAN (1976) and others. Supertethys, a central belt within the Tethyan realm proposed by KAUFFMAN & JOHNSON (1988) will be discussed by KOLLMANN (this volume).

Tethys or Mésogée

The Mésogée concept is undoubtedly very useful when applied to Cretaceous shallow marine environments. Nevertheless it cannot be upheld in its original context as its distribution does not correspond to the Cretaceous Tethys of NEUMAYR and SUESS as DOUVILLÉ thought.

While DOUVILLE's Mésogée is the total area of distribution of rudists throughout the Cretaceous, the concept of the Tethyan realm as it is used now by many paleontologists is a dynamic one taking into account the fluctuation of realm boundaries during geologic times. The differences in the distribution of Lower Aptian and Campanian to Maastrichtian rudists have been shown by MASSE (1985) and PHILIP (1985).

It is therefore difficult to decide, which term to use: Mésogée, which can't be upheld in its original static conception but has the advantage of having an uncompromised name. Or Tethys as a dynamic paleo-biogeographic concept which is acceptable from a scientific point of view but does not agree with the original content of this term. Which term to use is not so much a matter of philosophy but of convention. IGCP Project 262 was named Tethyan Cretaceous Correlation because Tethyan is used all over the world in a paleo-biogeographic sense and did not need much of an explanation. But nothing should be static in geology and therefore J. P. MASSE's support of Mésogée is of great importance for further discussions.

Another premise of the project is that accepting a boundary deduced exclusively from a single shallow water fossil group such as rudists is too limiting for further work. It will be necessary to achieve a broader understanding of this realm by collecting and interpreting data on as many fossil groups as possible. The concept of the Tethyan realm must be kept open for discussion in order to improve our knowledge of Cretaceous paleo-biogeography.

References

AGER, D. V. (1967). Some Mesozoic brachiopods in the Tethys region. – Systematics Ass. Publ. **7,** Aspects of Tethyan biogeography (ed. C. G. Adams & D. V. Ager): 135–151.

BERGGREN, W. A., & HOLLISTER, C. D. (1974): Paleogeography, paleobiography and history of circulation in the Atlantic Ocean. – Studies in Paleo-Oceanography, Soc. Econ. Paleont. Mineralog., Spec. Publ., **20:** 126–186.

BIJU-DUVAL, B., DERCOURT, J., & PICHON, X. LE (1977): From the Tethys ocean to the Mediterranean seas: a plate tectonic model of the evolution of the Western alpine system. – Int. Symp. on the structural history of the Mediterranean Basins, Split. B. Biju-Duval and L. Montadert, Eds. Editions Technip, pp. 143–164.

BITTNER, A. (1896): Bemerkungen zur neuesten Nomenclatur der alpinen Trias. – pp. 1–32.

DACQUÉ, E. (1926): Paläogeographie. pp. 1–196.

DANA, J. D. (1875): Manual of Geology, 2nd edition. pp. I–XVI, 1–828.

DEWEY, J. F., PITMAN III., W. C., RYAN, W. B. F., BONNIN, J. (1973): Plate tectonics and the evolution of the Alpine system. – Geol. Soc. Am. Bull., **84/10:** 3137–3180.

DIENER, K. (1925): Grundzüge der Biostratigraphie. – pp. 1–304.

DONOVAN, D. T. (1967): The geographical distribution of Lower Jurassic ammonites in Europe and adjacent areas. – Systematics Ass. Publ. **7,** Aspects of Tethyan biogeography (ed: C. G. Adams & D. V. Ager): 111–134.

DOUVILLÉ, H. (1900): Sur la distribution géographique des rudistes, des orbitolines et des orbitoides. – Bull. Soc. Géol. Fr. (3), **28:** 222–235.

GIGNOUX, M. (1950): Géologie Stratigraphique. – pp. I–VII, 1–735.

GUEMBEL, F. (1888): Geologie von Bayern 1: Grundzüge der Geologie. pp. I–XVI, 1–1142.

HALL, J. (1859): Geological Survey of New York. Palaeontology, vol. III. – pp. 1–532.

HAUG, E. (1900): Les géosynclinaux et les aires continentales, contribution á l'étude des transgressions et des régressions marines. – Bull. Soc. Géol. Fr. (3). **28:** 617–711. Paris.

JENKYNS, H. C. (1980): Tethys: past and present. – Proc. Geol. Ass., **91:** 107–118.

KAUFFMAN, E. G. (1973): Cretaceous Bivalves. In: Atlas of Palaeobiogeography (A. Hallam, ed.). pp. 353–383.

KAUFFMAN, E. G., & JOHNSON, C. (1988): The morphological and ecological evolution of the Middle and Upper Cretaceous reef-building rudistids. – Palaios, **5/3:** 194–216.

KENNEDY, W. J., & COBBAN, W. A. (1976): Aspects of Ammonite biogeography, and biostratigraphy. – Spec. Pap. Palaeont., **17:** 1–94.

KOSSMAT, F. (1936): Paläogeographie und Tektonik. – pp. I–XXIII, 1–413.

LAUBSCHER, H., & BERNOULLI, D. (1973): Mediterranean and Tethys. – The ocean basins and margins (A. E. M. Nairn and W. H. Kanes, ed.) 4A: 1–28.

MASSE, J.-P. (1985): Paléobiogéographie des Rudistes du domaine péri-méditerranéenne à l'Aptien inférieur. – Bull. Soc. géol. France (8), **1/5:** 715–721.

MASSE, J.-P. (this volume): The Lower Cretaceous Mesogee: The state of the art.

NAIDIN, D. P. (1986): Tetis: Termin i ponjatie (Tethys-terminus and meaning). – Vestn. Moskv. Un-ta, ser. 4. (Geol.), **6:** 3–18.

NEUMAYR, M. (1883): Über klimatische Zonen während der Jura- und Kreidezeit. – Denkschr. Akad. Wiss., mathem.-naturw. Kl., **57:** 277–310.

NEUMAYR, M. (1887): Erdgeschichte Vol. 2: Beschreibende Geologie. pp. I–XI, 1–879.

PHILIP, J. (1985): Sur les relations des marges téthysiennes au Campanien et au Maastrichtien déduites de la distribution des Rudistes. – Bull. Soc. géol. France (8), **1/5:** 723–731.

RAKUS, M., DERCOURT, J., NAIRN, A. E. M. (1989): Evolution of the Northern Margin of the Tethys. Vol. 1–3. Mém. Soc. géol. France, N. S., **154.** Paris.

SCHUCHERT, Ch. (1910): Paleogeography of North America. – Bull. Geol. Soc. Am., **20:** 427–606.

SCHUCHERT, Ch. (1935): Historical Geology of the Antillean-Caribbean region. – pp. I–XXVI, 1–811.

SMITH, A. G. (1971): Alpine deformation and the oceanic areas of the Tethys, Mediterranean and Atlantic. – Geol. Soc. Am. Bull., **82/8:** 2039–2070.

SOHL, N. F. (1971): North American Cretaceous Biotic Provinces delineated by gastropods. – Proc. North Am. Paleont. Convention, part L: 1610–1638.

SOHL, N. F. (1987): Cretaceous gastropods: Contrasts between Tethys and the Temperate provinces. – Journ. Paleont., **61/6:** 1085–1111.

STAUB, R. (1924): Der Bau der Alpen. Versuch einer Synthese. – Beitr. Geol. Karte d. Schweiz, N. F., **52:** 1–272.

STAUB, R. (1928): Der Bewegungsmechanismus der Erde. – pp. I–VIII, 1–270.

SUESS, E. (1893): Are great ocean depths permanent? – Nat. Sci., **2:** 180–187.

SUESS, E. (1901): Das Antlitz der Erde, vol. III/1, pp. 1–508.

UHLIG, V. (1911): Die marinen Reiche des Jura und der Unterkreide. – Mitt. Geol. Ges. Wien, **4:** 329–448.

New Aspects on Tethyan Cretaceous Fossil Assemblages.
Band 9 Schriftenreihe der Erdwissenschaftlichen Kommissionen der Österreichischen Akademie der Wissenschaften. 15–33.

1.2. The Lower Cretaceous Mésogée: A state of the art

By Jean-Pierre MASSE*)

Abstract

The Mésogée as defined by DOUVILLÉ (1900a) is both a paleobiogeographic and a paleogeographic concept. The configuration is revised here for the Lower Cretaceous by applying advances in knowledge over the last 80 years to the original concept. The global distribution of rudist-orbitolinid assemblages support the broad configuration of Mésogée proposed by DOUVILLÉ. Their distribution is more extended than in the former definition, especially concerning the latitudinal range: the circumterrestrial extension is confirmed.

The Tethys sensu SUESS (1893) is more limited longitudinally but is wider latitudinally: Mésogée and Tethys have two distinct meanings and are not to be confused.

1. Introduction

Mésogée was originally defined by DOUVILLÉ (1900a) as:
– the Cretaceous phase of the "Tethys" sensu SUESS (1893, 1900) or the "Central Mediterranean" sensu NEUMAYR (1885)
– the sea inhabited by rudists and associated faunas.

Therefore Mésogée involves both, paleogeographic and paleobiogeographic aspects.

In fact the boundaries of the Mésogée as mentioned by DOUVILLÉ are not indentical to those of the Tethys of SUESS. First, the Mésogée corresponds only to the "equatorial zone" of the Tethys, i. e., its southern part. Secondly, Tethys is limited in longitude by Central America to the West and the Bengali delta to the East (the "Sino-Australien continent" acts as a barrier between the Te-

*) Centre de sédimentologie et Paléontologie – URA 1208 du C.N.R.S. Dynamique des plates-formes carbonatées – Université de Provence – Place V. Hugo F-13331 Marseille Cedex 03.

thys and the Pacific Ocean), whereas Mésogée is inferred to have been circumterrestrial.

The contradiction between the geographic configuration proposed for Mésogée and the simple assimilation of its configuration to a particular phase of Tethys was misleading. This is probably the explanation for the extensive usage of "Tethys" in the sense of "Mésogée", i. e., as a paleobiogeographic entity (see for example DOUGLASS, 1960; COATES, 1973; KAUFFMAN, 1973; RAMSON, 1973; MIDDLEMISS, 1973; STEVENS, 1980; ENAY, 1980; SOHL, 1987) in spite of the original, essentially paleogeographic meaning.

Furthermore, as remarked by PHILIP (1982), Mésogée was used mistakenly for example as an oceanic domain limited to the Eastern Mediterranean (BIJU-DUVAL et al., 1977, DERCOURT et al., 1986) or in a too restricted sense for example as the perimediterranean area only (PELISSIE et al., 1982).

As a consequence of the author's experience and knowledge, the present paper is limited to the Lower Cretaceous phase of Mésogée, its objectives are:

– to recall the conceptual framework of DOUVILLÉ based on the paleobiogeographic reference taxa, their broad paleoclimatic significance and the inferred paleogeographic boundaries of their corresponding area of distribution.

– to test DOUVILLÉ's model against our present knowledge concerning the geographic distribution of rudist-orbitolinid assemblages, and to discuss their paleogeographic distribution using global plate-tectonic reconstructions,

– to present briefly the main guidelines for improvement of the model by a better definition of biotic assemblages using paleogeographic entities linked to specific paleoenvironments and their general paleoclimatic framework.

2. The Mesogée sensu Douvillé

The Lower Cretaceous Mésogée of DOUVILLÉ (1900a) was essentially the sea inhabited by rudists and large foraminifera, especially orbitolinids. The boundaries of this sea were derived from the gross geometry of the plotted localities in which both reference taxa have been recognized. In fact, the Mésogée area defined by DOUVILLÉ deals either with both reference taxa or only with one, i. e., many localities are characterized only by orbitolinids. Furthermore, ammonites and even some echinoids are also used as mesogean indicators.

In spite of Valanginian to Hauterivian (?) rudist developments in Western Europe (although Orbitolinids are lacking) the history of Mésogée is mainly recorded since the Barremian. Its extension at that time was from Colombia to Iran (or Pakistan?); but the central part was situated in the Mediterranean area to which rudists seemed to be limited. Orbitolinids appeared during the Late Barremian and were limited to the SE of France. Thus during this stage, outside Western Europe, Mésogée was essentially defined by ammonites.

Rudists and orbitolinids of Aptian age are known from North America (Texas), North-Africa (Algeria), the southern edge of the Black Sea, the Caucasus and several localities in Iran; they also existed in Western Europe, up to Southern England. The contemporaneous presence of mesogean ammonites from Venezuela to Iran and North India has been noted. The spreading of Mesogée to North Pakistan was proved in 1926 after the study of a collection of rudists and orbitolinids from the Indus upper valley (Kohistan).

The Albian stage was characterized by important paleogeographic changes in Europe, but rudist-orbitolinid assemblages still existed. The mesogean ammonites and some echinoids are found from Chile through Peru and Venezuela to Texas, while they are also reported from North America to Syria. At the same time this mesogean fauna extended to the South Atlantic (Brazilian and Angolan coastal basins), some ammonites are recorded from Roumania to the Cameroons and even to India. In his paper on American rudists Douvillé (1900b) was not convinced about the precise age of the Middle Cretaceous fauna from Mexico: thus the record of the reference mesogean taxa in this region was doubtful for the Albian stage.

This brief summary of Douville's knowledge shows that the Mésogée was poorly defined during the Berriasian to Hauterivian interval.

During the Barremian-Albian interval the Mésogée is better defined by ammonites rather than by the rudist-orbitolinid assemblage, which is mainly known in the perimediterranean area with some extensions into North America (Texas) and the Middle East (Iran and Pakistan). Consequently, there is no evidence that Mésogée was a circumterrestrial entity during the Lower Cretaceous because no mesogean fauna was recorded in Eastern Asia as it was in Upper Cretaceous times.

No map was provided by Douvillé but his text is precise enough to draw the broad configuration of his Mésogée considered as a circumterrestrial entity during the whole Cretaceous. It corresponds to a latitudinal band about 20° wide in latitude; its median line coincides with the present situation of the equator in SE Asia (Borneo), while it shifts in the Euro-African area to about 20° N and in America to about 10° N (Fig. 1). The southern boundaries of Mésogée are related to shorelines and are considered as the northern border of the Brazilian and Saharian shields. The northern boundaries are not related to shorelines but to the distribution of the rudist-orbitolinid assemblage followed northwards by the "boreal communities" in the sense of Neumayr (1895). In spite of the reported coincidence between the paleogeographic and paleobiogeographic boundaries on its southern side "southern faunal extensions" (see above) show that at least during the Albian, the Mésogee probably spread south of the present equatorial line, especially on South Atlantic margins.

3. The present knowledge of Mésogée

Improved knowledge concerning the distribution of rudist-orbitolinid assemblages since Douville's work requires some modifications of his results concerning the geographic extension of Mésogée. Moreover, plate-tectonic concepts also require that the paleogeographic framework of the paleobiogeographic reconstructions be modified.

Furthermore, the development of orbitolinid paleontologic studies require that DOUVILLE's orbitolinid conceptions be rediscussed. Actually this group is now divided into two "subfamilies" (see MOULLADE, 1965): The Orbitolinidae correspond with the forms to which DOUVILLÉ referred as "orbitolinids" (large, flat; frequently discoidal). They are only known from Barremian to Albian times (ARNAUD-VANNEAU, 1982; ARNAUD-VANNEAU & al., 1985). The Dictyoconinae (relatively smaller, essentially conical) occurred during the entire Lower Cretaceous (ARNAUD-VANNEAU, 1982; ARNAUD-VANNEAU & al., 1985).

Regarding the paleo-environmental significance of the rudist-orbitolinid assemblage on the one hand and the ammonite assemblage on the other, and based on the present stage of the investigation, the rudist-orbitolinid assemblage is considered here as the main reference because of its unequivocal shallow water evidence.

A – New Data (Fig. 2)

1 – Within the geographic area originally indicated as mesogean, many localities with rudists and/or orbitolinids have been added to those mentioned by DOUVILLÉ. This confirms the proposed broad configuration. These localities are as follows:

America

– Texas-Arizona. New information has put rudist-orbitolinid assemblages mainly into the Albian stage (COOGAN, 1977; PERKINS, 1974; SCOTT, 1979; 1981) while DOUVILLÉ (1900b) believed them to be of Aptian age. Numerous rudist taxa have been described from here since 1900 (see ADKINS, 1930; PERKINS, 1960; COOGAN, 1973), all of Albian age. Nevertheless, Lower Aptian rudists were recently found in the Sligo formation from the subsurface of Texas (SKELTON, 1982).

– Caribbean area. Lower Aptian (to Barremian?) rudists were recorded in Trinidad and Venezuela (HARRIS & HODSON, 1922; IMLAY in COOGAN, 1977; MASSE & ROSSI, 1987) as well as in Jamaica (CHUBB, 1971; COATES, 1977). Albian Caprinids were reported from Cuba, the Dominican Republic, Puerto Rico and their presence in the Virgin Islands is also possible (SOHL, 1976).

The Mexican Lower Cretaceous rudists, first recognized by FELIX (1891), then by PALMER (1928) and MUELLERRIED (1933), have been described recently in more detail (ALENCASTER & PANTOJA ALOR, 1986). Orbitolinid occurrences from Arizona, New Mexico, Texas, and from the subsurface of Mississippi and Florida are well documented by DOUGLASS (1960a). Later Lower Cretaceous orbitolinid-bearing limestones are also mentioned from the Dominican Republic, Puerto Rico, Barbados and Trinidad (DOUGLASS, 1961) as well as from Guatemala and Venezuela (DOUGLASS, 1960b).

Europe

Extensive data have been provided on rudist communities from Switzerland, France (PAQUIER, 1903–1905; DOUVILLÉ, 1918; MASSE, 1976) Italy (TORRE, 1965; MAINELLI, 1975–1983; CAMOIN, 1982; MASSE & al., 1984; LUPERTO-SINNI & MASSE, 1982; MASSE, 1985), Spain (BATALLER,1974; PASCAL, 1984) and Portugal (REY, 1972; BERTHOU & al., 1983 and unpublished observations from MASSE, REY & SKELTON). Orbitolinids have also been described from these localities especially in the Barremian-Albian interval, as characteristic components of "Urgonian limestone" formations. Dictyoconids are present from Valangian to Albian in the European Province, whereas they are mainly known from Hauterivian to Albian in the African one (PELISSIE & al., 1982; MOULLADE & al., 1984), while rudists occur subcontinuously. Large amounts of data about rudists have been provided from Bulgaria (PAQUIER & ZLATARSKI, 1901; TZANKOV, 1960; ATANASSOVA-DELTCHEVA, 1978), where these bivalves are mainly of Barremian? to Lower Aptian age. Rudists have been mentioned from different parts of Roumania: Carpathic zone (DRAGASTAN, 1975) as well as Dobrogea (NEAGU & al., 1977). Lower Aptian caprotinids and caprinids were reported from the West Carpathians in Czechoslovakia (KÜHN & ANDRUSOV, 1942) and also from the Pienine Klippen Belt (MESIK, 1966). Barremian to Aptian Urgonian limestones with rudists have been described from the High Tatric Massif in South Poland (PASSENDORFER, 1930; LEFELD, 1968) as well as from Hungary (BENKÖ-CZABALY, 1970). In nearly all these localities orbitolinids are associated with rudists. Monopleurid and requienid limestones are well documented from Eastern Serbia (JANKICEVIC, 1978, 1984); some caprotinids and caprinids have been recoreded from Croatia (POLSAK, 1970), Slovenia (PLENICAR & BUSER, 1967) and Bosnia (MASSE & al., 1984), where some Aptian radiolitids have been recently described (SLISKOVIC, 1982, 1984). Barremian to Albian rudists were also reported from the Parnassus zone of Central Greece (COMBES & al., 1981) as well as from Albania (PEZA & al., 1981).

Urgonian formations are well known in the Southern USSR: in Crimea (PCHELINTSEV, 1959; YANIN, 1975a–b, 1985) and in the Caucasus (RENGARTEN, 1950; YANIN, 1985) with a well developed rudist-orbitolinid association within the Barremian-Aptian, whereas rudists have also been found in the Berriasian to Hauterivian interval, especially in Crimea. Similar Urgonian rudists were also described from the Ukrainian Carpathians (YANIN & TCHERNOV, 1979).

North Africa

Rudists associated with Urgonian limestones are known from the Aptian of Algeria (eastern part) (BLAYAC, 1908; EMBERGER, 1954; VAN DE FLIERT, 1952; CHIKHI-AOUIMEUR, 1980, 1983; MASSE & CHIKHI-AOUIMEUR, 1982) and Tunesia (PERVINQUIÈRE, 1903; MASSE, 1984). At the same time, caprotinids and requienids are present in Hauterivian limestones both in Algeria and Tunisia (MASSE, unpublished). Requienids have also been mentioned from Aptian limestones of Morocco (Canerot & al., 1981). In all these localities orbitolinids (both Orbitolininae and Dictyoconinae) are associated with rudists.

Near and Middle East

From Sinai, Lebanon and Syria orbitolinids were reported by Henson (1948), Said & Barakat (1957) and Saint-Marc (1970).

A typical mesogean fauna is documented from shallow shelf carbonates extending from Yemen to Iran through Saudi Arabia and Oman. Thus, rudists are known from Hauterivian to Aptian, orbitolinids are recorded mainly during the latter stage (Masse & al., 1984; Alsharhan & Nairn, 1986; Simmons & Hart, 1987; Moshier & al., 1988). Urgonian limestones with unidentified rudists but with a clear Lower Cretaceous age were also reported from Central Turkey (Gutnic & Moullade, 1967).

Around the Persian Gulf, orbitolinid occurrences have been mainly reported by Henson (1948).

Central and Southern Asia

Following the first observations of Douvillé (1926), detailed studies in Pakistan (Rossi-Ronchetti, 1965; Pudley & al., 1985) and Afghanistan (Montenat & al., 1982) have shown the importance of orbitolinid-rudist asemblages of Aptian-Albian age in this region. Orbitolinids were also reported from Burma (Sahni, 1937), but the corresponding paleontologic and sedimentologic framework is not well known.

Many new sites have been recorded, of which locations were predictable from the Douvillé model whereby previous data were definitely lacking. The corresponding locations are mainly the result of offshore and/or deep-sea drilling investigations.

North Atlantic Sites

On the Grand Banks of Newfoundland, shelf petroleum wells discovered orbitolinids (Gupta & Grant, 1971). D.S.D.P. recorded a rudist-orbitolinid assemblage of Barremian?-Lower Aptian age (Perkins, 1979; Schroeder & Cherchi, 1979) at the southern toe of the Bank. At Meriadzek Bank, caprinids were identified in association with a typical Urgonian assemblage of microfossils (Pastouret & al., 1974).

Western Pacific Sites

An Aptian orbitolinid-rudist assemblage (with caprinids) was dredged off Japan (Geisha guyots) and between the Marshall and Hawaiian Islands (Pacific Mountains) (Heezen & al., 1973).

From many localities where Douvillé only mentioned ammonites, orbitolinids have been described in the meantime. This is the case for Colombia and Venezuela (GERHARDT, 1897, and KARSSEN, 1858–1886), both references in DOUGLASS (1960a). At least the occurrence of orbitolinids in Borneo reported by FRITSCH (1878) (in DOUGLASS, 1960a), which is not included in DOUVILLÉ's localities, falls within his Mésogeé.

Many new sites have been recorded outside the proposed Mesogean realm of DOUVILLÉ.

On the Pacific side of South America, Lower Cretaceous requienids and/or monopleurids have been mentioned in Chile at 30° S present latitude and Peru. These rudists are associated with calcareous sponges and corals (FRITZSCHE, 1924). Coral biostromes were recently described in the Hauterivian-Barremian of Peru (SCOTT & ALEMAN, 1984) and have been interpreted as typical Tethyan (= Mesogean) assemblages, as well as similar faunas from the Middle Jurassic of Chile (PRINZ, 1986).

In Eastern Africa, rudist-orbitolinid assemblages have been described from Somalia (TAVANI, 1947; SILVESTRI, 1932 in PEYBERNES, 1982) and from Tanzania (HENNIG, 1916; DIETRICH, 1925 in PEYBERNES, 1982). The Somalian fauna is dated as Albian (a possible Aptian age is also possible for part of the Mid-Cretaceous formations). The Tanzanian fauna recently revised by PEYBERNES & FORSTER (1987) belongs to the Aptian. Orbitolinids have also been figured from Kenya and Ethiopia (PEYBERNES, 1982).

In the Japanese archipelago several orbitolinid-rudist-bearing beds have been found since 1920 (YABE & NAGAO, 1926; YABE & HANZAWA, 1926; OKUBO & MATSUSHIMA, 1959). All these beds are apparantly of Aptian age.

B – Mésogée Extension: A Discussion

New data since the publication of DOUVILLÉ's work show that the Lower Cretaceous Mésogée auct. needs to be extended:

– southward on the Pacific side of South America and on the eastern side of Africa.

– northward on the Pacific side of Eastern Asia.

A problem arises concerning the significance of the northern part of the West African, central and southern Atlantic margins considered by DOUVILLÉ as "Mesogean extensions" of Mid-Cretaceous age. Similarly, Brazilian marginal basins pose the same question.

No rudist-orbitolinid assemblages have yet been found in any of these regions. However, shallow water carbonates with calcareous green algae and Foraminifera with Mesogean significance are known from the following localities:

– offshore of Senegal, dasyclads are associated with orbitolinids in limestones of Aptian-Albian age (PEYBERNES, 1982; MICHAUD, 1984),

– Southeast of Nigeria, dasyclads are associated with Trocholina in limestones of Albian age (DESSAUVAGIE, 1968; FORSTER, 1978; POIGNANT & LOBITZER, 1982).

– offshore of the Brazilian coast (Sergipe basin), Albian carbonate platforms have been recognized (unpublished results of Petrobras) along with shallow water foraminifera (BENGTSON & BERTHOU, 1982).

Platform carbonates have also been recognized offshore of Marocco and Mauritania, where their age is thought to be Aptian (JANSA & WIEDMANN, 1982; VON RAD & WISSMANN, 1982; RANKE & al., 1982), although no precise information has been given on their paleontologic content. In all these localities, rudists, and/or orbitolinids are to be expected and further investigations are needed. "Mesogean", but pelagic, foraminifera have been reported by CHEVALIER and FISCHER (1982) from Gabon. Here, Gargasian forms similar to the perimediterranean ones support the idea of a pre-Albian communication between the Southern and Central Atlantic and the extension of Mésogée to the meridional part of Western Africa. This point of view is not shared by ANGLADA & RANDRIANASOLO (1985), for whom the Central Atlantic pelagic foraminifera were mainly controlled by migrations coming from the South Atlantic. Consequently, the question of a Mesogean extension southward of Nigeria is still open.

On the Pacific side of North America, no Lower Cretaceous rudist-orbitolinid assemblages have been found, but JONES (1972) refers to a "Tethyan faunal assemblage" near 45° N, whose "warm shallow water" significance is not clearly demonstrated. Furthermore, this area belongs to a terrane complex whose geographic origin is probably more meridional (BLAKE & JAYKO, 1986; AUBOUIN & al., 1986). Therefore the precise significance of this Tethyan assemblage is not quite clear, neither from palaeoecologic nor from palinspastic points of view.

The configuration of proposed Mésogée is different from PHILIP's (1982), which was based mainly on Upper Cretaceous data. Regarding the "Tethys" as a paleobiogeographic concept closely similar to Mésogée, some differences are also noticeable to the "Tethyan realm" of KAUFFMAN (1973), KAUFFMAN & JOHNSON (1988). In this model the North-West Pacific area (the Japanese and Hawaiian seamounts), the Atlantic coastal zone of North America (including the Grand Banks area) and the Atlantic margin of France and Great Britain are all placed in the North temperate realm. This reconstitution also ignores the occurrence of Central-European (Poland, Czechoslovakia) sites. COATE's (1973) reconstitution using rudists and/or hermatypic corals as well as SOHL's (1987) using gastropods are very similar to those of KAUFFMAN.

The configuration proposed here is similar to this by GUPTA & GRANT (1971) using the distribution of the orbitolina group and the LLOYD model (1982) based on "warm water fauna including bivalves, hermatypic corals and orbitolinas". But all of these reconstitutions fail to take into account the Pacific side of South America, the West Pacific seamounts and the West Africa-Brazilian localities.

Consequently, the southward extension of Mésogée is far greater than it was previouly thought. The present day shift of this realm between the South American and East African latitudes is about 20°. The northward shift between

East Asia and West Europe is only about 10°. The total latitude extension appears now to be of about 55° both on the Pacific side of America and between Central Europe and Eastern Africa.

C – Mésogée Configuration in Palinspastic Reconstructions

The following is evident for the Lower Cretaceous prior to the communication of the southern and the northern part of the Atlantic by the opening of the central part of this ocean by regarding palaeotectonic reconstructions, especially those of BARRON & al. (1981) with some complements of DERCOURT & al. (1986) for the Mediterranian and the Middle East:

On palinspastic paleogeographic maps (Fig. 3) DOUVILLÉ's Mésogée configuration shows a strong discrepancy between the location of Mésogée and the position of the paleo-equator. This is also indicated by earlier attempts of KAUFFMAN (1973), PHILIP (1982) as well as PARRISH & BARRON's (1986) reconstructions of warm seas/carbonate platform relationships. According to this, Mésogée is mainly developed in the northern paleohemisphere. In the configuration proposed here (Fig. 4), the paleolatitudes of the Mésogée boundaries near the oceanic margins are very similar in the southern hemisphere, i. e. about 35° S. In the northern hemisphere they are about 35° N in Western Europe and on the Atlantic side of North America; on its Pacific side the paleolatitude is close to 40° N, although this value needs further confirmation. On the East Asiatic margins it ranges close to 60° N.

It appears from this reconstruction that the latitudinal extension of Mésogée is wider than proposed by previous models: probably near 100°, i. e., far more like the present intertropical situation of warm seas. Mésogée is nearly symmetrical on both sides of the paleo-equator, especially in the Neo-Tethysian area and in the Eastern Pacific. This situation fits well with the paleo-equator position proposed by the palinspastic plate-tectonic reconstructions (see above), whereas this fit was anomalous in previous models. An important asymmetric pattern is nevertheless observed on the East Asiatic margin. Similarily, as the result of possible paleogeographic barriers in the Central Atlantic region, Mésogée has here a limited extension southward, at least in pre-Albian times. In this region the limited extension southward in the Albian poses some questions which need further investigations.

Conclusions

The Lower Cretaceous Mésogée as proposed by DOUVILLÉ (1900a–1926) is supported by many data obtained since his work. Nevertheless, the following findings, tentative conclusions, and remarks are to be stressed.

1. In spite of the lack of many oceanic settings DOUVILLÉ's Mésogée was essentially circumterrestrial, a hypothesis supported by modern data concerning oceanic areas which fit with plate tectonic reconstructions.

2. The latitudinal extension of Mésogée is wider than postulated by DOUVILLÉ; the main regions which need to be added to DOUVILLÉ's Mésogée are: Eastern Asia, the Central Pacific side of South America and North America counterparts (north of California), and the North Atlantic. Moreover, the southward shifting of Mésogée boundaries around the South America-Africa blocks fits with the paleo-equator position proposed by paleotectonic reconstructions. That is to say, Mésogée has a nearly symmetrical extension on both sides of the paleo-equator.

3. As mentioned by DOUVILLÉ, the Mésogée boundaries in the South American-African block (i. e., the Brazilian-Ethiopian continent of NEUMAYR) are formed by shorelines, whereas its northern boundaries are climatically and/or paleobiologically controlled. This configuration was modified during the Albian as a result of the opening of the central Atlantic. Orbitolinid-rudist assemblages have not yet been recorded here.

4. Consequently, Tethys differs at least from Mésogée by its limited longitudinal extension and wider latitudinal spreading, especially in northwestern Europe. The two words have significantly different meanings, both from the conceptual point of view (PHILIP, 1982; MONOD, 1985) and from the objects they tend to describe. As proposed by MONOD (1985), Tethys must be used essentially in a paleogeographic and paleostructural sense, while Mésogée should be used essentially in a paleobiogeographic sense even if, as a paleobiologic entity, it also has a paleogeographic meaning.

5. Palinspastic reconstructions of the Mésogée need to be discussed and tested with paleoclimatic and ocean circulation/atmosphere relationship models. A reappraisal of Mesogean biota in order to obtain a better definition of bioprovinces and communities and to establish accurate models of biotic community/environment relationships should be undertaken.

References

ADKINS, W. S. (1930): New Rudistids from the Texas and Mexican Cretaceous. *Texas Univ. Bull.*, **3001:** 77–100.

ALENCASTER, G. (1973): Una nueva especie de Toucasia en el Cretacico medio de los estados de Oaxaca y Puebla. *Paleont. Mexicana,* **36:** 3–37.

ALENCASTER, G. (1986): Nuevo Rudista (Bivalvia-Hippuritacea) del Cretacico inferior de Pihuamo, Jalisco. *Bol. Soc. Geol. Mexicana,* **XLVII, 1:** 47–61.

ALENCASTER, G., PANTOJA ALOR, J. (1986): *Coalcomana ramosa* (Boehm) (Bivalvia-Hippuritacea) del Albiano temprano del Cerro de Tuxpan, Jalisco. Ibid., 33–46.

ALSHARHAN, A. S., NAIRN, A. E. M. (1986): A review of the Cretaceous formations in the Arabian Peninsula and Gulf: part 1. Lower Cretaceous (*Thamama* Group) Stratigraphy and Paleogeography. *J. of Petrol. Geol.,* **9/4:** 365–392.

ARNAUD-VANNEAU, A. (1980): Micropaléontologie, paléoécologie et sédimentologie d'une plate-forme carbonatée de la marge passive de la Téthys. *Geol. Alpine,* Grenoble, **11:** 874 p.

Arnaud-Vanneau, A. (1986): Variations dans la composition et dans la diversité des faunes de Foraminifères benthiques du Crétacé inférieur sur quelques plates-formes carbonatées téthysiennes de l'Europe et du Moyen-Orient. *Bull. Soc. Géol. France*, (8) **II/2:** 245–253.

Arnaud-Vanneau, A., Cherchi, A., Chiocchini, M., De Castro, P., Fourcade, E., Hamaoui, M., Luperto-Sinni, E., Neumann, M., Prestat, B., Schroeder, R. (1985): Les grands Foraminifères du Crétacé Moyen de la région méditerranéenne. *Géobios*, Mém. spec., **7:** 160 p.

Atanassova-Deltcheva, M. (1978): Représentants du genre *Matheronia* Munier-Chalmas, 1873, dans l'Aptien inférieur du Prébalkan. *Ann. Univ.* Sofia, **70/1:** 135–160.

Aubouin, J., Blanchet, R., Roure, F., Tardy, M. (1986). Traits généraux des cordillères de l'ouest des Etats-Unis. *Bull. Soc. Géol. France*, II, **5:** 741–754.

Barron, E. J., Harrison, C. G. A., Sloan, J. L., Hay, W. H. (1981): Paleogeography, 180 millon years ago to the present. *Eclogae Geol. Helv.*, **74/2:** 443–470.

Bataller Catalayud, J. (1947): Sinopsis de las especies nuevas del Cretacico de Espana. *Mem. Real Acad. Cienc. Art.* Barcelona, 440 p.

Bengtson, P., Berthou, P. Y. (1982): Microfossiles et Echinodermes Incertae sedis des dépôts Albiens à Coniaciens du bassin de Sergipe-Alagoas, Brésil. *Cahiers de Micropaleont.*, **3:** 13–14.

Benkö-Czabalay, L. (1970): Les biofaciès des formations récifales du Crétacé. *Act. Géol. Acad. Sc. Hungaricae*, **14:** 271–286.

Berthou, P. Y., Correia, F., Prates, S., Taugourdeau, J. (1983): Essai de synthèse du Crétacé de l'Algarve: biostratigraphie, paléogéographie, sédimentation argileuse. Première partie. *Bull. Géol. Bassin de Paris*, **20/2:** 1–18.

Biju-Duval, J., Dercourt, J., Pichon, X. le (1977): From the Tethys Ocean to the mediterranean seas: a plate tectonic model for the evolution of the western alpine system. Biju-Duval & Montadert edit., Structural history of the Mediterranean Basins. Technip, Paris, 143 p.

Blayac, J. (1908): Note sur l'existence de formations récifales à la base du Barrémien inférieur du Djebel Taya et du Djebel Debar, près Guelma (Algérie). *Bull. Soc. Géol. France*, (8) **8:** 71–77.

Camoin, G. (1982): Plates-formes carbonatées et récifs à Rudistes du Crétacé de Sicile. Thèse 3ème cycle, Univ. de Provence, 244 p.

Chevalier, J., Fischer, M. (1982): Présence de *Colomiella* Bonet (Calpionellidea) dans le Crétacé inférieur (Madiela) du Gabon. *Cahiers de Micropaleont.*, **2:** 29–34.

Chikhi-Aouimeur, F. (1980): Les Rudistes de l'Aptien supérieur du Djebel Ouenza (Algérie, NE). Thèse 3ème cycle, Univ. Sc. & Tech. Houari Boumédiene, Alger, 111 p.

Chikhi-Aouimeur, F. (1983): Etude paléontologique de quelques Rudistes de l'Aptien supérieur du Djebel Ouenza (Algérie Nord-Orientale). *Géologie Méditerranéenne*, **X/1:** 33–48.

Chubb, L. J. (1971): Rudists of Jamaica. *Paleont. Americana*, **7/45:** p. 162–257.

Coates, A. G. (1973): Cretaceous tethyan coral-rudist biogeography related to the evolution of the atlantic ocean. *In:* Organisms and continents through time. *Paleontol. Ass.*, London, p. 169–174.

Combes, P. J., Fourcade, E., Masse, J. P., Philip, J. (1981): Observations stratigraphiques et paléontologiques sur le Crétacé de la zone du Parnasse. *Acad. Yougoslave Sc. et Arts*, ICSOBA-AIM conference, Cagliari, Italy, sept. 26/28, 1979, p. 347–365, Zagreb.

Coogan, A. H. (1973): New Rudists from the Albian and Cenomanian of Mexico and South Texas. *Rev. Inst. Mexicano Petrol.*, **V/2:** p. 51–82.

Coogan, A. H. (1977): Early and Middle Cretaceous Hippuritacea (Rudists) of the Gulf Coast. *In* BEBOUT and LOUCKS edit. Cretaceous carbonates of Texas and Mexico, p. 32–70.

Dercourt, J., Zonenshain, L. P., Ricou, L. E., Kazmin, V. G., Pichon, X. le, Knipper, A. L., Grandjacquet, C., Sbortshikov, I. M., Geyssant, J., Lepvrier, C., Pechersky, D. M., Boulin, J., Sibuet, J. C., Savostin, L. A., Sorokhtin, O., Westphal, M., Bazhenov, M. L., Lauer, J. P., & Biju-Duval, B. (1986): Geological evolution of the Tethys Belt from the Atlantic to the Pamirs since the Lias. *Tectonophysics,* **123:** p. 241–315.

Desauvagie, T. F. J. (1968): Cenomanian *Trocholinas* from Nigeria. *Micropaleont.,* **14/1:** 64–72.

Douglass, R. C. (1960a): Revision of the family Orbitolinidae. *Micropaleont.,* **6/3:** 249–270.

Douglass, R. C. (1960b): The Forminiferal genus *Orbitolina* in North America. *Geol. Surv. prof. pap.,* **333:** 1–52.

Douglass, R. C. (1961): *Orbitolinas* from Caribbean Islands. *J. of Paleont.,* **35/3:** 475–479.

Douvillé, H. (1900a): Sur la distribution géographique des Rudistes, des Orbitolines et des Orbitoïdes. Bull. Soc. Géol. France, (3) **XXVIII:** 222–235.

Douvillé, H. (1900b): Sur quelques Rudistes Américains. *Bull. Soc. Géol. France,* (3) **28:** 205–221.

Douvillé, H. (1918): Le Barrémien supérieur de Brouzet. Part. 3, Les Rudistes. *Mém. Soc. Géol. France,* Paléont., **XXII/1:** 1–28.

Douvillé, H. (1926): Fossiles recueillis par Hayden dans le Kashmir en 1906 et les Pamirs en 1914: leur description. *Rec. Geol. Surv. India,* **58/4:** 349–357.

Dragastan, O. (1975): Upper Jurassic and Lower Cretaceous microfacies from the Bicaz Valley Bassin (East Carpathians). *Mem. Inst. Geol. Geophys.* Bucarest, **XXI:** 87 p.

Emberger, J. (1954): Présence d'*Agria marticensis* d'Orb. et de *Praeradiolites pulchellus* Vidal dans les terrains crétacés du bassin du Hodna (Algérie). *C. R. som. Soc. Géol. France,* **3:** 49–51.

Enay, R. (1980): Evolutions et relations paléobiogéographiques de la Téthys mésozoïque et cénozoïque. in «Géologie des chaînes alpines issues de la Téthys». *Mém. Bur. Rech. Géol. Min.,* **115:** 276–283.

Forster, R. (1978): Evidence for an open seaway between Northern and Southern protoatlantic in Albian time. Nature, **272:** 158–159.

Fritzche, C. H. (1924): Neue Kreidefaunen aus Südamerika (Chile, Bolivia, Peru, Columbia). *N. Jahrb. Mineralog.:* 313–334.

Gupta, S. B. K., Grant, A. C. (1971): *Orbitolina,* a Cretaceous Larger Foraminifera, from Flemish Cap: Paleoceanographic implications. *Science,* **173:** 934–936.

Gutnic, M., Moullade, M. (1967): Données nouvelles sur le Jurassique et le Crétacé inférieur du Barla Dag au Sud de Senirkent (Taurus de Pisidie, Turquie). *Bull. Mineral. Res. Explor. Inst. Turkey,* **69:** 60–78.

Heezen, B. C., Matthews, J. L., Catalano, R., Natland, J., Coogan, A., Tharp, M., Rawson, M. (1973): Western Pacific guyots. *In:* Heezen & al. *Init. Rep. of the Deep Sea Drill. Proj.,* **XX:** 653–723.

Henson, F. R. S. (1948): Larger imperforate Foraminifera of southwestern Asia. *British Mus.,* London, 127 p.

Jankicevic, J. (1978): Barrémien et Aptien des parties moyennes des Carpatho-balkanides dans la Serbie orientale au point de vue du développement Urgonien. *Ann. géol. péninsule balkaniq.,* **XLII:** 103–194.

JANKICEVIC. J. (1984): Urgonian limestones of Moravia near Ljig. *Ann. Geol. Peninsule Balkan.*, **XLVIII:** 53–58.
JONES. D. L. (1972): Structural elements and biostratigraphical framework of Lower Cretaceous rocks in Southern Alaska. *In:* Casey & Rawson edit. The Boreal Lower Cretaceous. p. 1–18. Seel House Press, Liverpool.
KAUFFMAN. E. G. (1973): Cretaceous bivalves. *In:* Atlas of paleobiogeography. A. Hallam. édit. Elsevier Sc. Publish. Company. p. 363–383.
KAUFFMAN. E. G.. JOHNSON. C. C. (1988): The morphological and Ecological Evolution of Middle and Upper Cretaceous Reef Building Rudistids. *Palaios,* **3:** 194–216.
KEMPER. E. (1987). Das Klima der Kreide-Zeit. *Geol. Jb.,* **A 96:** 5–185.
KUHN. O.. ANDRUSOV. D. (1942): Stratigraphie und Palaogeographie der Rudisten. III. Rudistenfauna und Kreideentwicklung in den West-Karpathen. *Neues Jahrb. Mineralog. geol. Paleont. Beiträge.* **86:** 450–480.
LEFELD. J. (1968)· Stratigraphy and palaeogeography of the High-Tatric Lower Cretaceous in the Tatra Mountains. *Stud. Geol. Polonica.* **XXIV:** 115 p.
LLOYD. C. R. (1982): The mid-Cretaceous Earth: paleogeography. ocean circulation and temperature. atmospheric circulation. *J. of Geol..* **90:** 393–413.
LUPERTO-SINNI. E.. MASSE. J. P. (1982): Contributo della paleoecologia alla paleogeografia della parte meridionale della piattaforma Apula nel Cretaceo inferiore. *Geologica Rom..* **21:** 859–877.
MAINELLI. M. (1975): *Requienia pietrarojensis* n. sp. del Barremiano di Civita di pietraroja (BN). *Boll. Soc. Paleont. Ital..* **14/1:** 3–10.
MAINELLI. M. (1975): *Requienia tortilis* nov. sp. nell albiano inferiore del Matese centro-meridionale e dei Monte d'Ocre (Apennino centro-meridionale). *Boll. Soc. Paleont. Ital.,* **14/2:** 203–215.
MAINELLI. M. (1983): Nuove Rudiste del Cretacico inferiore medio di Monte La Costa (S. Polo Matese. Campobasso). *Boll. Soc. Paleont. Ital..* **22/3:** 189–208.
MASSE. J. P. (1976): Les calcaires urgoniens de Provence (Valanginien-Aptien inférieur). Thèse doct. Sc.. Univ. Aix-Marseille II, 445 p.
MASSE. J. P. (1985): Paléobiogéographie des Rudistes du domaine périméditerranéen à l'Aptien inférieur. *Bull. Soc. Géol. France,* (8) **I/5:** 715–721.
MASSE. J. P.. ROSSI. T. (1987): Le provincialisme sud-caraïbe à l'Aptien inférieur. Sa signification dans le cadre de l'évolution géodynamique du domaine caraïbe et de l'Atlantique Central. *Cret. Res.,* **8:** 349–363.
MASSE. J. P.. SKELTON. P., SLISKOVIC. T. (1984): *Glossomyophorus costatus* n. gen. n. sp., Rudiste (Caprotinidae) nouveau de l'Aptien du domaine méditerranéen central et oriental. *Géobios.* **17/6:** 723–732.
MICHAUD. L. (1984): Les milieux de sédimentation crétacés du Sénégal et leur évolution diagénétique (Etude de subsurface). Thèse de 3° cycle, Univ. de Droit. d'Economie et des Sciences d'Aix-Marseille. 286 p.
MIDDLEMISS. F. A. (1973): The geographical distribution of Lower Cretaceous Terebratulacea in Western Europe. *In:* Casey & Rawson. edit. The Boreal Lower Cretaceous, Seel House Press. p. 111–120.
MISIK. M. (1966): Microfacies of the Mesozoic and Tertiary limestones of the West Carpathians. *Vydav. Slovensk. Akad. Vied.* Bratislava, 269 p.
MONOD. O. (1985). Tethys. in "Encyclopedia universalis" Edit. Paris, p. 990–995.
MONTENAT. C.. MOULLADE, M.. PHILIP. J. (1982): Le Crétacé inférieur à Orbitolines et Rudistes d'Afghanistan central. *Géologie Méditerranéenne.* **IX/2:** 109–122.
MOSHIER. S. O.. SCOTT. R. W.. HANDFORD. C. R.. BOUTELL. R. D. (1988): Giant gas

accumulation in a "chalky" textured micritic limestone, Lower Cretaceous Shuaiba Fm. Eastern United Arab Emirates. In Lomando & Harris "Giant Oil and Gas Fields – a core workshop". *Soc. Econ. Paleont. Mineral.*, Core Workshop n° 12, Houston, p. 229–272.

MOULLADE, M. (1965): Contribution au problème de la classification des Orbitolinidae (Foraminiferida, Lituolacea). C. R. Acad. Sci. Paris, **260:** 4031–4178.

MOULLADE, M., PEYBERNES, B., REY, J., SAINT-MARC, P. (1984): Intérêt paléobiogéographique de Grands Foraminifères benthiques (Orbitolinidae) du Crétacé inférieur/moyen des marges européennes et africaines à l'articulation Atlantique-Téthys. *Trav. lab. Strat. Paléoécol.*, Univ. Provence, Marseille, **3:** 107–112.

MULLERRIED, F. K. G. (1947): Paleobiologia de la caliza de Cordoba y Orizaba, Vera Cruz. *An. Inst. Biol.*, **18:** 361–462.

NEAGU, T., PANA, I., DRAGASTAN, O. (1977): Biostratigraphie de la série des calcaires éocrétacés de l'aire Cernavoda-Alimanu-Ostrov. *Rev. Roum. géol. géophysiq et géogr.*, **21:** 137–144.

NEUMAYR, M. (1885): Die geographische Verbreitung der Juraformation. *Denkschr. K. Akad. Wiss. Wien, Math.-Natur.*, **50:** 57–142.

OKUBO, M., MATSUSHIMA, N. (1959): On a new species of *Pachyodonta* from the Akaishi Mountains, Central Japan. Earth Sci., *J. An. Geol. collab.*, Japan, **42:** 1–4.

PALMER, R. M. (1928): The Rudistids of Southern Mexico. *Occ. pap. Californ. Acad. Sc.*, **14:** 132 p.

PAQUIER, V. (1903): Les Rudistes urgoniens. Part I. *Mém. Soc. Géol. France, Paléont., Mém.*, **29:** 1–46.

PAQUIER, V. (1905): Les Rudistes urgoniens. Part II. Ibid., 49–95.

PAQUIER, V., ZLATARSKI, M. (1901): Sur l'âge des couches urgoniennes de Bulgarie. *Bull. Soc. Géol. France,* (4), **I:** 287.

PASCAL, A. (1984): Les systèmes biosédimentaires urgoniens (Aptien-Albien) sur la marge nord-ibérique. Thèse Univ. Dijon, 561 p.

PASSENDORFER, E. (1930): Etude stratigraphique et paléontologique du Crétacé de la série haut-tatrique dans les Tatras. *Prace Panst. Inst. Geol.*, **II/4:** 351–677. Warszawa.

PASTOURET, L., MASSE, J. P., PHILIP, J., AUFFRET, G. A. (1974): Sur la présence d'Aptien inférieur à faciès Urgonien sur la marge continentale armoricaine. Conséquences paléogéographiques. *C. R. Acad. Sc. Paris,* t. **278:** 2011–2014.

PELISSIE, T., PEYBERNES, B., REY, J. (1982): Tectonique des plaques et paléobiogéographie des Grands Foraminifères benthiques et des algues calcaires du Dogger à l'Albien sur le pourtour de la Mesogée. *Bull. Soc. Géol. France,* **7: 24/5–6,** 1069–1076.

PERKINS, B. F. (1960): Biostratigraphic studies in the Comanche Cretaceous series of Northern Mexico and Texas. *Geol. Soc. Amer.*, **83:** 138 p.

PERKINS, B. F. (1974): Paleoecology of a Rudist reef-complex in the Comanche Cretaceous Glen Rose limestone of Central Texas. *Geosc. and Man,* **VIII/1:** 131–173.

PERKINS, B. F. (1979): Rudists from D.S.D.P. Leg 43, site 384. *Init. Repts. DSDP,* **43:** 5909.

PEYBERNES, B. (1982): Les Orbitolinidés crétacés d'Afrique. Essai de Synthèse. *Cahiers de Micropal.*, **2:** 13–28.

PEYBERNES, B., FORSTER, R. (1987): Présence d'une association micropaléontologique de la zone à *Praeorbitolina cormyi* (Aptien inférieur) dans l'Urgonien de Tanzanie méridionale. Implications paléogéographiques au niveau de la marge indienne de l'Afrique. *Géol. Méditer.*, **XIV/2:** 143–151.

PEZA, L. H., MARKU, D., PIRDENI, A. (1981): Biostratigraphie et paléogéographie des dépôts crétacés de la région de Munelle. *Permbled. Stud. Stratigr. Paleont.*, Tirana, **2:** 95–108.

PHILIP, J. (1982): Paléobiogéographie des Rudistes et géodynamique des marges mésogéennes au Crétacé supérieur. *Bull. Soc. Géol. France*, (7), **XXIV:** 995–1006.

PLENICAR, M., BUSER, S. (1967): The Cretaceous macrofauna in the Western part of the Trnovski Gozd. *Geologija*, **10:** 147–160, Ljubljana.

POLSAK, A. (1970): Offneria rhodanica du Crétacé inférieur de la Lika en Croatie. *Geol. Vjesn. Just. za geol. Istraz.* Zagreb, **23,** p. 145–149.

PRINZ, P. (1986): Middle Jurassic corals from Northern Chile. *N. Jb. Geol. Palaont. Mh.*, **1986/12:** 736–750.

POIGNANT, A. F., LOBITZER, H. (1982): Les algues de l'Albien supérieur du Nigéria. *Cahiers de Micropal.*, **2:** 35–40.

PTCHELINTSEV, V. F. (1959). Rudisti mezozoja gornovo Krima. *Ak. Nauk. SSSR*. Geol. Muz. A. P. Karpinszkovo, p. 1–178, Moskwa.

PUDSEY, C. J., SCHROEDER, R., SKELTON, P. (1985): Cretaceous (Aptian-Albian) Age for Island Arc Volcanics, Kohistan, N. Pakistan. *In*. Gupta & al., Contributions to Himalaya Geology, **3:** 150–168.

VON RAD, U., WISSMANN, G. (1982): Cretaceous-Cenozoic history of the West Saharan. Continental Margin (NW Africa), development, destruction and gravitational sedimentation. Geology of the Northwest African Continental Margin. Eds. VON RAD, U., HINZ, K., SARNTHEIN, M., SEIBOLD, E., 106–131.

RANKE, U., VON RAD, U., WISSMANN, G. (1982). Stratigraphy, facies and tectonic development of the on-and offshore of the Northwest African Continental Margin. Eds. VON RAD, U., HINZ, K., SARNTHEIM, M., SEIBOLD, E., 86–105.

RAWSON, P. F. (1973): Lower Cretaceous (Ryazanian-Barremian) marine connections and cephalopod migrations between the Tethyan and Boreal realms. *In:* CASEY & RAWSON, edit.: the Boreal Lower Cretaceous. 131–144.

RENGARTEN, V. P. (1950): Rudisztovie fauni melovih ottozsenijii Zakavkazja. *Trud. inszt. geol. nauk.*, **180:** (51), 1–92, Moskwa.

REY, J. (1972): Recherches géologiques sur le Crétacé inférieur de l'Estremadura (Portugal). *Mem. Serv. Geol. Portugal*, n. s. **21:** 477 p.

ROSSI-RONCHETTI, C. (1965): Rudiste e Nerine del Cretaceo di Yasin (Pakistan nord-occidentale). Ital. Exped. Karakorum (K2) and Hindu Kush, 229–272, Brill Edit.

SAHNI, M. R. (1937): Discovery of Orbitolina-bearing rocks in Burma with a description of *Orbitolina birmanica*. *India Geol. Surv. Recs.*, **71, 4:** 360–375.

SAID, R., BARAKAT, M. G. (1957). Lower Cretaceous Foraminifera from Khashm el Mistan, Northern Sinaï, Egypt. *Micropaleont.*, **3/1:** 39–47.

SAINT-MARC, P. (1970): Contribution à la connaissance du Crétacé basal du Liban. *Rev. Micropaleont.*, 12/4: 224–233.

SCHROEDER, R., CHERCHI, A. (1979): Upper Barremian-Lowermost Aptian Orbitolinid Foraminifera from the Grand Banks continental rise, northwestern Atlantic (D.S.D.P. Leg 43, site 384). *In: Initial Reports of the Deep Sea Drilling Project*, **43:** 575–583.

SCOTT, R. W. (1979): Depositional Model of Early Cretaceous Coral-Algal-Rudist Reefs, Arizona. *Am. Assoc. Petrol. Geol. Bull.*, 63/7: 1108–1127.

SCOTT, R. W. (1981): Biotic relations in Early Cretaceous Coral-Algal-Rudist reefs. Arizona. *J. of Paleont.*, 55/2, 463–478.

SCOTT, R. W., ALEMAN, A. (1984): *Stylina columbaris* n. sp. in a Lower Cretaceous Coral Biostrome, Peru. *J. of Paleont.*, **58/4:** 1136–1142.

SIMMONS, M. D., HART, M. B. (1987): The biostratigraphy and microfacies of the Early to mid-Cretaceous carbonates of Wadi Mi Aidin, Central Oman Mountains. In Hart edit. Micropaleontology of carbonate environments. Ellis Horwood Ltd, Chichester, **10:** 176–207.

SKELTON, P. (1982): Aptian and Barremian Rudist Bivalves of the New World: Some Old World similarities. *Cretaceous Research,* **3:** 145–153.

SLISKOVIC, T. (1982): New radiolitidae (Hippuritacea) from the Cretaceous deposits of Bosnia and Herzegovina. *Pos. Ot. Glmasn. Zemalj. Muz.,* N. S. **21:** 1–19.

SLISKOVIC, T. (1984): *Eoradiolites jovanovici* n. sp. from the Lower Cretaceous beds of Eastern Bosnia. *Zborn. Radov. Akad. Nauk. i Umjefn.* Bosne i Hercegovine, **LXXV:** 257–262.

SOHL, N. F. (1976): Notes on Middle Cretaceous macrofossils from the Greater Antilles. *Ann. Mus. Hist. Nat. Nice,* **IV:** XXXI-I-XXXI-6.

SOHL, N. F. (1987): Cretaceous gastropods: contrasts between Tethys and the Temperate Provinces. *J. of Paleont.,* **61/6:** 1085–1111.

STEVENS, G. R. (1980): Tethyan faunal influences in the Indo-Pacific and South Pacific regions. *In:* Geologie des Chaînes alpines issues de la Téthys. *Mém. Bur. Rech. Geol. Mines,* **115:** 285–287.

SUESS, E. (1893): Are the great Oceans depths permanent? *Nat. Sci.,* **II:** 183.

SUESS, E. (1900): La face de la terre. 3° part: Les Mers. A. COLIN & Cie edit., Paris, 873 p.

TAVANI, G. (1947): Fauna malacologica cretacea della Somalia e dell Ogaden. *Paleont. Ital.,* XLII, n. s. **XIII:** 83–153.

TORRE, D. (1965): Rudiste presenoniane delle Murge baresi. *Paleontographia Italica,* 59, n. s. **29:** 63–83.

TZANKOV, V. (1960): Sur le niveau stratigraphique de quelques Pachyodontes urgoniens en Bulgarie. *Rev. of the Bulgarian Geol. Soc.,* **XXI/1:** 38–73.

VAN DE FLIERT, J. R. (1952): Liste des Rudistes du Crétacé du Constantinois. Monogr. rég., XIX, Congr. géol. Alger, **13:** 31–45.

WIEDMANN, J., BUTT, A., EINSELE, G. (1982): Cretaceous environment and subsidence history at the Moroccan Continental margin. Geology of the Northwest African Continental Margin. Eds. VON RAD, U., HINZ, K., SARNTHEIM, M., SEIBOLD, E., 366–395.

YABE, M., HANZAWA, S. (1926): Geological age of Orbitolina bearing rocks of Japan. *Sci. Rep. Tohuku Imp. Univ.,* **9:** 13–20.

YABE, H., NAGAO, T. (1926): Praecaprotina nov. gen. from the Lower Cretaceous of Japan. *Sci. Rep. Tohuku Imp. Univ.,* **9:** 21–24.

YANIN, B. T. (1985): Rudistae *Matheronia* from Lower Valanginian of the Crimea. *Recueil Paleont.,* **22:** 25–30.

YANIN, B. T. (1986): Parapachytraga, nouveau genre de Rudiste du Barrémien du Caucase. *J. Paleont.,* **1:** 113–115.

YANIN, B. T., TCHERNOV, B. G. (1979): Les Rudistes Requienia et Matheronia des sédiments urgoniens des Carpathes d'Ukraine. *Rev. Paleont.,* L'vov, **16:** 49–55.

YEHARA, S. (1920): A Pachyodont Lamellibranch from the Cretaceous deposits of Miyako in Rikuchu. *J. Geol. Soc. Tokyo,* **XXVII:** 39–44.

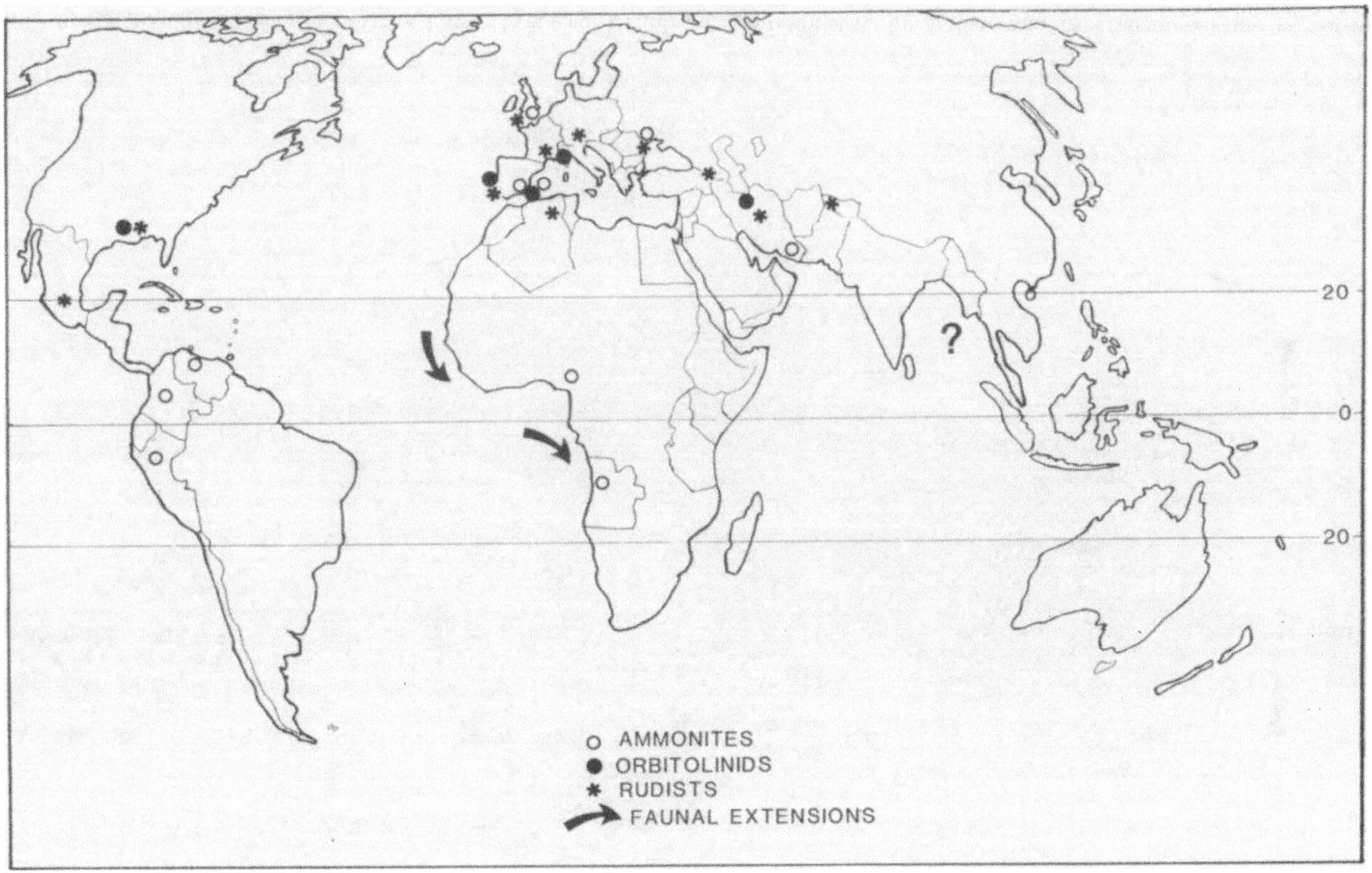

Fig. 1: Geographic distribution of the Lower Cretaceous Mesogean fauna after Douvillé (1900–1926).

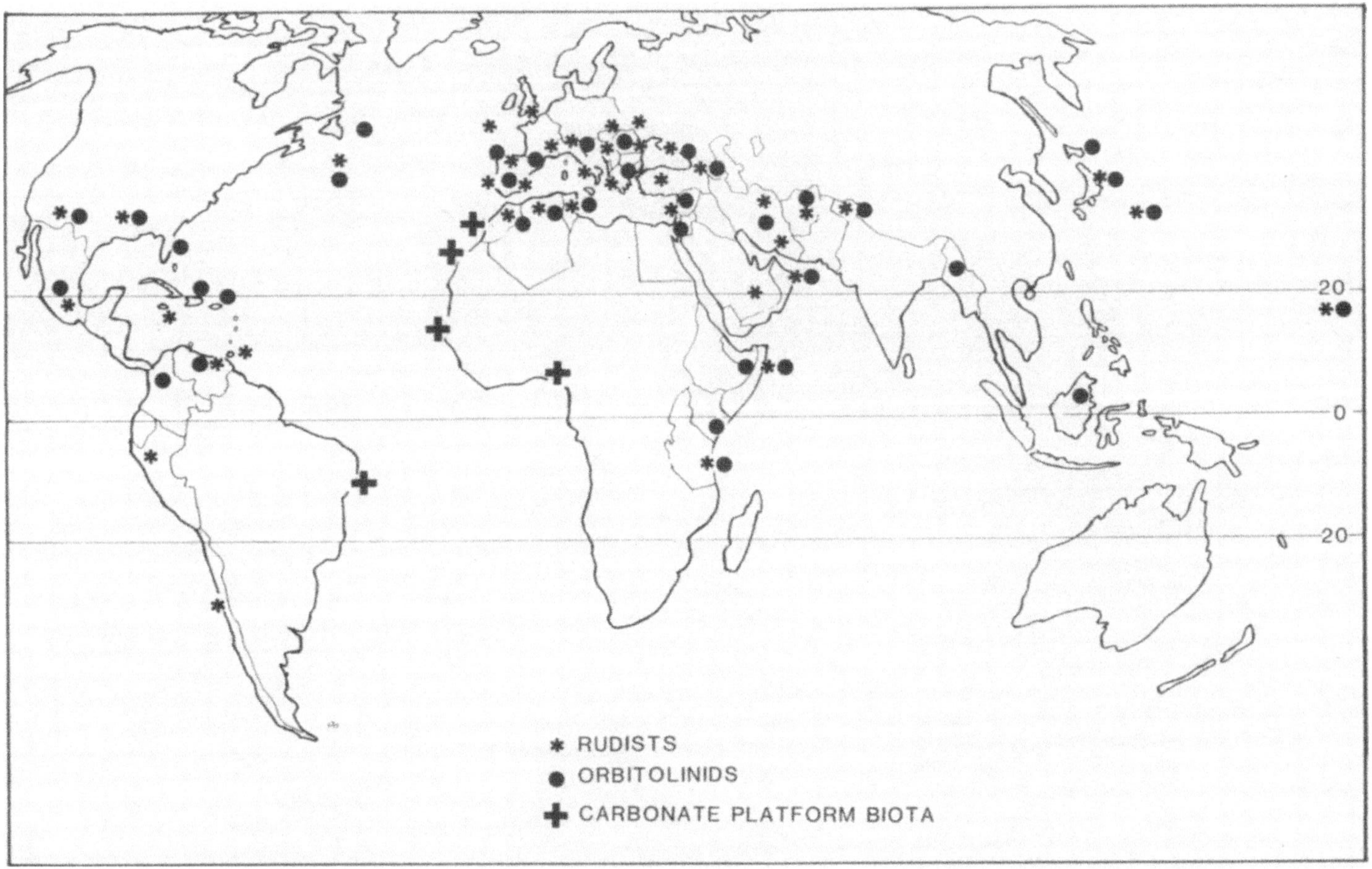

Fig. 2: Geographic distribution of the Lower Cretaceous Rudists, Orbitolinids (Orbitolininae and Dictyoconinae) and carbonate platform biota (Large Foraminifera – dasyclads aceans) according to present knowledge on Lower Cretaceous.

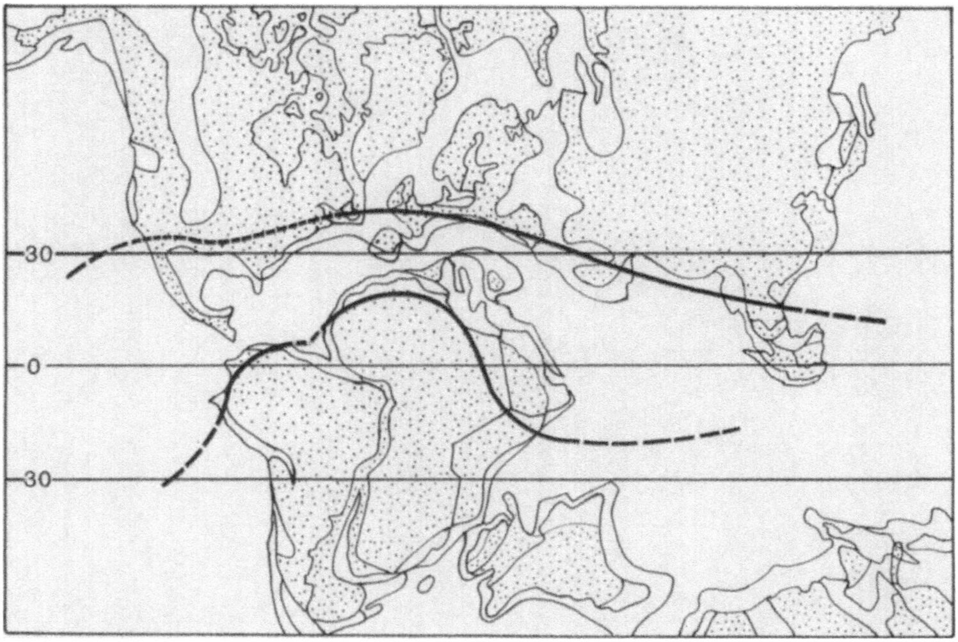

Fig. 3: Palinspastic configuration of the Lower Cretaceous Mésogée using DOUVILLÉ's data (paleogeographic reconstruction of continents after BARRON & al. [1981] and DERCOURT & al. [1986]).

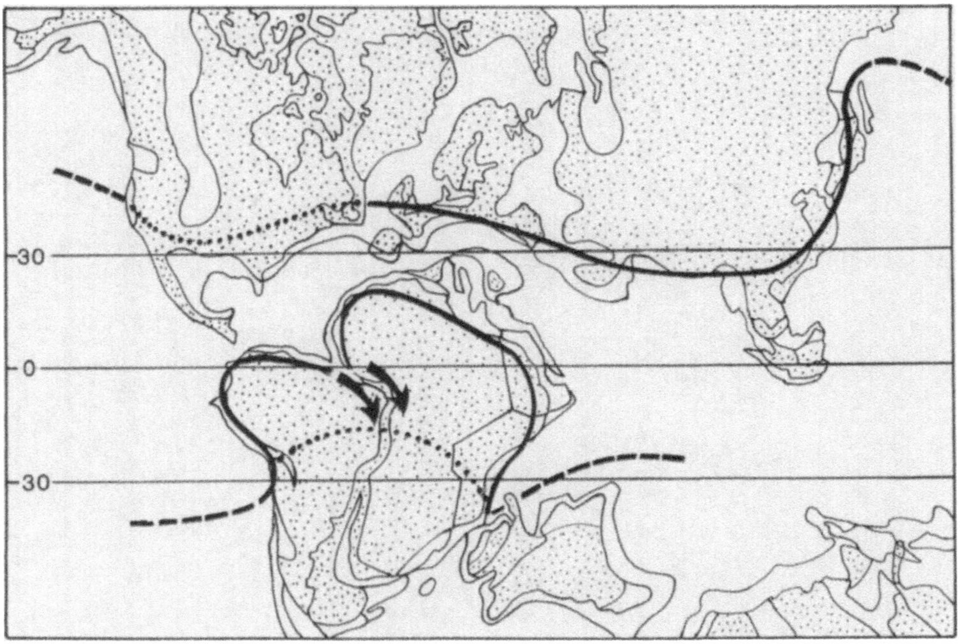

Fig. 4: Palinspastic configuration of the proposed Lower Cretaceous Mésogée. Arrows correspond to Albian Mesogean extensions on Central Atlantic.

2. Tethyan Cretaceous floral and faunal elements

New Aspects on Tethyan Cretaceous Fossil Assemblages.
Band 9 Schriftenreihe der Erdwissenschaftlichen Kommissionen der Österreichischen Akademie der Wissenschaften, 37–44.

2.1. Les Algues crétacées: tendances générales

Par Alain François POIGNANT*)

Résumé

Le Crétacé est une période importante pour les Rhodophycées et les Dasycladacées, en particulier le Crétacé inférieur. Durant cette période, les Dasycladacées diminuent d'importance, à l'inverse des Rhodophycées. Plusieurs chiffres illustrent cette tendance, l'accélération de l'expansion océanique semble pouvoir expliquer les observations.

Abstract

The Cretaceous and especially the Lower Cretaceous is an important period for the Rhodophyceae and Dasycladaceae. During this period the Dasycladaceae are loosing their predominance to the Rhodophyceae. This is shown by a comparison of the number of Lower and Upper Cretaceous species and genera. It is assumed that this evolutionary trend is caused by sea floor spreading.

1. Introduction

Le système crétacé représente un laps de temps (65 millions d'années) suffisamment long pour voir se dessiner des tendances générales, mais, pour le paléoalgologue, le début et la fin de la période posent un problème d'ordre stratigraphique.

En effet, la base, mal définie ne renferme pas, ou très peu, d'algues et les auteurs ont consciemment ou non associé le Berriasien et le Valanginien. On

*) Centre d'Etudes des Algues fossiles, Université P. & M. Curie – Boîte 2000 – 4, place Jussieu – F-75252 Paris cédex 05.

constate pour la limite supérieure un extraordinaire degré d'incertitude puisque pour certains auteurs le Sénonien comprend le Maestrichtien, pour d'autres, il en est exclu, le Crétacé supérieur inclut ou non le Danien, le Paléocène est mal défini ou encore intégré à l'Eocène. Il est évident que l'interprétation des données bibliographiques est entachée d'une relative imprécision.

2. Genres

Deux grands groupes ont été choisis: la totalité des Rhodophycées calcaires et les Dasycladacées calcaires (algues vertes). Il ne s'agit pas de taxons hiérarchiquement identiques, mais de groupes de même importance.

A) Rhodophycées

30 genres; 8 sont hérités du Jurassique et 20 se poursuivront au Paléocène.

1. Apparitions: 22 genres dont 17 au Crétacé inférieur et 5 au Crétacé supérieur. Pour l'ensemble, le taux de renouvellement est 0,73, ce qui est considérable.

2. Disparitions: 10 au total, soit 3 au Crétacé inférieur et 7 au Crétacé supérieur. Le taux de disparition est de 0,33.

3. La limite Jurassique-Crétacé n'est pas très marquée; 8 genres assurent la continuité; les genres nouveaux n'apparaissant que plus tard.

4. La limite Crétacé-Paléocène n'est pas mieux marquée: 19 genres sont ubiquistes et sur les 7 genres qui disparaissent au cours du Crétacé supérieur,2 seulement le font au Maestrichtien supérieur et 2 apparaissent à la base du Paléocène.

B) Dasycladacées

Sur les 24 genres de Dasycladacées, 11 sont hérités du Jurassique et 6 se poursuivront au Paléocène. On peut rappeler qu'une vision plus générale des Algues vertes comprend 35 genres de Dasycladales et 5 genres d'Udotéacées.

1. Apparitions: 13 uniquement au Crétacé inférieur, aucune au Crétacé supérieur, ce qui peut être interprété comme un signe de déclin. Le taux de renouvellement n'est que de 0,54, c'est-à-dire assez faible.

2. Disparitions: 18 au total soit 11 au Crétacé inférieur (61%) et 7 au Crétacé supérieur (39%), soit un taux élevé de disparition: 0,75.

3. La limite Jurassique-Crétacé n'apparaît pas ici non plus de façon nette puisque 11 genres sont issus du Jurassique et 13 apparaîtront mais au cours du Crétacé inférieur (Barrémien surtout).

4. La limite Crétacé-Paléocène n'est pas mieux marquée: 7 genres disparaissent au cours du Crétacé supérieur dont 2 seulement au Maestrichtien supérieur. La «crise» semble profitable aux Dasycladacées puisque 11 genres nouveaux feront leur apparition dès le début du Danien.

En résumé

	Rh.	Dasy.
– Genres apparus au *Crétacé inférieur*	57%	54%
– Genres apparus au *Crétacé supérieur*	17%	0
– Genres hérités du *Jurassique*	27%	46%
– Genres disparus au *Crétacé inférieur*	10%	46%
– Genres disparus au *Crétacé supérieur*	23%	29%
– Genres passant au *Paléocène*	63%	25%
– Taux d'apparition	0.73	0.54
– Taux de disparition	0.33	0.75

Il y a là des comportements inverses, un groupe semble vouloir remplacer l'autre.

3. Espèces

On compte environ 600 citations dont 41% pour le Crétacé inférieur.

A) Crétacé inférieur

	V.	H.	B.	Ap.	Al.	Tot.
Rh.	19	14	18	26	36	113
Dasy.	21	16	44	37	11	129
Tot.	40	30	62	63	47	242

Les Rhodophycées représentent 47% des espèces citées durant le Crétacé inférieur.

B) Crétacé supérieur

	Ce.	T.	C.	S.	Ca.	M.	T.
Rh.	54	41	34	47	49	75	300
Dasy.	17	11	6	5	5	10	54
Tot.	71	52	40	52	54	85	354

Les Rhodophycées représentent 85% des espèces citées dans le Crétacé supérieur; le renversement de tendance est net.

On peut exprimer ce renversement de tendance d'une autre façon.

1. Rhodophycées: sur la totalité des espèces citées, l'Hauterivien représente 0.3% et le Maestrichtien 18%.

2. Dasycladacées: sur la totalité des espèces citées le Barrémien représente 24% et le Santonien (ou le Campanien) 0.27%.

3. Au Crétacé inférieur, les Rhodophycées représentent 47% des citations, au Crétacé supérieur 85%.

On constate encore qu'il y a 1.25 fois plus de genres de Rhodophycées que de genres de Dasycladacées, mais il y a 2,26 fois plus d'espèces.

Le renversement de tendance est incontestable et il s'effectue durant l'Aptien-Albien. Le détail est en réalité plus complexe: au cours du Valanginien et de l'Hauterivien, on constate d'abord une diminution du nombre des espèces d'Algues rouges et vertes; c'est la fin du cortège jurassique.

*) Le renouvellement des Dasycladacées se fait avec éclat au Barrémien (34% des espèces du Crétacé inférieur) à partir duquel se produit le lent déclin.

*) Le renouvellement des Rhodophycées est plus discret, car plus lent. Il commence au Barrémien mais se poursuit en s'amplifiant à l'Aptien et à l'Albien (32% des espèces du Crétacé inférieur). Durant le Crétacé inférieur, on assiste à la mise en place des Corallinacées avec quelques genres, peu d'espèces et de nombreux individus.

*) Le Crétacé supérieur verra l'amplification du phénomène: déclin des Dasycladacées (jusqu'à 1% du cortège) et domination des Rhodophycées (21% du cortège au Maestrichtien).

4. Espèces/Unité de temps

On peut trouver par ce biais des ordres de grandeur, compte-tenu de la variété des chiffres proposés pour les datations.

	V.	H.	B.	Ap.	Al.
Espèces totales	4	6	31	13	4
Rh.	1,9	2,8	9	5,2	3
Dasy.	2,1	3,2	22	7,8	1

	Ce.	T.	Co.	S.	Ca.	M.
Espèces totales	18	17	20	17	5	11
Rh.	13,5	13,6	17	15,6	4,4	9,4
Dasy.	4,5	3,4	3	1,4	0,6	1,6

Ces chiffres apportent peu d'informations mais on doit retenir le comportement des Rhodophycées au Coniacien (17 espèces par million d'années) et des Dasycladacées au Barrémien (22 espèces par million d'années). Cela ne fait que confirmer ce qui a été dit plus haut.

5. Apparitions d'espèces nouvelles (Rhodophycées)

Durant l'ensemble du Crétacé , on enregistre 114 espèces nouvelles dont 67% au cours du Crétacé supérieur.

	V.	H.	B.	Ap.	Al.	Tot.
Nb. d'espèces nouvelles	8	1	6	8	15	38

Ce.	T.	Co.	S.	Ca.	M.	Tot.
30	2	4	14	5	21	76

Le Cénomanien représente à lui seul 40% des espèces nouvelles du Crétacé supérieur (ou 26% du Crétacé total), à l'inverse l'Hauterivien et le Turonien marquent le pas.

Le taux d'apparition par unité de temps est:

V.	H.	B.	Ap.	Al.
0,8	0,2	3	1,6	1,25

Ce.	T.	Co.	S.	Ca.	M.
7,5	6,7	2	4.7	0,45	2,6

6. Disparitions des espèces (Rhodophycées)

On compte 102 disparitions d'espèces durant le Crétacé dont 72% au Crétacé supérieur.

		V.	H.	B.	Ap.	Al.	Tot.
		6	3	1	5	14	29
Nb. d'espèces disparues							
	Ce.	T.	Co.	S.	Ca.	M.	Tot.
	15	12	1	3	5	37	73

Le Maestrichtien représente 51% des disparitions du Crétacé supérieur (36% du Crétacé total). La «crise» peut être invoquée. A l'inverse, les comportements du Barrémien et du Coniacien peuvent retenir l'attention.

A titre de comparaison, on peut calculer le taux de disparition par unités de temps.

V.	H.	B.	Ap.	Al.	Ce.	T.	Co.	S.	Ca.	M.
0,6	0,6	0,5	1	1,2	3,75	4	0,5	1	0,45	4,6

En résumé, le Crétacé inférieur a un solde positif de 9 espèces nouvelles et le Crétacé supérieur de 3, soit +12 pour l'ensemble du Crétacé . Le détail est alors:

V.	H.	B.	Ap.	Al.	Ce.	T.	Co.	S.	Ca.	M.
+2	-2	+5	+3	+1	+15	-10	+3	+11	0	-16

Le Barrémien et le Cénomanien sont nettement positifs, Turonien et Maestrichtien sont nettement négatifs.

Cela confirme l'installation de l'association et son développement au Barrémien et au Cénomanien, mais aussi l'influence de la «crise» au Maestrichtien. Les conclusions sont moins claires, en ce qui concerne le Turonien dont la définition précise, les faciès et les limites sont restées longtemps floues.

7. On peut exprimer la vitalité du renouvellement d'une association par le rapport

$$\frac{\text{nombre d'espèces nouvelles}}{\text{nombre d'espèces disponibles}}$$

V.	H.	B.	Ap.	Al.	Ce.	T.	Co.	S.	Ca.	M.
0,4	0,07	0,3	0,3	0,4	0,5	0,04	0,1	0,1	0,1	0,3

ou encore par le rapport $\frac{\text{nombre d'espèces nouvelles}}{\text{nombre d'espèces disparues}}$

V.	H.	B.	Ap.	Al.	Ce.	T.	Co.	S.	Ca.	M.
1,3	0,3	6	1,6	1	2	0,16	4	4,6	1	0,6

là encore la réponse est claire.

8. Enfin, le rapport $\frac{\text{Algues rouges}}{\text{Algues vertes}}$ permet de quantifier un aspect de la paléoécologie.

Dans l'actuel, ce rapport peut aller jusqu'à 6 (ou plus) exprimant alors un mode battu. Les chiffres proposés devraient être modifiés pour tenir compte des formes calcaires non comptabilisées mais on peut retenir l'ordre de grandeur:

1)

V.	H.	B.
0,9	0,8	0,4

2)

Ap.	Al.	Ce.	T.	Co.	S.	Ca.	M.
0,7	3,3	3,2	3,7	5,7	9,4	9,8	7,5

Dans un premier temps, ce taux baisse; on peut suggérer que l'on assiste d'abord à la mise en place de milieux abrités, calmes, type plate forme interne. Puis les taux augmentent de façon considérable jusqu'au Campanien inclus. On peut imaginer que les conditions changent, le mode battu domine nettement (expansion océanique?).

La baisse du rapport au Maestrichtien est plus délicate à interpréter; c'est peut-être à la signature de la «crise» mais bien des hypothèses peuvent être proposées.

9. Répartition géographique

Les Algues crétacées ont dans leur ensemble une très vaste répartition géographique dépassant les limites habituelles de la Téthys.

Au Crétacé inférieur, avant la véritable éclosion des Rhodophycées, les Algues sont fréquentes dans les faciès urgoniens diachrones et sont plus ou moins circonscrites aux limites de la Téthys.

Au Crétacé supérieur, et même dès l'Albien inférieur, les Rhodophycées deviennent dominantes et sortent très largement des limites téthysiennes:

latitudinalement on les trouve actuellement entre le tropique du Capricorne et 55°N, c'est à dire du S de la Suède, du Schleswig-Holstein-Basse-Saxe-Mecklembourg à la Namibie ou Madagascar. Les Algues accompagnent très bien l'ouverture de l'Atlantique S.

longitudinalement on les trouve actuellement des Rocheuses (au delà de 120°W) jusqu'en Inde (au delà de 90° E) et probablement bien plus à l'Est.

La répartition générique ou spécifique n'apporte aucun argument nouveau concernant le domaine téthysien. Il n'en reste pas moins vrai qu'on a davantage d'informations sur la Téthys proprement dite, d'où un point de vue faussé.

Les éventuelles barrières gênant ou limitant l'expansion des Algues sont sédimentologiques (faciès détritiques wealdiens, craie, flysch ...).

10. Conclusions générales

Durant la longue histoire des algues, le Crétacé est une étape importante. C'est, en effet, une période d'inventions:

– invention du sporange isolé,

– invention du sore,

– invention du conceptacle multiperforé, le conceptacle monoperforé étant déjà connu,

– invention de l'articulation (l'existence d'Archamphoroa au Jurassique supérieur n'étant pas confirmée).

Il restera à inventer le mégacyte et à faire réapparaître les cellules polygonales. Il est important de souligner qu'aucune disparition de structure n'est constatée.

En héritage, le Jurassique a laissé deux grandes familles: Solénoporacées et Gymnocodiacées qui vont peu à peu disparaître. On a là un très bel exemple d'une compétition, d'une lutte d'influence entre Solenoporacées-Gymnocodiacées et les Corallinacées, curieusement absentes pendant plus de 120 millions d'années. Les Corallinacées sont mieux adaptées (présence de tissus différenciés et surtout d'un hypothalle de fixation, protection des organes reproducteurs, articulation, moindre sensibilité aux variations de température, de salinité, de profondeur ...) et vont donc occuper les niches écologiques.

Les nouveaux genres apparaissent lentement au Crétacé inférieur; ils sont tous, au début du moins, monospécifiques mais les individus sont nombreux. Au Crétacé supérieur, ce sera la multiplication rapide des espèces (exemple: Archaeolithothamnium: 2 espèces au Crétacé inférieur, 26 espèces au Crétacé supérieur).

Cette rivalité va quasiment exterminer les Solenoporacées et les Gymnocodiacées au bénéfice des Corallinacées mais se fait aussi de façon très nette au détriment des Dasycladacées.

Le Crétacé est pour les Dasycladacées une période difficile, le relais entre les formes aspondyles et euspondyles ne semblent pas leur conférer un avantage; leur déclin est très net.

Il y a dans tout cela des raisons phylogénétiques certaines et difficiles à cerner, mais il y a peut-être et surtout des causes paléogéographiques. Il semble que l'expansion océanique et les bouleversements qu'elle apporte soit une des explications des comportements contradictoires des Algues crétacées. Les variations de l'expansion océanique contribuent à de profondes transformations: transgressions, régressions, modifications des marges, hydrodynamisme, réduction des plates formes, approfondissement des milieux marins, modifications du thermalisme sous-marin, évolution du chimisme de l'eau de mer, changements dans la répartition des courants marins, variations des températures océaniques. C'est ainsi que certaines périodes verraient la quasi-disparition des plates-formes et l'approndissement du milieu marin favorable aux Corallinacées. D'autres périodes seraient favorables à la formation de l'aragonite (d'où prolifération des Dasycladacées) ou de la calcite (prolifération des Corallinacées). Il est évident que les formes les mieux adaptées, ou les mieux adaptables vont occuper efficacement les nouveaux territoires; les Corallinacées sont le type même de ces formes conquérantes. D'autres organismes marins (Rudistes ?) semblent avoir des réactions synchrones.

Il paraît nécessaire de replacer les Algues crétacées dans un contexte biologique et géodynamique général.

11. Bibliographie

Bassoullet, J. P., Bernier, P., Conrad, M. A., Deloffre, R., et Jaffrezo, M. (1978): Les Algues dasycladales du Jurassique et du Crétacé. – Géobios, **2:** 330 p.

Deloffre, R., et Genot, P. (1982): Les Algues dasycladales du Cénozoïque. – Bull. Cent. Rech. Expl. Prod. Elf Aqu., **4:** 247 p.

Deloffre, R., et Poignant, A. F. (1978): Détermination générique d'algues mésozoïques: Floridées et Dasycladacées. – Bull. Centre Rech. Expl. Prod. Elf Aqu., **2/1:** 39–60.

Poignant, A. F. (1979): Les Algues rouges crétacées: relations mer boréale-Téthys. – Aspekte der Kreide Europas. I.U.G.S., Series A, **6:** 273–278.

Poignant, A. F. (1981): Sur des formes nouvelles d'Algues rouges crétacées. Cret. Research, **2:** 187–195.

Poignant, A. F. (1981): Les Algues calcaires au Crétacé moyen. – Cret. Res., **2:** 405–408.

Poignant, A. F. (1981): Les Algues du Crétacé moyen; tendances générales. – Cret. Res., **2:** 503–504.

Poignant, A. F. (1982): Les Algues turoniennes. – Mém. Mus. Nat. Hist. Nat., **XLIX:** 197–202.

Poignant, A. F. (1983): Les Algues crétacées (Barrémien à Coniacien). – Zitteliana **10:** 309–312.

Poignant, A. F. (1983): Les Algues sénoniennes: aspects évolutifs. – Géol. Méditerr., **X/3–4:** 239–241.

Poignant, A. F., & Lobitzer, H. (1982): Les Algues de l'Albien supérieur du Nigéria. – Cahiers de Micropaléontologie, **2:** 35–38.

Poignant, A. F., & Michaud, F. (1985): *Lithophyllum berriozabalense* et *Lithothamnium subguabairense;* deux nouvelles espèces de Mélobésiées du Crétacé supérieur mexicain. – Bull. Centres Rech. Expl. Prod. Elf-Aquitaine, **9/1:** 127–135.

Ramakavelo, G. (1989): Etude micropaléontologique du Crétacé supérieur et du Paléocène de la région de Toliary (Tuléar), Bassin de Morondava (Madagascar). – Thèse Université P. & M. Curie, Paris VI, 174 p.

New Aspects on Tethyan Cretaceous Fossil Assemblages.
Band 9 Schriftenreihe der Erdwissenschaftlichen Kommissionen der Österreichischen Akademie der Wissenschaften, 45–55.

2.2. A review of low-latitude "Tethyan" calcareous nannoplankton assemblages of the Cretaceous

By Michael WAGREICH*)

Abstract

Distinctive low-latitude ("Tethyan") and high-latitude (boreal-austral) nannofossil assemblages can be recognized in Early Cretaceous, Mid-Cretaceous and Late Cretaceous times. Early and Mid-Cretaceous Tethyan assemblages are situated within a broad equatorial belt about 40° N to S. During the Late Cretaceous nannofossil bioprovincialism increases resulting in the existence of typical Tethyan/tropical, subtropical and high-latitude nannofossil species.

1. Introduction

Coccoliths, the minute calcareous skeletal elements of unicellular marine, planktonic protists (coccolithophorids), form a major component of marine sediments. Coccolithophores are predominantly autotrophic and therefore photosynthisizing living cells are restricted to the upper 200 meters of the water column. Studies of production, transportation and thanatocoenosis of modern coccoliths show that species diversity and distribution patterns are largely controlled by surface-water temperature (e. g. ROTH & BERGER, 1975), although assemblages change markedly during the sinking process through the water column (HONJO, 1976). Besides the strong latitudinal control of Recent calcareous nannoplankton distribution, other factors such as salinity, surface water fertility (especially nutrient advection caused by upwelling water masses), oceanic water circulation and proximity to the shore are known to influence nannofloral distributions (e. g. BUKRY, 1974; OKADA & HONJO, 1975; REID, 1980; ZHANG & SIESSER, 1986). Another important factor governing the distribution

*) Dr. Michael WAGREICH, Institut für Geologie der Universität Wien, Universitätsstraße 7/III, A-1010 Wien, Austria.

of coccoliths in oceanic sediments is dissolution during transportation and deposition, especially in depths below the calcite lysocline (THIERSTEIN, 1980).

It can be assumed that the controlling factors of Ancient coccolith distributions are more or less the same as for Recent calcareous nannoplankton. As a consequence studies of coccolith assemblages have proven to be a powerful tool in Cenozoic palaeoceanographic reconstructions as indicators of temperature and to a lesser degree fertility or salinity of oceanic surface waters (e. g. HAQ & LOHMANN, 1976; HAQ, 1980; RUDDIMAN & MCINTYRE, 1976). For the interpretation of Mesozoic coccolith assemblages the problem of the increasing influence of postdepositional, diagenetic alterations of assemblages arises. Dissolution and overgrowth result in an increase of the most resistant species such as *Watznaueria barnesae* and *Micula decussata* in the Cretaceous (THIERSTEIN, 1981). Samples with poorly preserved coccoliths are a major problem in palaeoceanographic reconstructions for the Mesozoic (e. g. ROTH, 1981).

The existence of Late Cretaceous latitudinal nannofossil provinces were for the first time demonstrated by WORSLEY & MARTINI (1970). They showed that occurences of *Micula murus* are strictly confined to low latitudes and *Nephrolithus frequens* only to high latitudes. Lateron, investigations of Mid-Cretaceous and Late Cretaceous coccolith assemblages also indicated latitudinal, water-temperature control for several other species and groups (e. g. ROTH & BOWDLER, 1981; THIERSTEIN, 1981).

This paper concentrates on various aspects of "Tethyan" coccolith assemblages. In nannofossil terms, "Tethyan" can be regarded as a synonym to "low-latitude", indicating the existence of a more or less broad, worldwide equatorial belt of compareable "tropical" nannofossil assemblages. No significant differences in nannofossil assemblage compositions can be demonstrated between the Pacific and the "Tethys" in a geological/tectonic sense, which is the former ocean lying along the later Alpine-Himalayan-Indonesian orogenic belt.

2. Early Cretaceous

(Berriasian, Valanginian, Hauterivian, Barremian)

The Early Cretaceous was characterized by the rise of calcareous nannoplankton to become the major source of calcareous material to the oceanic system, resulting in a gradual shallowing of the calcite lysocline during the Early Cretaceous. Data on Early Cretaceous coccolith distribution and provincialism are scarce, although latitudinal differences are known to exist from the Late Jurassic onwards (COOPER, 1989). THIERSTEIN (1989) also indicated pronounced neritic-oceanic gradients. The first low-latitude, shallow-marine nannofloral element recognized for the Early Cretaceous was the genus *Nannoconus* (e. g. THIERSTEIN, 1973), although occurences of certain species of this group were later also reported from mid-latitudes and oceanic plateaus (PERCH-NIELSEN, 1979; ROTH, 1981).

Table 1: Taxonomic index of Cretaceous calcareous nannofossil species

Biscutum constans (GORKA 1957) BLACK 1959
Calcicalathina oblongata (WORSLEY 1971) THIERSTEIN 1971
Ceratolithoides aculeus (STRADNER 1961) PRINS & SISSINGH in SISSINGH 1977
Conusphaera rothii (THIERSTEIN 1971) JAKUBOWSKI 1986
Corollithion silvaradion FILEWICZ et al. in WISE & WIND 1977
Cretarhabdus surirellus (DEFLANDRE 1954) REINHARDT 1970
Crucibiscutum salebrosum (BLACK 1971) JAKUBOWSKI 1986
Cruciellipsis cuvillieri (MANIVIT 1966) THIERSTEIN 1971
Cyclagelosphaera margerelii NOEL 1965
Cylindralithus serratus BRAMLETTE & MARTINI 1964
Eprolithus floralis (STRADNER 1962) STOVER 1966
Helicolithus trabeculatus (GORKA 1957) VERBEEK 1977
Lithraphidites bollii (THIERSTEIN 1971) THIERSTEIN 1973
Lithraphidites quadratus BRAMLETTE & MARTINI 1964
Marthasterites furcatus (DEFLANDRE in DEFLANDRE & FERT 1954) DEFLANDRE 1959
Micula decussata VEKSHINA 1959
Micula murus (MARTINI 1961) BUKRY 1973
Nannoconus abundans STRADNER & GRÜN 1973
Nannoconus borealis PERCH-NIELSEN 1979
Nannoconus truitti BRÖNNIMANN 1955
Nephrolithus frequens GORKA 1957
Octocyclus magnus BLACK 1972
Quadrum trifidum (STRADNER in STRADNER & PAPP 1961) PRINS & PERCH-NIELSEN in MANIVIT et al. 1977
Rhagodiscus asper (STRADNER 1963) REINHARDT 1967
Repagulum parvidentatum (DEFLANDRE & FERT 1954) FORCHHEIMER 1972
Seribiscutum primitivum (THIERSTEIN 1974) FILEWICZ et al. in WISE & WIND 1977
Sollasites falklandensis FILEWICZ et al. in WISE & WIND 1977
Sollasites horticus (STRADNER et al. 1966) CEPEK & HAY 1969
Speetonia colligata BLACK 1971
Tegulalithus septentrionalis (STRADNER 1963) CRUX 1986
Tegumentum striatum (BLACK 1971) CRUX 1989
Thiersteinia ecclesiastica WISE & WATKINS in WISE 1983
Tubodiscus verenae THIERSTEIN 1973
Watznaueria barnesae (BLACK 1959) PERCH-NIELSEN 1968
Zeugrhabdotus elegans (GARTNER 1968) MUTTERLOSE 1988
Zeugrhabdotus embergeri (NOEL 1959) PERCH-NIELSEN 1984
Zeugrhabdotus sisyphus (GARTNER 1968) CRUX 1989
Zygodiscus spiralis BRAMLETTE & MARTINI 1964

A first overview about the general composition of Tethyan nannofloras of the Early Cretaceous was given by THIERSTEIN (1973, 1976). Recently, MUTTERLOSE & HARDING (1987), APPLEGATE et al. (1989), COOPER (1989), CRUX (1989), and MUTTERLOSE (1989) compared Tethyan nannofossil assemblages with those from the Boreal Early Cretaceous of NW Europe (Great Britain, North Sea, Northern Germany). The following conclusions can be drawn from Early Cretaceous nannofossil assemblages:

1. Typical low-latitude, Tethyan nannofloras are dominated by the genera *Nannoconus, Micrantholithus, Lithraphidites, Watznaueria* and *Rhagodiscus* (THIERSTEIN, 1973; WIND & CEPEK, 1979; APPLEGATE et al., 1989; MUTTERLOSE, 1989). The nannoconids show greater diversities than in the boreal realm (PERCH-NIELSEN, 1989).

2. Tethyan or warm water species, which are very rare or absent in high latitude samples are the nannoconids of the "*Nannoconus truittii*-assemblage" (e. g. MUTTERLOSE, 1989), *Lithraphidites bollii, Speetonia colligata, Cruciellipsis cuvillieri, Calcicalathina oblongata, Conusphaera rothii, Tubodiscus venerae* (CRUX, 1989: Tab. 8.1). Species with high abundances in the Tethyan Early Cretaceous, but also found within the Boreal realm are *Watznaueria barnesae, Zeugrhabdotus embergeri, Cretarhabdus (Stradneria) crenulatus,* and *Rhagodiscus asper.*

3. The Tethyan "*Nannoconus truittii*-assemblage" reaches as far north as the Irish Sea. The southernmost occurences are known from Surinam, but related nannoconids are found also in Mozambique (DERES & ARCHERITEGUY, 1980; MUTTERLOSE, 1989). Nannofloras from deep sea drilling cores offshore North Carolina (DSDP Leg 93) and Portugal (ODP Leg 103) represent assemblages of the northern Tethys (APPLEGATE et al., 1989), whereas assemblages of northern Germany and Great Britain have a Boreal character. This indicates a belt of Tethyan assemblages between 35° S and 45° N.

4. Transgressive events with influxes of warm water Tethyan elements into northern and southern mid- to high-latitude regions are characterized by extremely high abundances of *Rhagodiscus asper* up to 50%, together with nannoconids and *Cyclagelosphaera margerelii* (CRUX, 1989; MUTTERLOSE, 1989).

5. Typical Boreal (and Austral) species of the Early Cretaceous are *Crucibiscutum salebrosum, Tegulalithus septentrionalis, Sollasites arcuatus, Sollasites horticus, Corollithion silvaradion, Tegumentum striatum, Zeugrhabdotus sisyphus* (APPLEGATE et al., 1989; COOPER, 1989; CRUX, 1989). Endemic species of the Boreal NW Europe include *Nannoconus borealis* and *N. abundans.*

6. The diversities of the assemblages show no significant tendency to decrease from low-latitudes polewards. The highest diversities are reported from NW Europe (COOPER, 1989; CRUX, 1989), although this unexpected distribution pattern can be also explained by the lack of quantitative data for non-European sequences.

3. Mid-Cretaceous

(Aptian, Albian, Cenomanian, Turonian)

The Mid-Cretaceous oceanic circulation was largely driven by differences in salinity with relatively low temperature gradients and less vigorous than that of the Cenozoic (BRALOWER & THIERSTEIN, 1984). The reorganization of this "sluggish" circulation towards higher temperature gradients and cold, nutrient-rich bottom waters seems to have taken place around the Cenomanian-Turonian boundary (e. g. KUHNT et al., 1986). The Mid-Cretaceous palaeoceanic situation was characterized by a shallowing of the calcite compensation depth by about 2 kilometers to dephts between 2,5 and 3,5 km (ROTH & KRUMBACH, 1986). There are four general trends in the distribution of Mid-Cretaceous nannofossils recognized by ROTH (1981), ROTH & BOWDLER (1981), ROTH & KRUMBACH (1986), WISE (1988), APPLEGATE et al. (1989), and MUTTERLOSE (1989):

1. There exists a wide belt of more or less uniform "tropical-subtropical" nannofossil assemblages from about 40° N to 40° S. Tethyan nannofossil assemblages correspond to this broad low latitude belt but no typical Tethyan assemblage composition can be given because of other controlling factors such as oceanic upwelling. Samples from Tethyan regions show nearly the same overall composition as those from Boreal sections (BRALOWER, 1988). Nevertheless, Tethyan assemblages are characterized by a predominance of *Watznaueria barnesae* and common *Prediscosphaera, Cretarhabdus* and *Eiffellithus* (e. g. BRALOWER, 1988). Also most of the Nannoconids are considered to be indicators of warm surface waters from the Tethys (MUTTERLOSE, 1989: "*Nannoconus truittii*-assemblage") and are therefore a common element of Tethyan assemblages, especially from neritic environments.

2. Surface water fertility gradients caused by upwelling waters along the eastern margin of the Atlantic strongly influenced nannofossil distributions. This results in a pronounced east-west gradient in assemblage composition. Indicators of high fertility waters are *Biscutum constans* and species of *Zygodiscus, Vagalapilla* and *Braarudosphaera* (ROTH & BOWDLER, 1981; ROTH & KRUMBACH, 1986).

3. High latitude assemblages of northern (Boreal) and southern (Austral) affinities show a more or less similar composition with dominance of several species such as *Serbiscutum primitivum, Zygodiscus spiralis* and *Eprolithus floralis* (ROTH & KRUMBACH, 1986). Species like *Serbiscutum primitivum, Sollasites falklandensis, Octocyclus magnus* and *Repagulum parvidentatum* seem to be restricted to high latitudes (ROTH & BOWDLER, 1981; WISE, 1988).

4. Nearly monospecific "blooms" of nannofossil species (e. g. *Corollithion, Zeugrhabdotus elegans, Biscutum constans*) occur preferentially in assemblages of high- and mid-latitudes. These blooms are ascribed to strong upwelling of nutrient-rich, cold waters and possibly lower salinities due to partly restricted surface water conditions (WISE & WIND, 1977; ROTH & BOWDLER, 1981; NOEL, BREHERET & LAMBERT, 1987; WISE, 1988).

4. Coniacian-Santonian

So far no general overview about the distribution patterns of nannofossil assemblages of Coniacian-Santonian times does exist. Palaeoceanographic reconstructions by means of nannofossils are hindered by the lack of oceanic sections with carbonate deposition during this time span because of a global highstand of the calcite compensation depth (e. g. WISE, 1988). An increase in nannofossil provincialism since the Turonian can be deduced by comparing the magnitude of the Mid-Cretaceous oceanic circulation with the Late Cretaceous situation. The Coniacian-Santonian seems to be also relatively "quiet" in regard of nannofossil evolution because often only one nannozone can be recognized, especially in oceanic sections (*Marthasterites furcatus* total range zone, e. g. CEPEK & HAY, 1969; STRADNER & STEINMETZ, 1984). *Marthasterites furcatus* is a tropical to subtropical element only sporadically found within mid- to high latitudes (PERCH-NIELSEN, 1979; CRUX, 1982).

Tethyan assemblages are characterized by high abundances of *Watznaueria barnesae* and rare *Micula* species (percentages below 1%, WAGREICH, 1986). Other common genera include *Eiffellithus, Prediscosphaera, Glaukolithus, Cretarhabdus, Cribrosphaerella* and *Tranolithus*. Within nearshore environments holococcoliths such as *Lucianorhabdus, Calculites, Orastrum, Russelia* and *Pharus* show high diversification within the Tethyan realm compared with northern assemblages. Southern high latitudes are characterized by *Thiersteinia ecclesiastica* together with *Repagulum parvidentatum* and *Serbiscutum primitivum* (WISE, 1988). Blooms of *Helicolithus trabeculatus*, a rare species in Tethyan assemblages, are reported by WATKINS et al. (1989) from the Santonian of southern high latitudes of the Falkland and Kerguelen Plateaus.

5. Campanian-Maastrichtian

It is generally assumed that there was an increase in temperature gradients between low and high latitudes from the Mid- to the Late Cretaceous (e. g. BARRON, 1983). For the Maastrichtian a major temperature decrease has been suggested because of increased nannofossil provincialism (WORSLEY, 1974; ROMEIN, 1977), resulting in distinct low, middle and high latitude assemblages (e. g. THIERSTEIN, 1976, 1981). Also the diversity of assemblages increases from the Mid to the Late Cretaceous (e. g. DOEVEN, 1983; CEPEK et al., 1985). As a consequence Campanian-Maastrichtian nannofossil thanatocoenoses are about twice as diverse as modern ones (THIERSTEIN, 1981: 377).

Micula decussata is the most resistant nannofossil species of the Campanian-Maastrichtian (THIERSTEIN, 1980, 1981). Therefore strong dissolution leads to enrichment of this species and *Watznaueria barnesae* and dissolution of the more susceptible ones.

The following conclusions can be drawn about Campanian-Maastrichtian nannofossil distribution patterns:

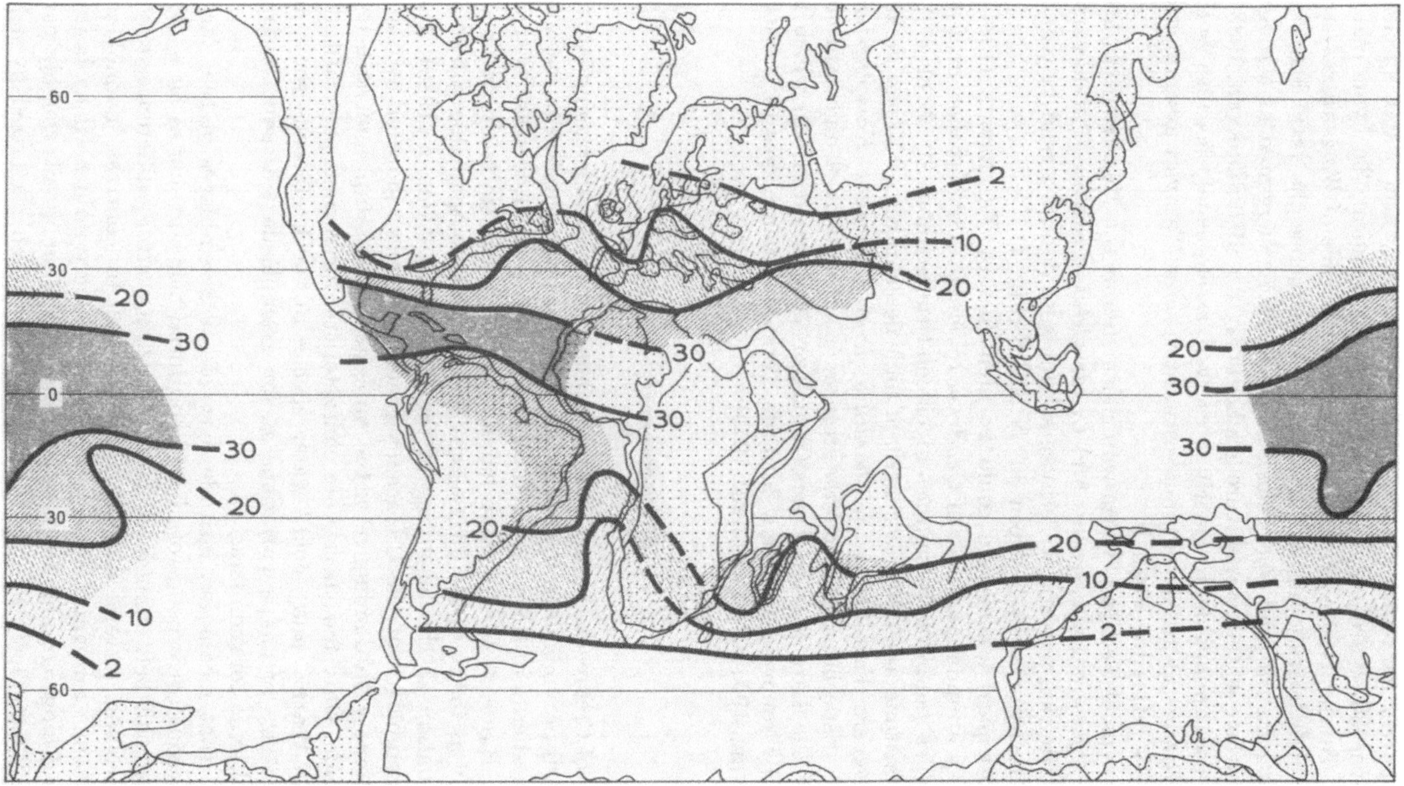

Fig. 1: Average percentage abundances of *Watznaueria barnesae* in Campanian-Maastrichtian assemblages after THIERSTEIN (1981: Fig. 15) (Mercator projection for 80 m. y. after BARRON et al., 1981).

1. *Watznaueria barnesae* is a typical low latitude, warm water species. Abundances over 30% are only reported from the tropical, Tethyan realm (comp. Fig. 1). Bukry (1973) even noted complete absence of *Watznaueria barnesae* in some high latitude assemblages. As a consequence the ratio between *Micula* and *Watznaueria* (and *Cyclagelosphaera*) has been suggested as a rough estimation of surface water temperature gradients (Wind, 1979; Doeven, 1983). This ratio is also applicable to poorly preserved assemblages, because both *Micula* and *Watznaueria* are dissolution resistant species. Tethyan assemblages display low M/W-ratios less than 1.

2. Other typical Tethyan, warm-water elements of the Maastrichtian are *Micula murus, Cylindralithus serratus* and *Cretarhabdus surirellus* (Thierstein, 1981). The distribution pattern of Tethyan assemblages defines a relatively narrow tropical "Tethyan" belt between 30° N and S (Fig. 1).

3. A subtropical assemblage could be identified by Thierstein (1981), characterized by abundance peaks of *Quadrum trifidum, Ceratolithoides aculeus* and *Lithraphidites quadratus* in samples of paleolatitudes about 30° N or S.

4. High latitude assemblages consist of high percentages of *Micula decussata* and the genera *Vagalapilla, Reinhardtites, Prediscosphaera, Kamptnerius,* and *Eiffellithus.* Especially, *Nephrolithus frequens* is an element of high latitude assemblages of the Late Maastrichtian (Worsley & Martini, 1970) which is only occasionally reported from low latitude Tethyan assemblages (e. g. Shafik & Stradner, 1971; Priewalder, 1973).

6. Discussion

A review of Cretaceous nannofossil assemblages indicates increasing latitudinal nannofossil provincialism probably as a consequence of increasing oceanic temperature gradients in the Late Cretaceous and of a general decrease of temperature (e. g. Barron, 1983). Weak nannofossil provincialism could be documented as early as for the Late Jurassic giving way to a compareable paleobiogeographic situation at the beginning of the Cretaceous (Cooper, 1989). The nannofossil assemblages of the Cretaceous generally display more or less uniform compositions with dominance of the same genera in samples of both low and high latitude. Only few species are characteristic for certain paleolatitudinal provinces. "Warm" and "cold" water species of the Cretaceous show no straightforward size or shape variations as for example dextral and sinistral coiling of planktonic foraminifera.

Tethyan, that means low-latitude, nannofossil assemblages of the Early and Mid-Cretaceous correspond to a broad equitorial belt reaching as far as 40° N and S. Within this belt nannofossil assemblages show strong differences resulting from upwelling waters and nearshore trends. In contrast, nannofossil assemblages of the Campanian-Maastrichtian are strongly influenced by latitudinal trends and biogeographic boundaries defined by nannofossils roughly parallel latitudes. Tropical, Tethyan assemblages are often characterized by high amounts of *Watznaueria barnesae* and certain other species, although influences apart from surface water temperature could be recognized (e. g. THIERSTEIN, 1981; DOEVEN, 1983). However, the lack of quantitative data especially for samples from Tethyan land sections hinderes a more precise examination of Tethyan nannofossil assemblages.

Acknowledgements

The critical reading of the manuscript by Dr. Kurt Decker (University of Tübingen) is gratefully acknowledged.

References

APPLEGATE, J. L., BERGEN, J. A., COVINGTON, J. M., & WISE, S. W. JR. (1989): Lower Cretaceous calcareous nannofossils from continental margin drill sites off North Carolina (DSDP Leg 93) and Portugal (ODP Leg 103): a comparison. – In: CRUX, J. A., & HECK, S. E. VAN (Eds.): Nannofossils and their Applications, pp. 212–222.

BARRON, E. J. (1983): A warm equable Cretaceous: the nature of the problem. – Earth-Sci. Rev., **19:** 305–338. Amsterdam.

BARRON, E. J., HARRISON, C. G. A., SLOAN, J. L., & HAY, W. W., 1981: Paleogeography, 180 million years ago to the present. – Ecglogae geol. Helv., **74:** 443–470. Basel.

BRALOWER, T. J. (1988): Calcareous nannofossil biostratigraphy and assemblages of the Cenomanian-Turonian boundary interval: implications for the origin and timing of oceanic anoxia. – Paleoceanography, **3:** 275–316. Washington.

BRALOWER, T. J., & THIERSTEIN, H. R. (1984): Low productivity and slow deep-water circulation in mid-Cretaceous oceans. – Geology, **12:** 614–618. Boulder.

BUKRY, D. (1973): Coccolith and silicoflagellate stratigraphy, Tasman Sea and Southwestern Pacific Ocean, Deep Sea Drilling. Project Leg 21. – Init. Repts. Deep Sea Drilling Proj., **21:** 885–893. Washington.

BUKRY, D. (1974): Coccoliths as paleosalinity indiators – evidence from the Black Sea. – Amer. Assoc. Petrol. Geol., Mem. **20:** 353–363. Tulsa.

CEPEK, P., & HAY, W. W. (1969): Calcareous nannoplankton and biostratigraphic subdivisions of the Upper Cretaeous. – Transact. Gulf. Coast Ass. Geol. Soc., **19:** 323–336.

CEPEK, P., KÖTHE, A., WEISS, W., & WOLFART, R. (1985): Paleogeographic evolution of the Atlantic Ocean during the Late Cretaceous. – Geol. Jb., **B 62:** 1–89. Hannover.

COOPER, M. K. E. (1989): Nannofossil provincialism in the Late Jurassic-Early Cretaceous (Kimmeridgian to Valanginian) period. – In: CRUX, J. A., & HECK, S. E. VAN (Eds.): Nannofossils and their Applications, pp. 223–246.

CRUX, J. A. (1982): Upper Cretaceous (Cenomanian to Campanian) calcareous nannofossils. – In: LORD, A. R. (Ed.): A Stratigraphical Index of Calcareous Nannofossils, pp. 81–135.

CRUX, J. A. (1989): Biostratigraphy and palaeogeographical applications of Lower Cretaceous nannofossils from north-western Europe. – In: CRUX, J. A., & HECK, S. E. VAN (Eds.): Nannofossils and their Applications, pp. 143–211.

DERES, F., & ARCHERITEGUY, J. (1980): Biostratigraphie des Nannoconides. – Bull. Centre Rech. Explor. Prod. Elf-Aquitane, **4:** 1–53, Pau.

Doeven, P. H. (1983): Cretaceous nannofossil stratigraphy and paleoecology of the Canadian Atlantic margin. – Bull. geol. Surv. Canada, **356:** 1–70. Ottawa.

Haq, B. U. (1980): Biogeographic history of Mioene calcareous nannoplankton and paleoceanography of the Atlantic Ocean. – Micropaleont., **26:** 414–443. New York.

Haq, B. U., & Lohmann, G. P. (1976): Early Cenozoic calcareous nannoplankton biogeography of the Atlantic Ocean. – Mar. Micropaleont., **1:** 119–197. Amsterdam.

Honjo, S. (1976): Coccoliths: Production, transportation and sedimentation. – Mar. Micropaleont., **1:** 65–80. Amsterdam.

Khunt, W., Thurow, J., Wiedmann, J., & Herbin, J. P. (1986): Oceanic anoxic conditions around the Cenomanian/Turonian boundary and the response of the biota. – Mitt. Geol.-Paläont. Inst. Univ. Hamburg, **60:** 205–246. Hamburg.

Mutterlose, J. (1989): Temperature-controlled migration of calcareous nannofloras in the northwest-European Aptian. – In: Crux, J. A., & Heck, S. E. van (Eds.): Nannofossils and their Applications, pp. 122–142, Chichester.

Mutterlose, J., & Harding, I. (1987): Phytoplankton from the anoxic sediments of the Barremian (Lower Cretaceous) of North-West-Germany. – Abh. geol. Bundesanst., **39:** 177–215. Wien.

Noel, D., Breheret, J. G., & Lambert, B. (1987): Enregistrement sedimentaire de floraisons phytoplanctoniques calcaires en milieu confine. Synthese de donnees l'actuel et observations geologiques. – Bull. Soc. geol. France, 1987 (8), **III/6:** 1097–1106. Paris.

Okada, H., & Honjo, S. (1975): Distribution of coccolithophores in marginal seas along the Western Pacific Ocean and in the Red Sea. – Mar. Biol., **31:** 271–285. Amsterdam.

Perch-Nielsen, K. (1979): Calcareous nannofossils from the Cretaceous between the North Sea and the Mediterranean. – IUGS Ser. A, **6:** 223–272. Stuttgart.

Perch-Nielsen, K. (1989): Tethyan and boreal Lower Cretaceous Nannoconus. – INA Newsl., **11:** 90. Firenze.

Priewalder, H. (1973): Die Coccolithenflora des Locus typicus von Pseudotextularia elegans (Rzehak), Reingruberhöhe, Niederösterreich; (Maastricht). – Jb. geol. Bundesanst., **116:** 3–34. Wien.

Reid, F. M. H. (1980): Coccolithophorids of the North Pacific Central Gyre with notes on their vertical and seasonal distribution. – Micropaleont., **26:** 151–176. New York.

Romein, A. J. T. (1977): Calcareous nannofossils from the Cretaceous/Tertiary boundary interval in the Barranco del Gredero I & II. – Proc. Kon. Ned. Akad. Wetensch., B, **80:** 256–279. Amsterdam.

Roth, P. H. (1981): Mid-Cretaceous calcareous nannoplankton from the Central Pacific: Implications for paleoceanography. – Init. Repts. Deep Sea Drilling Project, **62:** 471–489. Washington.

Roth, P. H., & Berger, W. H. (1975): Distribution and dissolution of coccoliths in the south and central Pacific. – Cushman Found. Foraminiferal Res., Spec. Publ., **13:** 87–113. Washington.

Roth, P. H., & Bowdler, J. L. (1981): Middle Cretaceous calcareous nannoplankton biogeography and oceanography of the Atlantic Ocean. – Soc. Econ. Paleont. Mineral. Spec. Publ., **32:** 517–546. Tulsa.

Roth, P. H., & Krumbach, K. R. (1986): Middle Cretaceous calcareous nannofossil biogeography and preservation in the Atlantic and Indian oceans: implications for paleoceanography. – Mar. Micropaleont., **10:** 235–266. Amsterdam.

Ruddiman, W. F., & McIntyre, A. (1976): Northeast Atlantic paleoclimatic changes over the past 600.000 years. – Mem. geol. Soc. America, **145:** 111–146. Boulder.

SHAFIK, S., & STRADNER, H. (1971): Nannofossils from the Eastern Desert, Egypt. – Jb. geol. Bundesanst., Sb. **17:** 69–104. Wien.

STRADNER, H., & STEINMETZ, J. (1984): Cretaceous calcareous nannofossils from the Angola Basin, Deep Sea Drilling Project Site 530. – Init. Repts. Deep Sea Drilling Project, **75:** 565–649. Washington.

THIERSTEIN, H. R. (1973): Lower Cretaceous calcareous nannoplankton biostratigraphy. – Abh. geol. Bundesanst., **29:** 1–52. Wien.

THIERSTEIN, H. R. (1976). Mesozoic calcareous nannoplankton biostratigraphy of marine sediments. – Mar. Micropaleont., **1:** 325–362. Amsterdam.

THIERSTEIN, H. R. (1980): Selective dissolution of Late Cretaceous and Early Tertiary calcareous nannofossils: experimental evidence. – Cretaceous Res., **2:** 165–176. Stuttgart.

THIERSTEIN, H. R. (1981). Late Cretaceous nannoplankton and the change at the Cretaceous-Tertiary boundary. – Soc. Econ. Paleont. Mineral., Spec. Publ., **32:** 355–394. Tulsa.

THIERSTEIN, H. R. (1989): Changes in the latitudinal structuring of nannofossil assemblages: Cretaceous-Holocene contrasts. – INA Newsl., **11:** 99–100. Firenze.

WAGREICH, M. (1986): Sedimentologische und stratigraphische Untersuchungen des tieferen Abschnittes der Gosauschichten von Gosau und Russbach (Oberösterreich-Salzburg). – Unpubl. Thesis Univ. Wien, 253 p. Wien.

WATKINS, D. K., WISE, S. W. JR., & POSPICHAL, J. J. (1989): Turonian-Maastrichtian nannofossil biostratigraphy and paleoceanography of the southern ocean. – INA Newsl., **11:** 106. Firenze.

WIND, F. H. (1979): Maestrichtian-Campanian nannofloral provinces of the Southern Atlantic and Indian Oceans. – In: TALWANI, M. (Ed.): Deep Drilling Results in the Atlantic Ocean Continental Margins and Paleoenvironment, Amer. Geophys. Union, M. Ewing Series, **3:** 123–137. Washington.

WIND, F. H., & CEPEK, P. (1979): Lower Cretaceous calcareous nannoplankton from DSDP Hole 379A (Northwest African Margin). – Init. Repts. Deep Sea Drilling Proj., **47:** 221–255. Washington.

WISE, S. W. (1988): Mesozoic-Cenozoic history of calcareous nannofossils in the region of the southern ocean. – Palaeogeogr., Palaeoclimat., Palaeoecol., **67:** 157–179. Amsterdam.

WISE, S. W., & WIND, F. H. (1977): Mesozoic and Cenozoic calcareous nannofossils recovered by DDSP Leg 36 drilling on the Falkland Plateau, southwestern Atlantic sector of the southern Ocean. – Init. Repts. Deep Sea Drilling Proj., **36:** 269–492. Washington.

WORSLEY, T. (1974): The Cretaceous-Tertiary boundary event in the ocean. – Soc. Econ. Paleont. Mineral., Spec. Publ. **20:** 94–175. Tulsa.

WORSLEY, T. R., & MARTINI, E. (1970): Late Maastrichtian nannoplankton provinces. – Nature, **225:** 1242–1243. London.

ZHANG, J., & SIESSER, W. G. (1986): Calcareous nannoplankton in continental-shelf sediments, East China Sea. – Micropaleont., **32:** 271–281. New York.

New Aspects on Tethyan Cretaceous Fossil Assemblages.
Band 9 Schriftenreihe der Erdwissenschaftlichen Kommissionen der Österreichischen Akademie der Wissenschaften, 57–74.

2.3. The structure and distribution of the European Cretaceous brachiopod assemblages with emphasis on the Tethyan fauna

By Jozef MICHALIK*)

Abstract

The Cretaceous was a time of decline for brachiopods, which tried unsuccessfully to re-occupy lost niches in the neritic communities. The structure of the Cretaceous brachiopod assemblages was much simpler than that of the Paleozoic ones. Nevertheless, the analysis of the quantitative composition of these assemblages provides an instructive view of their development, which is characterized by frequent opportunism and interruptions of the evolution of assemblages. Fluctuations in distribution are also typical for Cretaceous brachiopod paleobiogeography, making the exact delimitation of the mutual boundaries between a Tethyan, a Jura and a Boreal realm problematic, if not speculative in detail.

1. Introduction

Brachiopods are an ancient group of neritic benthic organisms which saw their maximum development during the Paleozoic. At the beginning of the Mesozoic, the brachiopods overcame several crises which diminished their importance in benthic faunas. The history of Mesozoic brachiopod development was an attempt to recolonize the shelves. However, many of the old brachiopod niches had been progressively substituted by other, more adaptable, ecologically more tolerant and flexible groups (MICHALIK, 1987a). Many of the Triassic and Jurassic brachiopod associations were dominated by opportunistic forms (AGER, 1971; VÖRÖS, 1987; MICHALIK, 1989 etc.).

Despite this fact, many specific faunal assemblages developed in various

*) Geological Institute, Slovakian Academy of Sciences, Dubravska cesta 9, CS-84226 Bratislava, Czecho-Slovakia.

regions. Since the basic works of AGER (1967; 1971; etc.) the European brachiopod faunas have been divided into the Mediterranean-, the Jura- and the Boreal types. However, the distribution of individual species changed in space and time, being affected by migration, destruction of populations, by unfavourable environmental conditions, and by geographic specialization. (The differences in individual taxonomic conceptions are not considered here, but they often represent a serious obstacle for the comparison of populations).

These factors make the exact delimitation of the boundaries between the paleobiogeographic zones rather uncertain. On the other hand, despite of the scarcity of reliable, complex and comparable data in the brachiopod literature the quantitative analysis of brachiopod associations could provide a better understanding of their paleobiogeography than some imaginary "index species".

2. Late Jurassic brachiopod development

At the end of the Jurassic, the system of basins with black clay and marl sedimentation originating in the Boreal epicontinental sea belt (London-Paris-, Rhone-, Swabian- and Polish Basins) was bordered by sponge and coral reefs. Here, the brachiopods played only a subordinate role in the benthic communities (FÜRSICH, 1977). The Sub-Mediterranean (Sub-Tethyan) realm, with its characteristic Jura-fauna, formed a belt from SE France through northern Switzerland, southern Poland, Dobrogea, Crimea, northern Caucasus to Kopet-Dagh (cf. MICHALIK, 1989). Brachiopod faunas were infrequent in the Tethyan realm during the Callovian-Oxfordian transgression, but they became more diversified and abundant during the Kimmeridgian and Tithonian. The coral reefs bordering the Mediterranean were inhabited by characteristic brachiopod associations. SMIRNOVA (1975) has described such an association consisting of *Weberithyris moravica, Tropeothyris cyclogonia, T. immanis, Cyclothyris astieriana, Lacunosella* sp., etc. from the Inwald locality of the former Stramberk limestone platform belt in the Outer Carpathians (Fig. 1).

Open marine shallow environments on the elevated, faulted block surfaces were inhabited by brachiopod associations dominated by *Camerothyris wahlenbergi, Karadagithyris bilimeki, Lacunosella hoheneggeri, L. zeuschneri* and by *Monticlarella* species. Nucleata and perforated pygopids, although subordinate, were stable components of these associations (BARCZYK, 1971; 1972). This can be shown in the Czorsztyn locality in the outer Carpathian Klippen Belt (Fig. 1).

The pygopids were the most significant (if not the only) elements of the Tethyan basinal brachiopod faunas (AGER, 1967; 1976). However, they also penetrated into northern seas as far as eastern Greenland. This means that the limits of the pygopid distribution (called the "Tethyan index") do not fit in the boundaries of the Tethyan realm. On the other hand, the distribution of the "Tethyan" species was not only controlled by bathymetry and temperature. The simple structure of the South Alpine pygopid associations (DIENI and

Middlemiss, 1981) is rather strange; it provokes some doubts as to wether the sampling methods used have been comprehensive and thorough enough for catching also the scarcely represented species of the brachiopod association (Fig. 1).

Ager (l.a.) suggested that the filtratory system of the pygopids was enormously effective allowing them to live in bathyal depths unfavourable for any other brachiopod groups. The deep-sea environment is poor in fluctuations, supporting the conservation of ancient animals living there. However, the pygopids interpreted as Upper Jurassic-Lower Cretaceous deep sea calm water benthos failed to survive the Lower Barremian and (with the exception of east Greenland) they never extended beyond the Tethyan deep marine basins. The mode of life, the relation to the environment, as well as the parameters governing the distribution of this obscure brachiopod group remain uncertain despite several attempts to elucidate them (Ager, 1978; Vogel, 1966; etc.).

3. Early Cretaceous brachiopod development

Many brachiopod taxa which arose during the Later Jurassic survived the Tithonian-Berriasian environmental changes and continued to evolve during the Valanginian and Hauterivian until the lowermost Barremian (Middlemiss, 1979; 1981; 1984; Fig. 2). Despite important paleogeographical changes connected with the Neo-Kimmerian orogenetic movements at the end of the Jurassic, the evolution of the principal paleo-biogeographical zones continued with only minor interruptions (Michalik et al., 1987). During the Berriasian, Boreal epicontinental seas yielded only few satisfactory conditions for brachiopod life (*Uralorhynchia* etc.). The absence of reefal and other pure limestones in this realm, along with the dominance of terrigenous clastic sedimentation was caused by a cold and relatively humid climate. Fluvial and brackish sedimentation prevailed in many areas. However, during the Valanginian, wide areas from eastern Greenland, Scotland, England to northern Germany were colonized by brachiopods. *Rouillieria, Rugitella* and *Rhynchonella,* supposed to be immigrants from the Russian Platform (Middlemiss, 1973; Owen, 1973) migrated to the west. *Ptilorhynchia* (syn. *Fusirhynchia,* cf. Owen, l. c.) migrated from western North America through Siberia to eastern Europe. Middlemiss (1973; 1979) reports a Valanginian mass migration of the Jura (or even Tethyan) faunas to the Boreal regions (Paris Basin, Northern Germany: *Oblongarcula, Sulcirhynchia, Sellithyris carteroniana, Musculina sanctaecrucis, Moutonithyris moutoniana*) and to Spain and Morocco. The Boreal realm was bordered by a dry-land belt to the south, with only shallow and narrow seaways (Middlemiss, 1976; 1979). Valanginian and Hauterivian Boreal cephalopods penetrated into the Mediterranean through the Polish Gate of the Danish-Polish Trough (Michalik and Kovac, 1982). However, no Lower Cretaceous brachiopods have yet been found in the adjacent Polish Basin. This is why the south Germany straits ("Rheinische Senkungszone") are considered as the main path of brachiopod exchange between the Boreal and the Sub-Tethyan realms. The bottom of the straits between the Paris and London Basins was probably too shallow and too agitated by the high water energy for rendering attachment opportunities for brachiopod larvae. The latter straits gradually lost their function as physiogeographic barriers in the Early Hauterivian (MIDDLEMISS, 1973). In contrast, the Polish Gate closed during the Late Hauterivian (MICHALIK and VASICEK, 1989).

The Sub-Mediterranean (Sub-Tethyan or simply Jura) realm was situated at the southern rim of the north European shelf. Its environment was characterized by well oxygenated water, by warm climate, by many substrates suitable for brachiopods and other benthic life, and by periodic oceanic influences. In this realm, the brachiopods formed a specific bioprovince, characterized by endemic brachiopod genera like *Sellithyris, Loriolithyris, Boubeithyris, Musculina, Glosseudesia, Lamellaerhynchia, Sulcirhynchia, Plicatirostrum* and others. Many brachiopods (*Lacunosella, Monticlarella, Belbekella, Plicatirostrum, Lamellaerhynchia, Belothyris*) came from middle Asia to Kopet Dagh, Caucasus, Crimea and spread into southern France and Morocco. Some even penetrated far north into Boreal basins (Fig. 3). (MIDDLEMISS, 1973; 1979; 1980; 1981; OWEN, 1973; SANDY, 1986; etc.).

No continuous barrier existed between the Subtethyan and the Tethyan realm. The Tethyan forms *(Monticlarella, Ismenia, Weberithyris, Peregrinella, Lacunosella)* were dominant in many places along the border of the North European shelf (DIENI et al., 1973). Despite their "conservative" distribution (MIDDLEMISS, 1984) they have spread between northern Caucasus, Crimea (YANIN and SMIRNOVA 1981), the Outer Carpathians (NEKVASILOVA, 1974–1978, cf. Fig. 4), the Pre-Alps of Switzerland and SE France (LE HEGARAT et al., 1973; SANDY, 1986), Sardinia (DIENI et al., 1973) and occurred even in eastern Greenland (AGER, 1963).

4. Mid-Cretaceous brachiopod distribution

All the typical "Jurassic" forms (*Pygope, Antinomia, Triangope, Pygites*, etc.) became extinct during the Lower Barremian. The distribution of the Tethyan and Sub-Tethyan forms diminished considerably. The Paris Basin was affected by a regression. *Moutonithyris* disappeared from Northern Europe, retreating deeply into the Tethyan realm. However, several species which developed during the Early Cretaceous migrated northwards and became dominant elements in the Late Barremian brachiopod assemblages (Fig. 5). This was the beginning of a new pulse of brachiopod migration.

Together with other "southern" forms (*Sulcirhynchia, Sellithyris*, etc.), *Moutonithyris* migrates during the Aptian through the German Basin as far as England. On the other hand, *Cyrtothyris* and other Boreal elements *(Platythyris, Oblongarcula, Rhombothyris)* spread into the Jura domain (Switzerland, SE France, Eastern Spain, SW Morocco) during the Late Albian. Other Tethyan elements penetrated through the Pyrenees and the opening Gulf of Biscaya into Spain. Sicily and Sardinia were inhabited by true Tethyan assemblages (according to MIDDLEMISS et al., 1973: *Sardope, Tropeothyris* etc., cf. Fig. 6). Extensive carbonate platforms of the "Urgonian" facies evolved on the Tethyan shelves, being characterized by rich assemblages of hermatypic corals, rudists, orbitolinids, bryozoans and algae. The lack of terrigeneous detritus is correlated with a lower percentage of rhynchonellids in the Tethyan brachiopod assemblages (Fig. 6). These facts indicate a hot and dry climate in the Tethyan Belt, contrasting with the more humid Boreal realm.

A rapid spreading of rhynchonellids during the Albian (*Cyclothyris, Orbirhynchia, Burmirhynchia,* etc., cf. LOBACHEVA, 1983; POPIEL-BARCZYK, 1977; MIDDLEMISS and Owen. 1980) and increased opportunism by *Ornatothyris, Musculina, Sellithyris;* (MIDDLEMISS, 1983; etc.) have been connected with a gradual substitution of carbonates by clastic sedimentation in wide areas, caused by important paleogeographical changes during the Austrian Phase of the Alpine Orogeny. This reorganization of environments led to substantial changes in brachiopod faunas. Virtually all the dominant Early Cretaceous types got extinct during the Cenomanian.

5. Early Upper Cretaceous brachiopod evolution

The end of the middle Cretaceous was characterized by a transgression affecting a broad belt of the European Craton. The rapid development of new life habitats led to a new brachiopod migration. However, the evolution and diversification of the brachiopod communities was limited by the predominantly terrigeneous character of the sedimentation. The rhynchonellids with a zig-zag commissure (*Burmirhynchia, Orbirhynchia, Cyclothyris, Grasirhynchia, Cretirhynchia* etc.), which were well adapted to silty and sandy substrates, became dominant elements of the brachiopod associations. Pelitic environments were inhabited by *Terebratulina, Kingena, Magas,* etc. (NEKVASIOLOVA, 1964 to 1974; POPIEL-BARCZYK, 1974; MIDDLEMISS and OWEN, 1980; etc.). Other Cenomanian associations consisting of *Concinnithyris, Gibbithyris, Capillithyris* and *Platythyris* (in marly facies), or of *Grasirhynchia, Orbirhynchia, Kingena,* etc. (glauconitic limestones) evolved in various environments of NW Europe (OWEN, 1978). Similar associations also inhabited Central and SE Europe. Several specific forms evolving on the Russian epicontinental shelf penetrated into adjacent areas (Don Basin etc., cf. KATZ, 1973) during the Turonian. The latter region was characterized by *Najdinothyris, Urbanirhynchia, Kaphirnigania, Orbignyrhynchia,* etc.

Subtethyan assemblages consisted mostly of calm-water forms like *Musculina, Sellithyris, Platythyris* (cf. DIENI et al., 1973; MIDDLEMISS, 1983; LOBACHEVA, 1983; etc.). This fauna is not diverse and shows a strong tendency towards opportunism (Fig. 6). A opportunistic Cenomanian association with *Trigonosemus* occurs in Turkmenia (VANCHUROV, 1966).

As Tethyan brachiopod assemblages of early upper Cretaceous age were extremely rare our knowledge about them is rather poor.

6. Late Cretaceous brachiopod development

The Santonian and Early Campanian was the time of a general retreat of brachiopods. The Cretaceous forms disappeared from all environments without being replaced by any other brachiopod taxa. This extinction event represents the most important crisis in the Mesozoic history of the phylum. The Campanian paleogeography (Fig. 7) differed from the middle Cretaceous one. Large areas were flooded by epicontinental seas and the source areas of clastic sediments diminished. On the other hand, surfaces in the Alpine folded belts which had fallen dry during the Austrian tectonic phase were the source areas for flysch sedimentations. The brachiopods *Neoliothyrina* and *Carneothyris* appeared along the coasts of the European continent in the Late Campanian and persisted until the earliest Palaeocene. They were accompanied by inarticulates *(Danocrania, Isocrania, Discinisca, Lingula)*, which occurred frequently on neritic clastic bottoms.

The association of the deeper neritic zone consisted of *Rhynchorina plena, Carneithyris suecica, C. uniplicata, C. circularis* (in the Don Basin, cf. KATZ, 1973), or of *Trigonosemus elegans, Terebratulina striata* and *Carneithyris carnea* (in Dobrogea; BARBULESCU et al., 1979). Some associations in Poland were dominated by *Terebratulina faujasi* and *Rugia tenuicostata* (BITNER and PISERA, 1979), or by *Carneithyris circularis, C. subpentagonalis, Chatwinothyris subcardinalis, Neoliothyrina obesa* and *Kingenella kongieli* (POPIEL-BARCZYK, 1968).

Several Maastrichtian associations of Denmark and their temporal variations were described in detail by SURLYK (1972). They were dominated by *Terebratulina chrysalis, T. faujasi, T. longicollis, Dalligas nobilis, Argyrotheca hirunda, A. bronii* and *Scumulus inopinatus* (Fig. 8). Upper Maastrichtian cancellothyrids *(Terebratulina, Gisilina, Rugia* and *Meonia)* were substituted by *Argyrotheca* which was dominant at the Maastrichtian-Danian boundary (JOHANSEN, 1987). This last author has described an abrupt extinction event at the Cretaceous/Tertiary boundary, which would be in a good agreement with the impact theory. However, she did not publish any data on the quantitative composition of successive stages of brachiopod assemblages.

Late Cretaceous brachiopods of the Kopet-Dagh in the Sub-Tethyan belt (VANCHUROV, 1967) were represented by craniids. From Danian deposits, VANCHUROV and KALUGIN (1966) have described some brachiopod associations strongly dominated by *Orbirhynchia rionensis* and *Gryphus orientalis.* A strong opportunism was also characteristic for other Sub-Tethyan areas.

Concerning the Tethyan Realm, MAKRIDIN & KATZ (1965) have drawn attention to an occurrence of *Trigonosemus, Basiliola* and craniids in the Caucasus area (without any literature citations). Moreover, they mentioned some craniids and rhynchonellids in the Balkan and Pyrenees. On the other hand, AGER

(1973) observed a "surprising lack of information" about brachiopod distribution in many highly fossiliferous Late Cretaceous sequences, especially in the Tethyan region. It seems that the Late Cretaceous crisis (certainly caused not only by the prevalence of fine terrigenous clastics and even flysch deposits in this area) had a more fatal effect on the Tethyan brachiopods.

7. Discussion and summary

The brachiopods represent a group suitable for solving a number of paleogeographical problems: their distribution depends on a narrow substrate tolerance, on small changes in marine currents, on little salinity fluctuations, and on the average water temperature. By studying their paleo-biogeographic distribution several faciological, paleoclimatological and paleoecological problems can be solved. However, such results cannot be achieved by investigating the occurrence of individual specimens. Even the distribution of species merely reflects their special (often irregularily patched) limitations. A more complex view can be obtained by the analysis of changes in assemblages. However, obtaining approximative knowledge of the original composition of fossil communities is not easy. A quantitative analysis of the associations seems to be an appropriate method not only for documenting the study material. Environmental changes are much better reflected by faunal successions than by the traditional "faunal list". Furthermore, these successions allow an estimation of how detailed and complete the knowledge of the individual association is.

In this context, our knowledge of the Cretaceous brachiopods is not adequate enough to give a definitive answer to many questions concerning their migration, the limits of their distribution, and the reasons for their decline. Many of these problems could perhaps be answered by a quantitative analysis of associations, based on sufficiently complete field data.

The mechanisms of the slowly proceeding substitution of "Jurassic forms" in Early Cretaceous brachiopod associations by more progressively evolving groups need to be explained. An interesting but still neglected problem is the relation between the development of the Early Cretaceous carbonate platforms and basins and the development of neritic brachiopod faunas. We also need an interpretation of the gradual retreat of brachiopods from the Tethyan domain during the Late Cretaceous. Tethyan faunas of this age are very inadequately known. Moreover, we know only a few, partly documented examples of brachiopod migrations at that time. Another problem is the insufficient knowledge of the cause, history and meaning of the Santonian crisis in brachiopod development. By recognizing all details of this process in fossiliferous, well sampled sections new light on the problems of animal extinction could emerge.

References

AGER, D. V. (1967): Some Mesozoic brachiopods in the Tethys Region. In: ADAMS, C. G. – AGER, D. V. (eds.): Aspects of Tethyan Biogeography. – Syst. Assoc. Publ., **7:** 135–151.

AGER, D. V. (1971): Space and Time in Brachiopod history. – In: MIDDLEMISS, F. A. (ed.): Faunal Provinces in Space and Time. – Geol. Journ. spec. iss., **4:** 95–110.

AGER, D. V. (1973): Mesozoic Brachiopoda. – In: HALLAM, A. (ed.): Atlas of Palaeobiogeography, pp. 431–436.

AGER, D. V. (1977): Mesozoic brachiopod migration and the opening of the North Atlantic. – Palaeog. Palaeocl., Palaeoecol., **21/2:** 85–100.

BĂRBULESCU, A., NEAGU, T., LĂZĂROIU, I., VODISLAV, C. (1975): Early Cretaceous brachiopods from Southern Dobrogea (in Roum.). – Cerc. Geol. Geof. Geogr., Geol., **20/1:** 111–141.

BĂRBULESCU, A., NEAGU, T., PICEREA, I., CIUPU, F. (1979): Contributions á l'étude des brachiopodes cretacés de la Dobrogea Méridionale. – Rev. Roum. Géol. Géoph., Géogr., Géol., **23/2:** 197–208.

BARCZYK, W. (1971): Upper Tithonian brachiopods Camerothyris wahlenbergi (Zejszner) from the Pieniny Klippen Belt. – Acta geol. Polon., **21/3:** 425–431.

BARCZYK, W. (1972): On the representatives of the genus Nucleata Quenstedt (Brachiopoda) from the Tithonian and Berriasian of the environs of Czorsztyn, Poland. – Prace Mus. Ziemi, **20:** 151–159.

BARCZYK, W. (1972): Some representatives of the family Pygopidae (Brachiopoda) from the Tithonian of the Pieniny Klippen Belt. – Acta geol. Polon., **22/3:** 507–513.

BARRON, E. J., HARRISON, C. G. A., SLOAN, J. L. II., HAY, W. W. (1981): Palaeogeography, 180 million years ago to the present. – Ecl. geol. Helv., **74/2:** 443–470.

BITNER, M. A., PISERA, A. (1979): Brachiopods from the Upper Cretaceous chalk of Mielnik (Eastern Poland). – Acta geol. Polon., **29/1:** 67–88.

DIENI, I., MIDDLEMISS, F. A., OWEN, E. F. (1973): The Lower Cretaceous brachiopods of East-Central Sardinia. – Boll. Soc. Paleont. Ital., **12/2:** 52.

DIENI, I., MIDDLEMISS, F. A. (1981): Pygopid brachiopods from the Venetian Alps. – Boll. Soc. Palaeont. Ital., **20/1:** 19–48.

JOHANSEN, M. B. (1987): Brachiopods from the Maastrichtian-Danian boundary sequence at Nye Klov, Jylland, Denmark. – Fossils and Strata, **20:** 1–99.

KAC, J. I. (1973): Brachiopody. – In: KRYMGOLTZ, G. J. (ed.): Atlas of the Cretaceous fauna of Donbas Area. (in Russ.). – Nedra, pp. 240–275.

LOBACHEVA, S. V. (1983): Lower Cretaceous brachiopods of SW foot of the Gissar Mts. Range – Biull. Mosc. obch. isp. prir., otd. geol., **58/5:** 100–110.

MAKRIDIN, V. P., KATZ, J. I. (1965): Importance of comprehensive palaeontological investigations for stratigraphy and palaeogeography. – Paleont. zhurn., **4/3:** 3–15.

MICHALÍK, J. (1987): Remarks to ecological relations of the Early Cretaceous organisms from pelagic deposits of the Krížna Nappe in Western Carpathians (in Slovak). – Spec. vol. from the 3rd work paleoecological seminary in Brno Univ. 1987, pp. 61–80.

MICHALÍK, J. (1987a): Development and structure of the Triassic and Liassic brachiopod communities. In: POKORNÝ, V. (ed.): Contribution of Czechoslovak paleontology to evolutionary science. Praha, Charles Univ., 39–53.

MICHALÍK, J. (1989): Development, changes and structure of Jurassic brachiopods. – In: Paleont. Conference of Slovakian Geological Society, Rižbašská Mílava 1988, Bratislava, 85–107.

Michalík, J., Borza, K., Snopková, P., Svobodová, M., Vašíček, Z. (1987): Contemporaneous trends in Lower Cretaceous stratigraphy and their application to the West Carpathian formations (in Slovak). – Geol. práce, Správy. **86:** 179–195.
Michalík, J., Kováč, M. (1982): On some problems of palinspastic reconstruction and Ceno-Mesozoic development of the Western Carpathians. – Geol. zborn. Geol. Carpath., **33/4:** 481–508.
Michalík, J., Vašíček, Z. (1989): Lower Cretaceous stratigraphy and paleogeography of the Czechoslovakian Western Carpathians. In: Wiedmann, J. (ed.): Cretaceous of the Western Tethys. Proc. 3rd Internat. Cret. Symp., Tübingen 1987. Schweitzerbart Verl. Stuttgart. 505–523.
Middlemiss, F. A. (1973): The geographical distribution of Lower Cretaceous Terebratulacea in Western Europe. – In: Casey, R., Rawson, P. F. (eds.): The Boreal Lower Cretaceous. Geol. Journ. spec. iss., **5:** 111–120.
Middlemiss, F. A. (1976): Lower Cretaceous Terebratulidina of Northern England and Germany and their geological background. – Geol. Jb., **A 30:** 21–104.
Middlemiss, F. A. (1979): Boreal and Tethyan brachiopods in the European Early and Middle Cretaceous. – Aspekte d. Kreide Europas. IUGS ser., **A 6:** 351–361.
Middlemiss, F. A., Owen, E. F. (1980): Condensed Cretaceous limestones in a section near Xiquena, Betic of Málaga, SE Spain. II. Albian-Cenomanian brachiopods. – Proc. Konink. Neder. Akad. Wetten., Geol., **B 83/2:** 201–208.
Middlemiss, F. A. (1980): Lower Cretaceous Terebratulida from SW Morocco and their biogeography. – Palaeontol., **23/3:** 515–556.
Middlemiss, F. A. (1981): Brachiopod events in the European Middle Cretaceous (Aptian-Cenomanian). – Cret. Res., **2:** 377–382.
Middlemiss, F. A. (1981): Lower Cretaceous Terebratulidae of the Jura Region, 1. Revision of some species described by Pictet and de Loriol (1872). – Ecl. geol. Helv., **74/3:** 701–733.
Middlemiss, F. A. (1983): Lower Cretaceous Terebratulidae of the Jura Region, 2. The Aptian fauna of La Presta and Noirvaux Dessus and the Upper Albian fauna of La Vraconne. – Ecl. geol. Helv., **76/3:** 669–687.
Middlemiss, F. A. (1984): Cretaceous terebratulid events in Western and Southern Europe and their relation to the stage boundaries. – Cret. Res., **5:** 345–348.
Middlemiss, F. A. (1984): Distribution of Lower Cretaceous brachiopods and its relation to climate. – In: Brenchley, P. (ed.): Fossils and Climate, pp. 165–170.
Middlemiss, F. A. (1984): Lower Cretaceous Terebratulidae of the Jura Region, 3. Terebratula salevensis de Loriol and the genera Praelongithyris and Tropeothyris. – Ecl. geol. Helv., **77/3:** 583–617.
Nekvasilová, O. (1964): Thecideidae (Brachiopoda) der böhmischen Kreide. – Sborn. geol. věd, **P 3:** 119–162.
Nekvasilová, O. (1969): New information on the occurrence of brachiopods at Štramberk (Stramberg), Moravia (Czechoslovakia, Upper Tithonian and Lower Cretaceous). – Věstn. Ústř. úst. geol., **44/4:** 261–263.
Nekvasilová, O. (1973): The brachiopod genus Bohemirhynchia gen. n. and Cyclothyris McCoy (Rhynchonellidae) from the Upper Cretaceous of Bohemia. – Sborn. geol. věd, **P 15:** 75–117.
Nekvasilová, O. (1974): Genus Cretirhynchia and Orbirhynchia (Brachiopoda) from the Upper Cretaceous of NW Bohemia. – Sborn. geol. věd, Paleont., **16:** 35–67.
Nekvasilová, O. (1977): Rhynchonellida (Brachiopoda) from the Lower Cretaceous of Štramberk (Czechoslovakia). – Paleontol., **19:** 45–76.

NEKVASILOVÁ, O. (1978): Cancellothyridacea (Brachiopoda) from the Lower Cretaceous of Štramberk (Czechoslovakia). – Čas. min. geol., **23/2:** 171–182.
OWEN, E. F. (1973): The distribution of Lower Cretaceous (Berriasian-Barremian) rhynchonellid and terebratelloid brachiopods in the northern hemisphaere. – In: CASEY and RAWSON (eds.): The Boreal Lower Cretaceous. – Geol. Journ. spec. issue, **5:** 121–130.
OWEN, E. F. (1978): The distribution of brachiopods within the Cenomanian of Northwest and Central Europe. – Géol. Méditerr., **5/1:** 147–154.
OWEN, H. G. (1983): Atlas of continental displacement, 200 million years to the Present. – Cambr. Earth Sci. ser., 159 p.
POPIEL-BARCZYK, E. (1968): Upper Cretaceous terebratulids (Brachiopoda) from the Middle Vistula Gorge. – Prace Muz. Ziemi, **12:** 3–86.
POPIEL-BARCZYK, E. (1972): Albian-Cenomanian brachiopods from the environs of Annopol on the Vistula with some remarks on related species from Cracow Region. – Prace Muz. Ziemi, **20:** 119–149.
POPIEL-BARCZYK, E. (1977): A further study of Albian-Cenomanian brachiopods from the environs of Annopol on the Vistula with some remarks on related species from the Cracow Region, Poland. – Prace Muz. Ziemi, **26:** 25–54.
SANDY, M. R. (1986): Lower Cretaceous brachiopods from Provence, France and their biostratigraphical distribution. – Bull. Brit. Mus. Nat. Hist., Geol., **40/4:** 177–196.
SMIRNOVA, T. N. (1975): Revision of the Upper Tithonian brachiopods from Inwald. – Acta geol. Polon., **25/1:** 115–139.
SURLYK, F. (1972): Morphological adaptations and population structure of the Danish chalk brachiopods (Maastrichtian, Upper Cretaceous). – Biol. Skr. Dan. Vid. Selsk., **19/2:** 1–57.
VANCHUROV, I. A., KALUGIN, V. P. (1966): Stratigraphical importance of the Danian-Montian brachiopods (in Russ.). – Izv. AN Turkm. SSR, ser. fiz.-tech. chim., geol. nauk, 1966/4: 114–122.
VOGEL, K. (1966): Eine funktionsmorphologische Studie an der Brachiopodengattung Pygope (Malm bis Unterkreide). – N. Jb. Geol. Paläont., Abh., **125:** 423–442.
VÖRÖS, A. (1987): Pliensbachian brachiopod biogeography of the "Mediterranean microcontinent". – Acta geol. Hung., **30/1–2:** 59–80.
YANIN, B. T., SMIRNOVA, T. N. (1981): Stratigraphical distribution of the bivalve molluscs and brachiopods during Berriasian and Valanginian in Crimea in Russ. – Biull. Mosc. obch. isp. prir., otd. geol., **56/1:** 82–94.

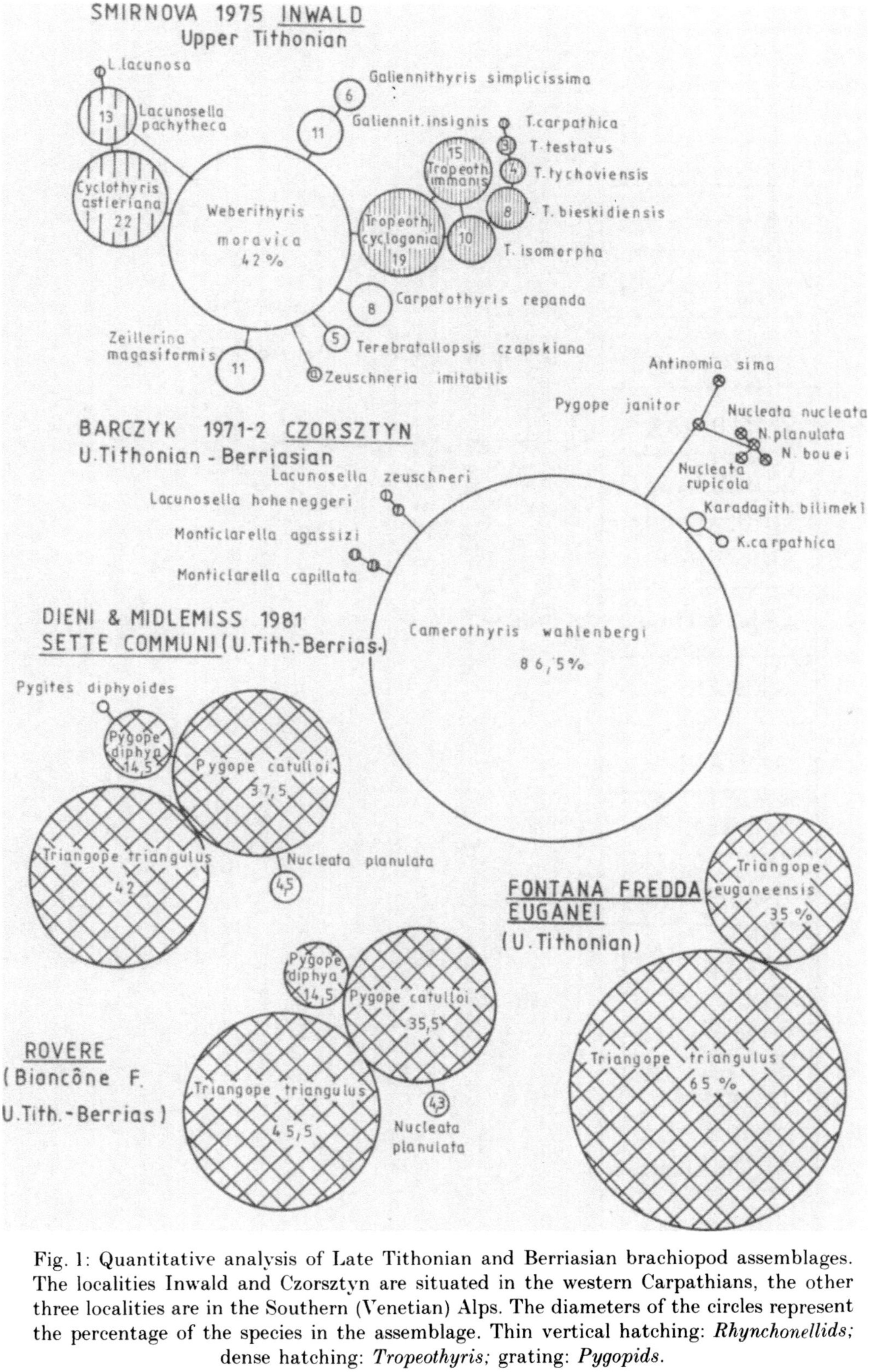

Fig. 1: Quantitative analysis of Late Tithonian and Berriasian brachiopod assemblages. The localities Inwald and Czorsztyn are situated in the western Carpathians, the other three localities are in the Southern (Venetian) Alps. The diameters of the circles represent the percentage of the species in the assemblage. Thin vertical hatching: *Rhynchonellids;* dense hatching: *Tropeothyris;* grating: *Pygopids.*

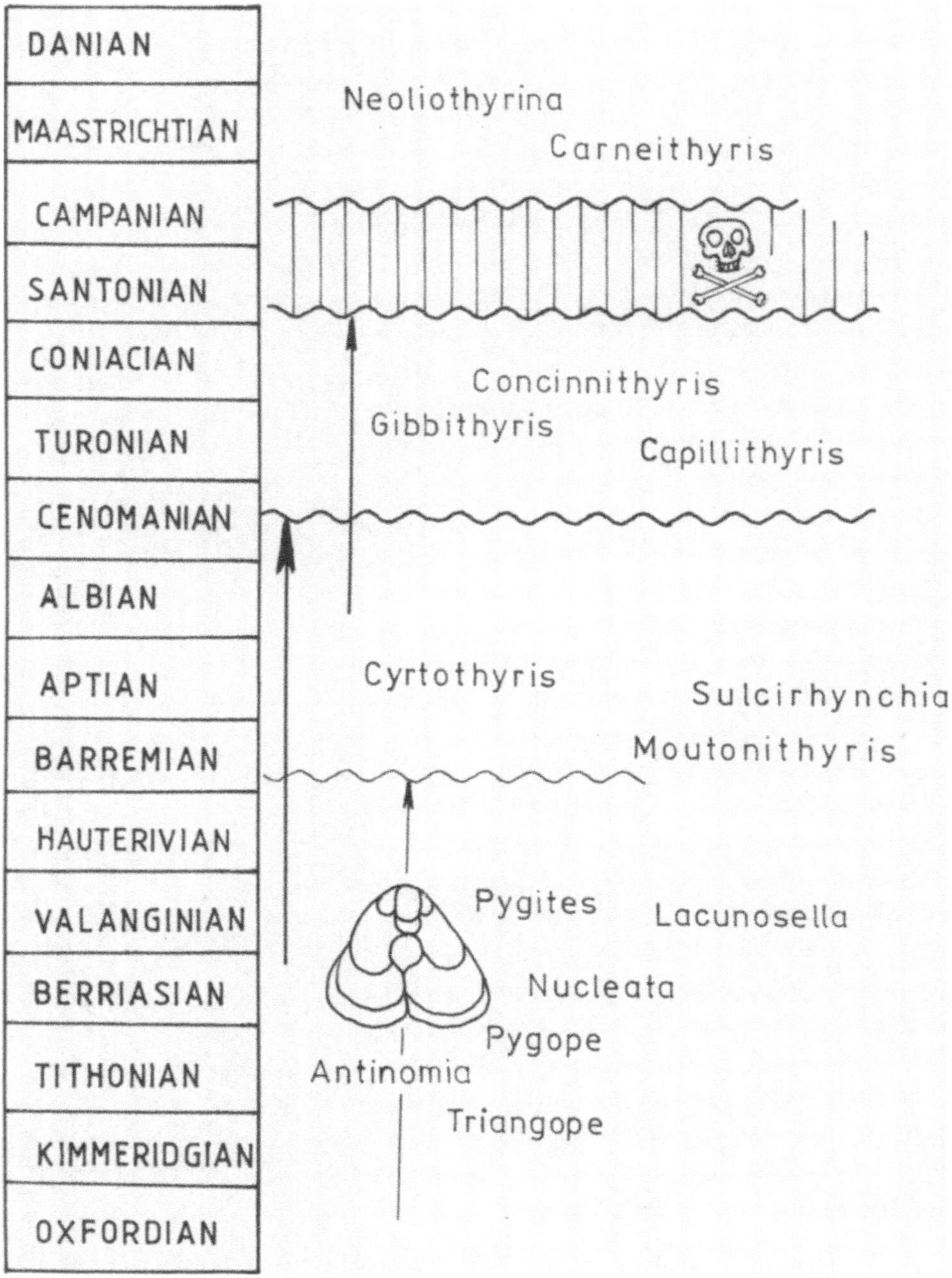

Fig. 2: The phases of Cretaceous brachiopod evolution (sensu MIDDLEMISS, 1984).

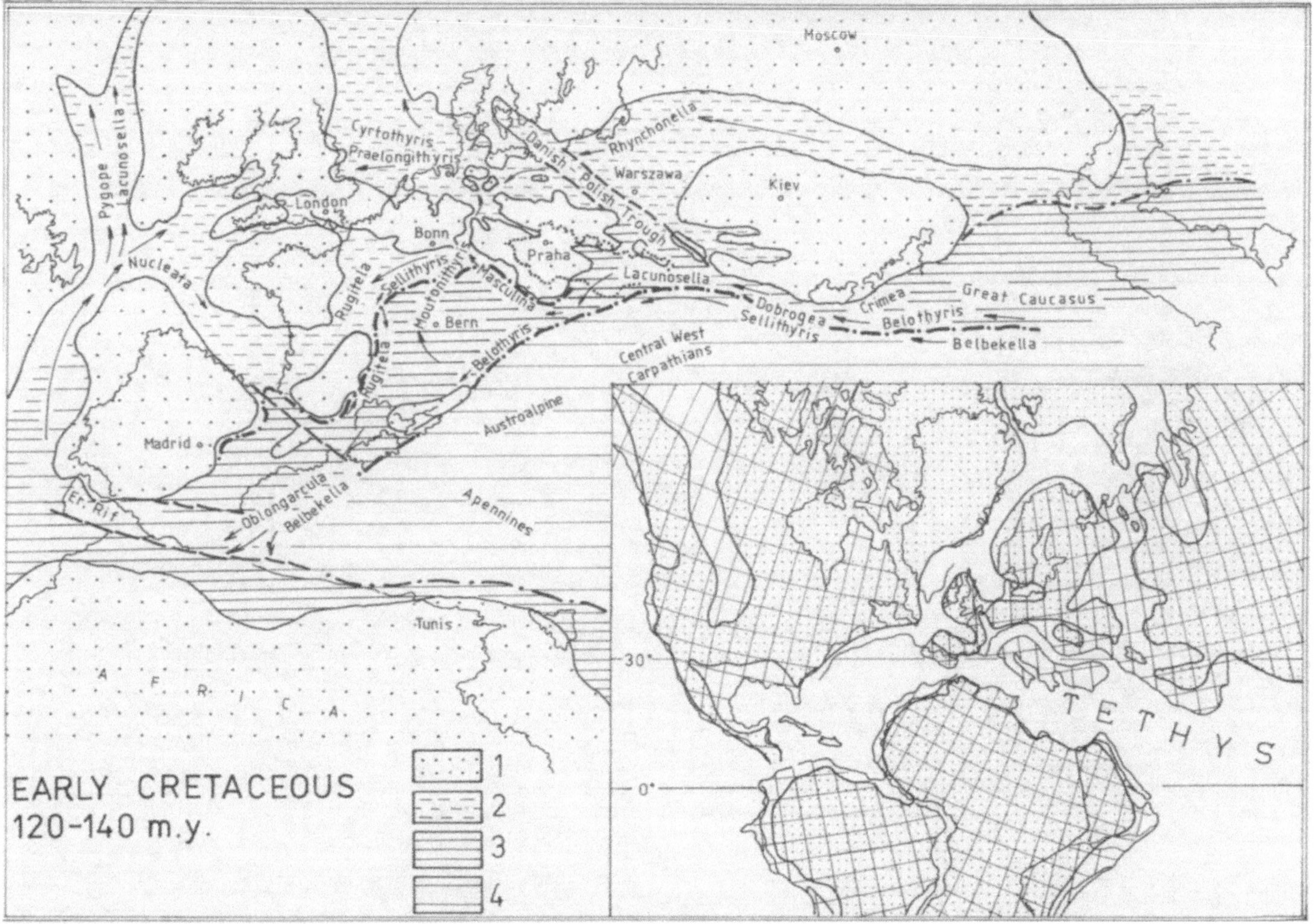

Fig. 3: Lower Cretaceous brachiopod provinces in Europe. Paleogeography after BARRON et al. 1981, OWEN, 1983; MAKRIDIN & KATZ, 1965. 1. Continental areas; 2. Boreal Realm; 3. Jura Subprovince; 4. Tethyan Realm. Arrows indicate supposed migration routes.

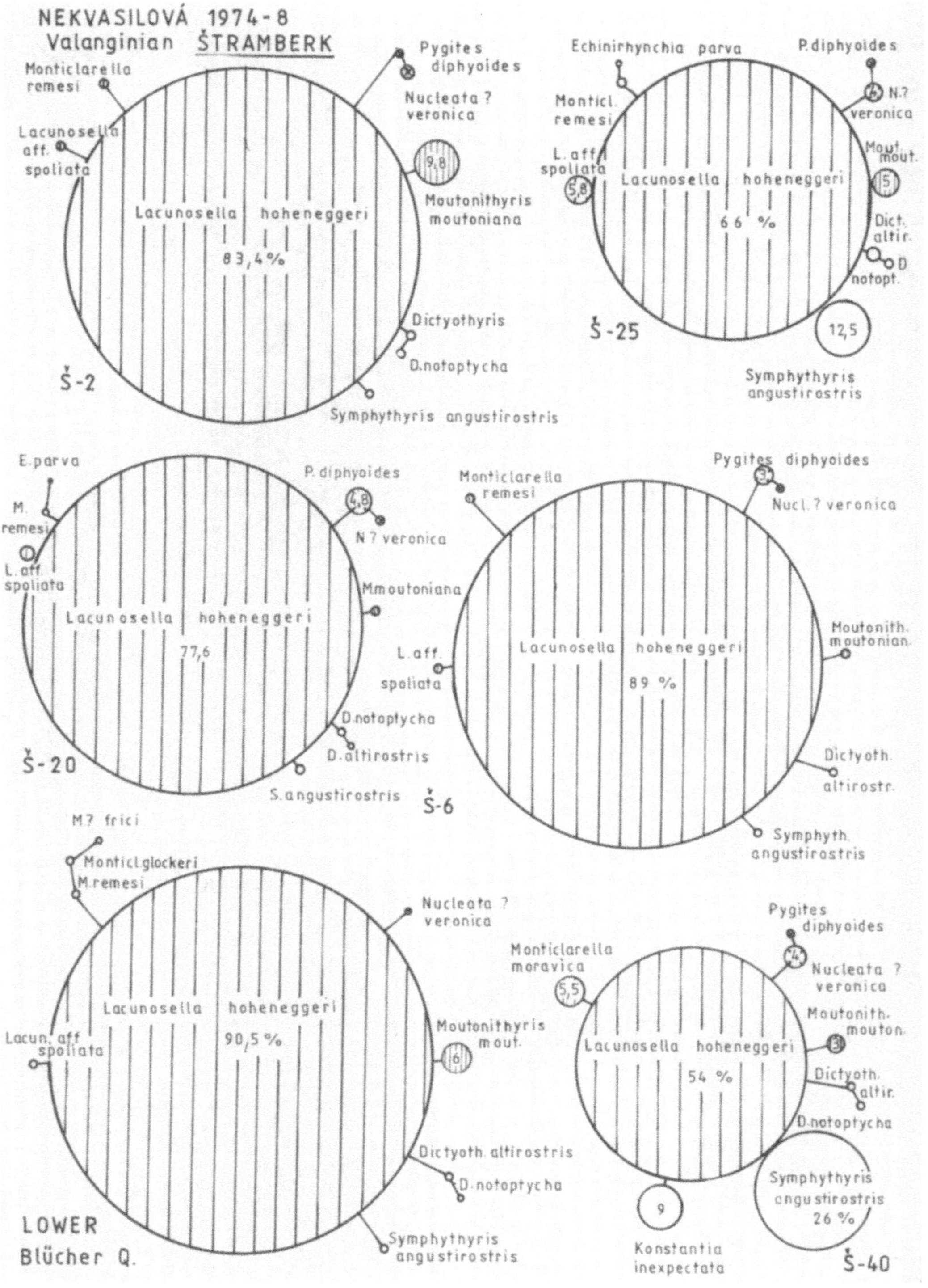

Fig. 4: Quantitative analysis of Valanginian brachiopod associations of the locality Štramberk (Outer Carpathians). Dense hatching: *Moutonithyris;* other symbols as in figure 1.

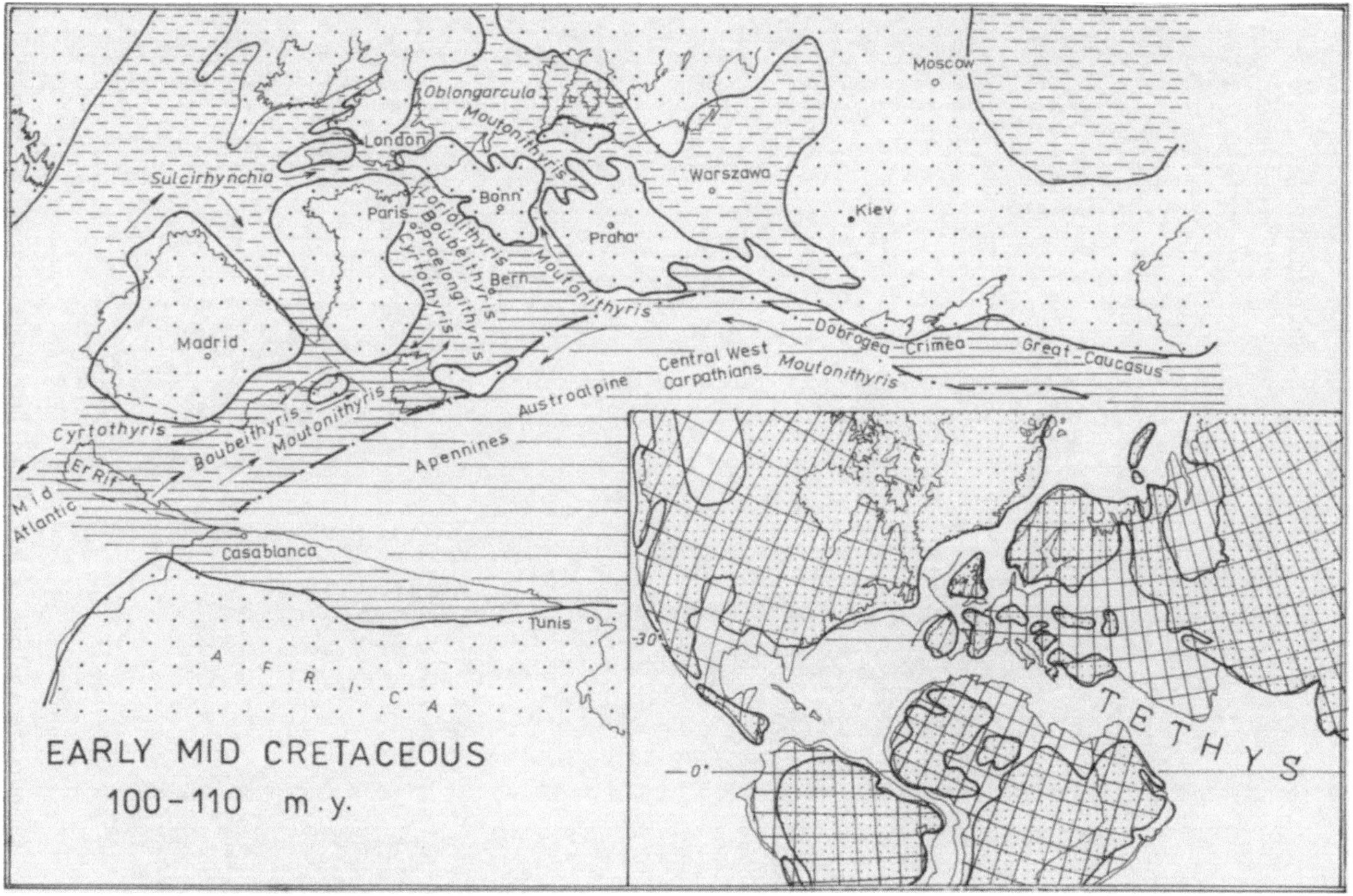

Fig 5. Distribution of selected middle Cretaceous brachiopod genera in Europe. Paleogeographic base as in figure 3.

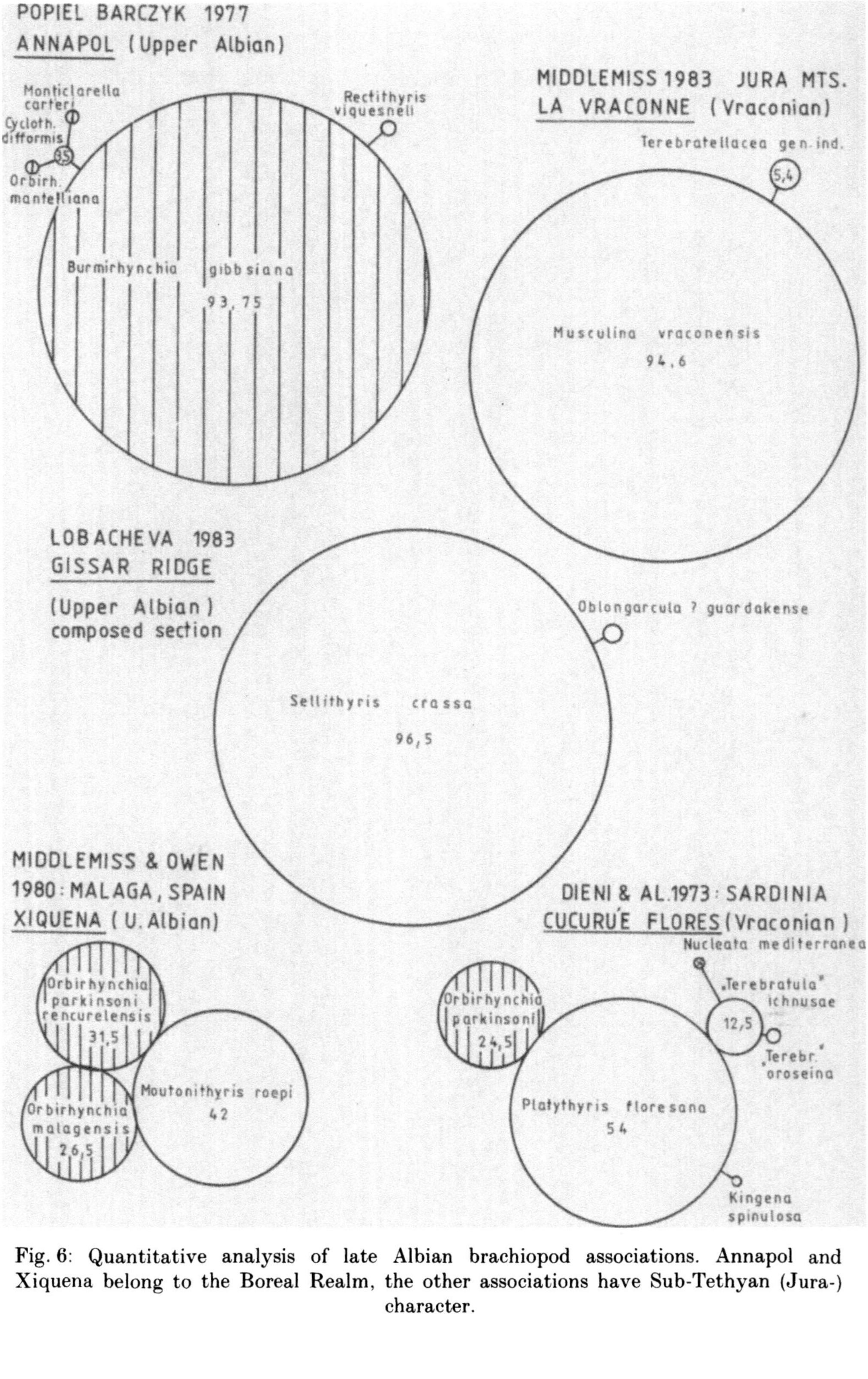

Fig. 6: Quantitative analysis of late Albian brachiopod associations. Annapol and Xiquena belong to the Boreal Realm, the other associations have Sub-Tethyan (Jura-) character.

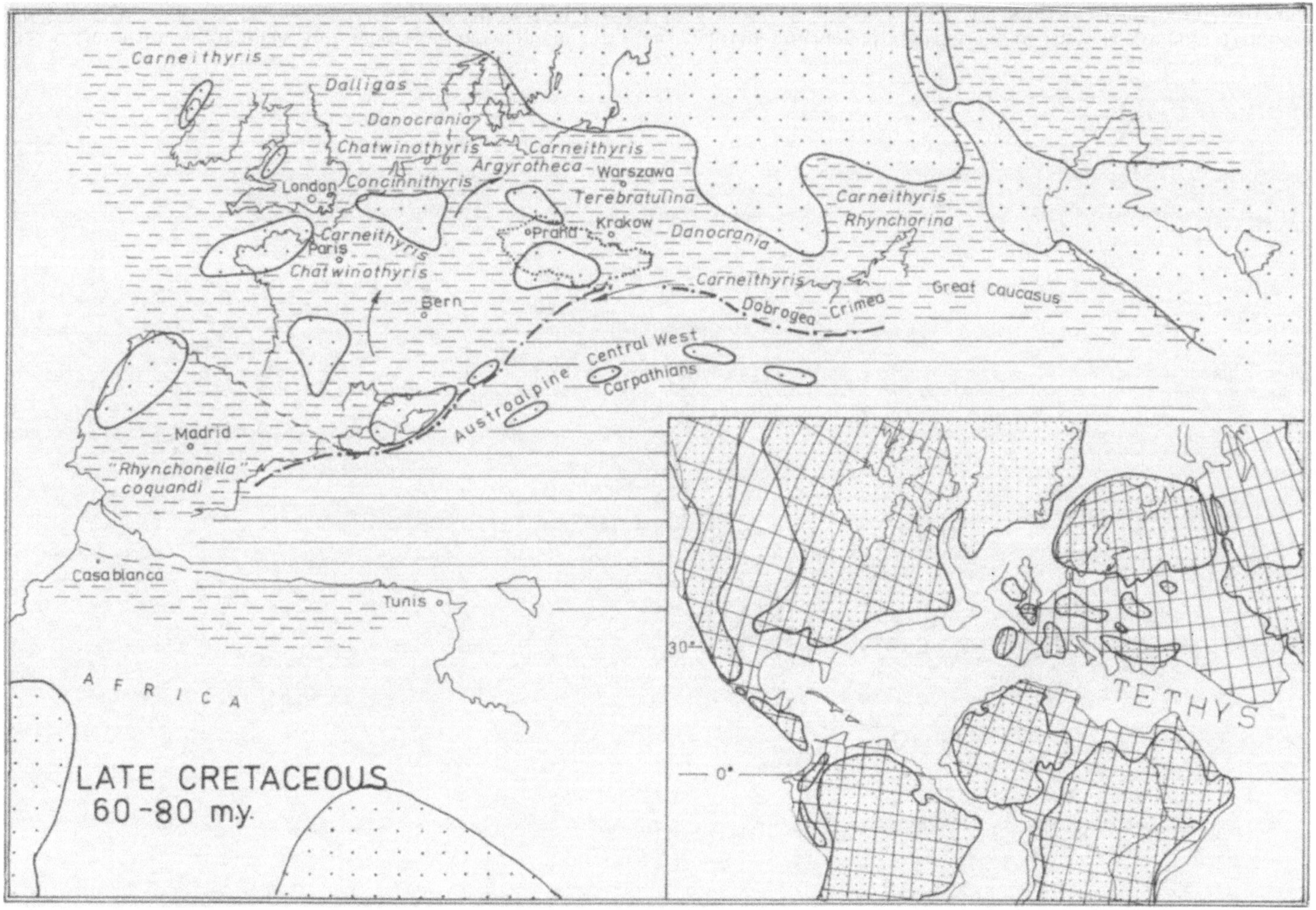

Fig. 7: The distribution of selected Late Cretaceous brachiopod genera in Europe. Paleogeography and symbols as in figure 3.

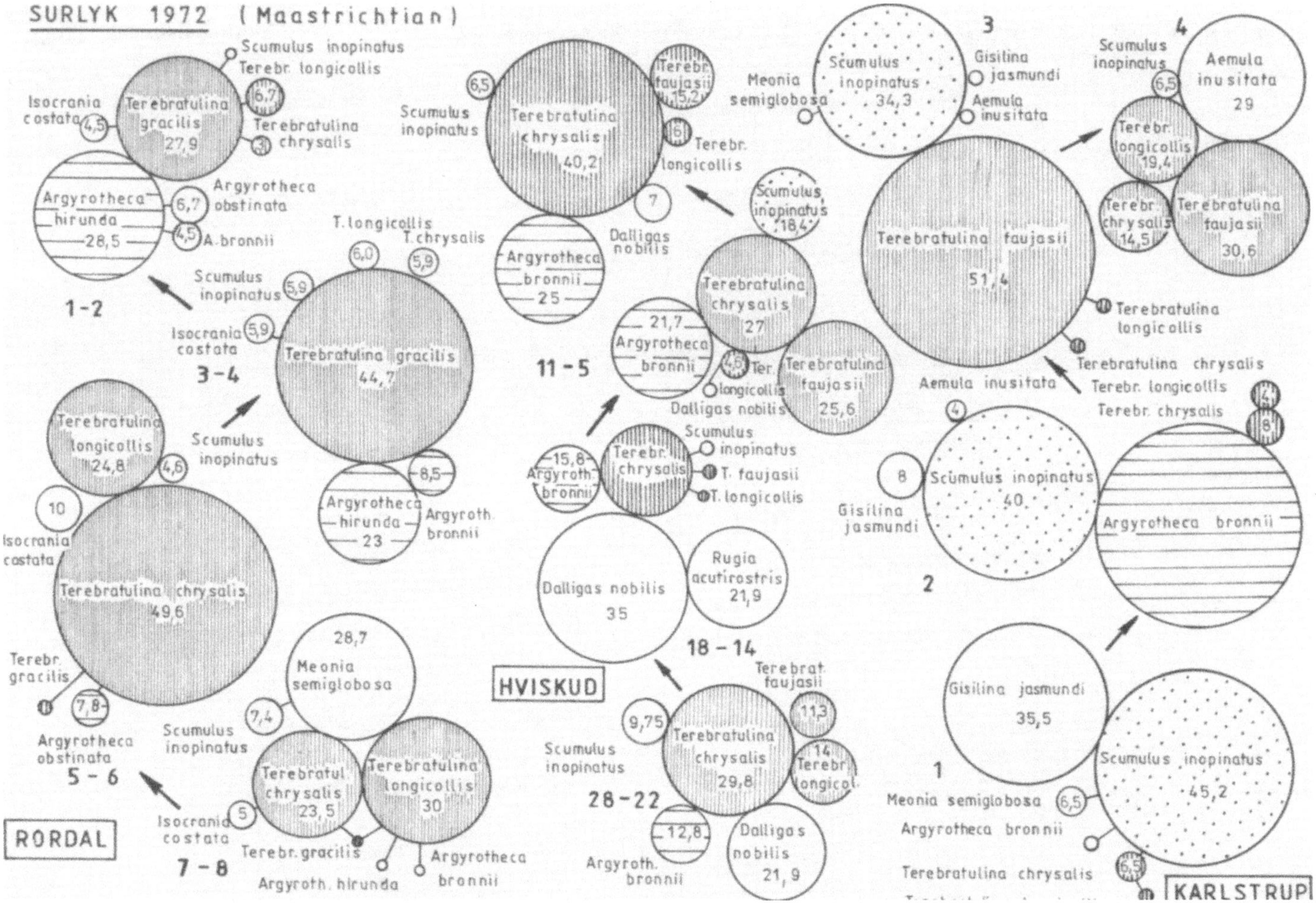

Fig. 8: Quantitative analysis and successive development of Maastrichtian brachiopod associations of three sections in Denmark (after SURLYK, 1972). The numbers refer to the original numbering of the fosiliferous levels by SURLYK. Horizontal hatching: *Argyrotheca;* vertical hatching: *Terebratulina:* dots: *Scumulus.*

New Aspects on Tethyan Cretaceous Fossil Assemblages.
Band 9 Schriftenreihe der Erdwissenschaftlichen Kommissionen der Österreichischen Akademie der Wissenschaften. 75–94.

2.4. Palaeogeographic distribution of Cretaceous Tethyan non-rudist bivalves

By Annie V. Dhondt*)

Abstract

In the non-rudist bivalve faunas of Valanginian-Hauterivian age two distinct realms can be distinguished: Boreal and warm-Temperate Tethyan. From the Barremian stage onwards with the spreading of the carbonate platform Tethyan faunas become more marked, with a rudist and a non-rudist facies. In the Albian a strong northern influence is visible in the northern Tethys. The successive Cenomanian transgressions result in a diversification of faunas; in the Tethys this is best seen in the oyster-facies. The Turonian-Santonian interval was relatively brief, and the faunas appear as being more or less homogeneous during that period, but with a differentiation between northern and southern sides of the Tethys. Campanian Tethys faunas are varied, but only occur on the southern side of the Tethys, and are mainly seen in the oyster-facies. In the Upper Campanian-Maastrichtian specific widespread Tethys faunas no longer exist: the Atlantic influence is more important than the Tethys influence.

Résumé

Parmi les faunes de bivalves d'âge Valanginien-Hauterivien on distingue deux unités paléobiogéographiques: le domaine boréal et le domaine tempéré (chaud) / téthysien. A partir du Barrémien la plateforme carbonatée s'étend et les faunes téthysiennes diffèrent d'après qu'elles sont oui ou non associées au faciès à rudistes. Pendant l'Albien une forte influence nordique est présente dans les faunes de la bordure nord de la Téthys. Les transgressions cénomaniennes successives sont à l'origine d'une diversification des faunes. Dans le

*) Department of Palaeontology, Koninklijk Belgisch Instituut voor Natuurwetenschappen, Vautierstraat 29, B-1040 Brussels, Belgium.

domaine de la Téthys ceci se remarque surtout dans les faciès à ostracées. L'intervalle Turonien-Santonien, relativement court, est caractérisé par des faunes assez homogènes, mais différentes sur les bordures nord et sud de la Téthys. Au Campanien de vraies faunes de bivalves téthysiens ne se retrouvent qu'en Afrique et en Asie (Proche Orient, et son extension vers l'Asie Centrale), et encore une fois surtout en faciès à ostracées. Au Campanien supérieur-Maastrichtien les faunes à large répartition téthysienne ont disparu: l'influence atlantique est devenue plus importante que celle de la Téthys.

Introduction

The ocean pattern changed drastically during the Cretaceous. At the beginning of the era the North Atlantic was barely open, the South Atlantic non existing and the major waterway between America and Eurasia a very wide Tethys. By the end of the Cretaceous most landmasses of the Northern Hemisphere occupied positions comparable to those of to-day, the North Atlantic was wide, the South Atlantic well established, and the Tethys much reduced and subdivided into smaller entities.

These changes in the oceans, and especially the transgressions which accompanied them and resulted in wide epicontinental seas, strongly influenced the paleobiogeographic distribution of the Cretaceous shallow marine faunas. This has been demonstrated beautifully by SOHL (1987) for the Cretaceous gastropods.

Aspects of the paleobiogeographic distributions of Bivalvia during the Cretaceous have been discussed by KAUFFMAN in HALLAM (1973); KAUFFMAN has taken the global distribution of all Bivalvia into consideration but has subdivided the Cretaceous stratigraphically only in Lower and Upper Cretaceous.

The most characteristic and best known representatives of Cretaceous bivalves are rudists and inoceramids, two groups which no longer occurred above the K/T boundary. Inoceramids were adapted to deeper seas: both cold water and warm water species have been recognised (KAUFFMAN, 1975).

Rudists are considered as indicating warm seas and they were present in large numbers in the Tethys. They lived in shallow environments. As discussed by SOHL (1987) the Cretaceous distribution of algal-coral, coral-rudist or rudist dominated organic framework buildings (SOHL, ibid., p. 1087, Fig. 1) shows the "... maximum extent of the Tethyan Realm during the Cretaceous" (SOHL, ibid., p. 1086).

As discussed during the meetings of the IGCP 262 the distribution of rudists could be used (and has been used by some authors) to delimit the distribution of Tethys in the Cretaceous. Yet, this distribution varies widely during the Cretaceous, reaching f. i. in the Late Maastrichtian, Maastricht (Limburg, The Netherlands). The term Tethys has been interpreted in many different ways and I would like to refer for details on this problem to NAIDIN (1986).

In the present paper distribution patterns and changes in geographic distribution of non-rudist bivalves throughout the Cretaceous shall be discussed.

Material and Methods

Bivalves were very numerous in the shallow Cretaceous seas. According to MOORE (1969) 80 families of marine bivalves occurred in the Cretaceous: of these 21 (8 of which are rudists), have not been found in strata younger than Cretaceous, 1 died out in the Miocene, and 58 still occur to-day. Thus the composition of the Cretaceous faunas was similar but not identical with those of to-day.

For paleogeographic purposes, the best data are given by those groups which (a) were widely distributed, (b) are easily recognizable and/or generally well-preserved. Among Cretaceous bivalves, these criteria are found in most groups with a calcitic shell (Isognomidae, Oxytomidae, Plicatulidae, Pectinidae, Spondylidae, Limidae, Ostreacea . . .) but also in groups with a thin shell and/or strong ornamentation (Inoceramidae, Lucinidae, Mactromyidae, Thyasiridae, Pholadomyidae, . . .). It excludes most of the trigoniids and many heterodonts.

Data have been assembled over many years concerning the distribution of bivalve species, based on type collections and general collections in many museums in Europe and North America. Emphasis has been on Pectinacea, Limacea and Ostreacea but other groups were also studied (list of collections as addendum).

Palaeobiogeographic Distributions of the species

At the beginning of the Cretaceous and up into the Hauterivian the faunal distribution resulted from a more or less stable ocean situation, similar to that known from the Upper Jurassic. Two distinct faunal biogeographic entities can be distinguished: a "Boreal realm" (CASEY & RAWSON, 1973) with bivalve faunas very similar to those from the Oxfordian-Volgian in the same regions, and a "warm temperate-Tethyan province" (KAUFFMAN in HALLAM, 1973) with old and new faunal elements. As an example of these different distributions I have chosen the typically Cretaceous pectinid genus *Neithea*, with the species *N. atava* (ROEMER) and *N. valangiensis* (PICTET & CAMPICHE) typical for the Valanginian-Hauterivian of the Tethys (s. l.), and the pectinid *Mclearnia* with the species *Mc. cinctus* (SOWERBY) and related species typical for the Boreal realm (Text-Figure 1) (KELLY, DHONDT & ZAKHAROV, 1984). Similar distinctive distributions could be shown using oysters [*Aetostreon* and *Rastellum* (DHONDT & DIENI, 1988) versus *Deltoideum*, or with the presence/absence of *Buchia* (ZAKHAROV, 1981)].

The Valanginian-Hauterivian interval in the Tethys is characterized by a very rich and diversified bivalve fauna which occurs over wide areas. This distribution of faunas was already recognised f. i. by LANGE (1914), and a similar distribution is accepted as a matter of course by ammonite workers today (KAKABADZE, 1981, 1983; KOTETISHVILI, 1970, 1983; THIEULOY, 1973, 1977; WIEDMANN, 1973). DHONDT & DIENI (1988, 1989) have discussed aspects of the bivalve distributions and shown that most species which are well-defined have a distribution ranging from South-America to Central Asia, and occasionally to Japan [Examples of such widely distributed taxa: *Grammatodon (Nanonavis) securis* (LEYMERIE), *Cucullaea (Noramya) gabrielis* LEYMERIE, *Pinna (Pinna) robinaldina* D'ORBIGNY, *Gervillaria alaeformis* (SOWERBY), *Gervillaria sowerbyana* (MATHERON), *Isognomon (Isognomon) ricordeanus* (D'ORBIGNY), *Mimachlamys robinaldina* (D'ORBIGNY), *Neithea (Neithea) atava* (ROEMER), *Aetostreon latissimum* (LAMARCK), *Ceratostreon boussingaulti* (D'ORBIGNY), *Trigonia carinata* (AGASSIZ), *Quadratrigonia nodosa* (SOWERBY), *Sphaera corrugata* (SOWERBY), *Ptychomya plana* AGASSIZ, *Panopea gurgitis* (BRONGNIART in CUVIER), *Pholadomya gigantea* (SOWERBY)]. These authors have shown that on the northern margin of the Tethys almost identical faunas can be found between the Paris Basin and the Caucasus, in strata of Valanginian-Hauterivian age, with about 80% of the species in common (Text-Figure 2).

From the Barremian onwards the spreading of the carbonate platforms and the presence of rudist build-ups is characteristic of the Tethys fauna. Among the non-rudist bivalves a differentiation can be made between those which lived in and around the rudist frameworks and those which lived in environments where rudists were rare. Rudist-associated bivalves in the Barremian-Aptian are f. i. the pectinid *Neithea deshayana* (MATHERON), the bakevelliid species *Turkmenia balkhanensis* KRIMHOLZ (Text-Figure 3).

The faunas which are not associated with rudists are very diverse (PELLAT & COSSMANN, 1907, COSSMANN, 1916) and contain new elements such as *Neithea (Neithea) syriaca* (CONRAD), *Pycnodonte* sp., *Gryphaeostrea* sp., etc.

Up to the Aptian the distribution of the Tethys fauna extends from the Caribbean to Japan via southern Europe and the Middle East (Text-Figure 4) [a few examples of such widely distributed taxa: *Grammatodon (Nanonavis) securis* (LEYMERIE), *Pinna (Pinna) robinaldina* D'ORBIGNY, *Gervillaria alaeformis* (SOWERBY), *Isognomon (Isognomon) ricordeanus* (D'ORBIGNY), *Neithea (Neithea) atava* (ROEMER), *N. (N.) syriaca* (CONRAD) etc. Occasionally extensions of the distribution of these species reach temperate regions, such as the Isle of Wight (UK).

The middle Cretaceous is stratigraphically well known, but because the type-localities of the stages are outside the Tethys some of the Tethyan faunas are not as well dated as most of the temperate faunas. Albian and Cenomanian were long stages, and transgressions and regressions, or at least changes in the sea level, occurred several times during that period, but the overall result was a progressive inundation of the land from the earliest Albian to the latest Cenomanian (HANCOCK & KAUFFMAN, 1979).

During the Albian the Northern Atlantic Ocean had become wider, and the sea level began to rise (HANCOCK & KAUFFMAN, 1979). In European Tethys faunas northern elements such as *Aucellina* are found (MORTER & WOOD,

1983) (Text-Figure 5). In Texas, Oklahoma and Mexico specific faunas are recognised which contain numerous pectinids [*Neithea* species and oysters (Böse, 1911; Kniker, 1918; Stanton, 1947; Scott, 1970)]. An unusual pattern in which endemic faunas alternate with more widely spread faunas, can be seen (Young, 1972) in the Texas basin.

A relatively widely distributed Tethyan fauna of Vraconian age has been recognised, but it also reached temperate parts of the Paris Basin, especially in localities with greensand facies of Albian and Cenomanian age [examples of species *Pseudoptera anomala* (Sowerby), *Acesta subovalis* (Sowerby), *Limaria ? elongata* (Sowerby) (and closely allied species) . . .] (Text-Figure 5).

In the Albian and Cenomanian as in the Barremian a few species are associated with rudist frameworks – *Neithea roemeri* (Hill) from the Upper Albian and lowermost Cenomanian of Texas, *Neithea fleuriausiana* (d'Orbigny) from the Albian-Cenomanian of France, Spain, Portugal (possibly both species are identical!), *Chondrodonta* spp. in Texas and Southern Europe (Text-Figure 5 & 6).

The Cenomanian stage in its type-locality and in the area around Le Mans is stratigraphically very well known: the many transgressions of the stage are precisely documented. Le Mans lies outside the Tethys, but some of the typical bivalves from the Upper Cenomanian faunas are found over wide areas, partly in the temperate regions and partly on the northern margins of the Tethys. Faunas seem to be different on both sides of the Tethys (s. l.) among oysters, pectinids and plicatulids f. i. the Upper Cenomanian is characterized by *Pycnodonte biauriculatum* (Lamarck), *Ceratostreon flabellatum* (Goldfuss), *Rhynchostreon suborbiculatum* (Lamarck), *Neithea aequicostata* (Lamarck), *N. alpina* (d'Orbigny) [= *N. texana* (Roemer)] on the northern side of the Tethys and in warm Temperate areas; *Costagyra olisiponensis* (Sharpe), *Ilymatogyra africana* (Lamarck), *Gyrostrea delettrei* (Coquand), *Actinostreon syphax* (Coquand), *'Curvostrea' rouvillei* (Coquand) (and related species), *Neithea hispanica* (d'Orbigny), *N. dutrugei* (Coquand), *N. coquandi* (Péron), *Plicatula auressensis* Coquand, on the southern side of the Tethys (Bobkova, 1961) (Text-Figure 6). It must be stressed that in Portugal and in Spain both southern and northern side Tethys species occur, and this is also true to a lesser degree for localities around Marseille: it might well be that the Tethys in the Late Cenomanian was relatively easily crossed by bivalve larvae between the Iberian Peninsula and Northwest Africa. Some of the same species occurring in Texas (and occasionally reaching into the Western Interior) have sometimes a slightly different stratigraphic age.

Kollmann (1978) gave explanations for the varying gastropod distributions during the Cenomanian of western Europe, stressing the Tethys influences in the Upper Cenomanian.

A differentiation between northern and southern sides of the Tethys is also documented in ostracods of the same age (Babinot, 1988).

The Cenomanian-Turonian transition is the time of strong regressions (Hancock & Kauffman, 1979) and, probably related to this, of faunal changes (Dhondt, 1981). Yet, in the Tethys deposits these changes are not so obvious as in the Temperate provinces, possibly:

a. because the stratigraphic boundary between the Cenomanian-Turonian stages was chosen in the Temperate Realm and the Tethyan equivalency was not easily found (result: the stratigraphic data indicated in collections for Tethyan bivalves are not always very precise);

b. the faunal break which is found between the Cenomanian and Turonian faunas in the Temperate province is probably not strongly present in the Tethys [f. i. many species continue into the Turonian and beyond in the Tethys (Dhondt, 1981), but this is relatively rare in the Temperate region].

As far as I know, specific Turonian Tethys non-rudist bivalves have not clearly been documented. Some of the uppermost Cenomanian Tethyan bivalves reach the Coniacian and are mainly Turonian – an interesting example is *Costagyra olisiponensis* (Sharpe) with a very wide Tethyan distribution (Reeside, 1929; Moroni & Ricco, 1968; Lefranc, 1983) (Text-Figure 7). More species might have similar distributions but few are as well documented.

From the same period a rudist-associated fauna is known containing *Neithea zitteli* (Pirona), and *Chondrodonta joannae* (Choffat) (Text-Figure 7).

In the Coniacian and Santonian the shallow oyster facies of North Africa continues to be well developed; a few examples of species occurring in it: *Pycnodonte flicki* Pervinquière, *P. (Costeina) costei* (Coquand), ? *Acutostrea acutirostris* (sensu Coquand, non Nilsson), *Rastellum dichotomum* (Bayle), *Plicatula flattersi* Coquand, *Pl. ferryi* Coquand, *Veniella* spp. etc.

In north eastern Spain and in southern France similar species are found, but in the western and northwestern European white chalk facies – a totally different environment – another fauna existed (Text-Figure 8). The northern margin Tethys is not completely documented, but several consecutive faunas of Coniacian and Santonian age are known from the 'Gosau' deposits in Austria (Summesberger, 1985): they contain Tethys elements (mainly species known from SE France) combined with shallow more northern Temperate elements (Dhondt, 1987).

During the Campanian the wide transgressions over Europe and Africa reached a maximum extension (Hancock & Kauffman, 1979). This influenced the faunas in as far that several somewhat deeper water faunas were present, which can clearly be seen in the faunas from near Bordeaux [northern Tethys margin- southern Temperate province (Dhondt, 1985)].

In the Tethys, oyster facies were present in North Africa and the Middle East. On an extensive material, with good stratigraphic control, from the Wadi Quena region (Egypt), Malchus (1990) has demonstrated that these faunas were changing quite rapidly, and that most species were short-lived. A different distribution in Upper Campanian and Maastrichtian Tethys strata is shown by *Merklinia perornata* (Cottreau): Cuba to Oman and Madagascar (Text-Figure 9). Similar distributions are found for certain inoceramids, even if they reached somewhat further (*Trochoceramus* spp., Text-Figure 9) (Dhondt, 1983; Etayo-Serna, 1985).

Among the fossils from the Upper Campanian strata on the northern side of the Tethys, specific Tethyan elements are difficult to recognize: these faunas seem to contain more elements of the warm Temperate faunas. The extension of the warm Temperate faunas went even further in the Maastrichtian and the spreading of faunas from the north of Europe towards the Near East is shown f. i. in the distribution of the pectinid *Microchlamys acuteplicata* (ALTH) (Text-Figure 9).

Thus, towards the K/T boundary the palaeogeographic distribution of the bivalves in the Tethys has lost its homogeneous aspect. With the opening of the Atlantic Ocean, the Tethys and its faunas have lost their continuity. This is clearly visible in the Upper Maastrichtian, and the terminal Cretaceous regressions resulted in accentuating this fact in the Paleocene.

Discussion

The distributions indicated for the Tethys bivalve faunas are – because of the nature of the material used – only approximative at the stratigraphic level. To improve on this it would be necessary to have new collections of all the consecutive faunas, and then study all systematic groups from these collections together (including stratigraphically relevant groups such as ammonites, inoceramids and Foraminifera). In recent years in the Tethys such work has been done for the Upper Cretaceous in Eastern Egypt (Wadi Quena) by a group from German universities organized by BANDEL, in NE Spain by the group working in Barcelona organized by PONS, in the Gosau in Austria by a group in Vienna organized by KOLLMANN and SUMMESBERGER; for the Lower Cretaceous such work is under way for deposits near the carbonate platform in SE France organized by A. and H. ARNAUD from Grenoble in France, and has been done for the E. Sardinian outcrops by DIENI from Padova, to name but a few. When all the data of these and other projects in similar direction shall be assembled, the precise reconstruction of the faunas at each exact moment shall be possible, for the complete Tethys. The stratigraphic control of the Cretaceous has improved so much in the last 20 years that a very good resolution (reaching often the level of an ammonite zone) could be reached for these paleobiogeographic reconstructions.

At present the data on Tethyan non-rudist bivalves suffer from occasional stratigraphic inaccuracies at the Barremian/Aptian, Cenomanian/Turonian, Campanian/Maastrichtian, Maastrichtian/Danian boundaries (as result of interpretations often dating back to the beginning of this century), and from the problem of the correlation/position of the Turonian. Thus, except for the Cenomanian, it is difficult to reach beyond the stage level.

Acknowledgments

Grants for research from the Nationaal Fonds voor Wetenschappelijk Onderzoek (Brussels), from the Ministerie van Onderwijs (Brussels), exchange grants from the Internationale Culturele Betrekkingen (Brussels), a postdoctoral fellowship at the Smithsonian Institution (Washington DC) have made possible the collecting of data used and are gratefully acknowledged. Through the IGCP 262, I had the opportunity to study specimens of the Coquand collection in Hungary.

Collections were made available and information was generously provided by many colleagues. Sincere gratitude is expressed to M. AQRABAWI (Hamburg and Amman), K. BANDEL (Hamburg), T. N. BOGDANOVA (Leningrad), R. J. CLEEVELY (London), W. A. COBBAN (Denver), F. COLLIER (Washington DC), G. CSASZAR (Budapest), I. DIENI (Padova), C. DURDEN (Austin), J. Cl. FISCHER (Paris), C. FORBES (Cambridge), R. FOUCHER (Marseille), S. FRENEIX (Paris), J. M. HANCOCK (London), I. HAYAMI (Tokyo), J. HELMS and H. JAEGER (Berlin), E. G. KAUFFMAN (Boulder, Co.), S. KELLY (Cambridge), E. KEMPER (Hannover), W. J. KENNEDY (Oxford), H. A. KOLLMANN (Vienna), E. KOTETISHVILI (Tbilisi), N. MALCHUS (Berlin), T. MATSUMOTO (Fukuoka), N. J. MORRIS (London), D. P. NAIDIN (Moscow), J. M. PONS (Barcelona), V. A. PROZOROVSKII (Leningrad), H. REMY (Bonn), R. B. RICKARDS (Cambridge), F. SCHMID (Hannover), N. F. SOHL (Washington DC), J. SORNAY (Tain l'Hermitage), H. SUMMESBERGER (Vienna), Y. TAKAYANAGI (Sendai), L. VIA BOADA (Barcelona), J. WIEDMANN (Tübingen), K. YOUNG (Austin).

To M. HAEMELINCK (Brussels) sincere thanks for drawing the figures.

References

BABINOT, J.-F. (1988): Premières données sur les Ostracodes du Cénomanien de Yougoslavie (Istrie du Sud). – Géobios, **21:** 5–15. Lyon.

BARRON, E. J., HARRISON, C. G. A., SLOAN, J. L. II, & HAY, W. W. (1981): Paleogeography, 180 million years ago to the present. – Eclog. geol. Helv., **74:** 443–470. Basel.

BOBKOVA, N. N. (1961): Late Cretaceous Oysters from the Tadjik Depression. – Trudy VSEGEI NS, **50:** 139 pp. Leningrad (in Russian).

BOESE, E. (1910–1911): Monografia geologica y paleontologica del Cerro de Muleros cerca de Ciudad Juàrez, Estado de Chihuahua y descripcion de la Fauna cretàcea de la Encantada, Placer de Guadalupe, Estado de Chihuahua. – Boletin Instituto geologico de Mexico, **25:** 193 pp. Mexico.

CASEY, R., & RAWSON, P. (Eds.) (1973): The Boreal Lower Cretaceous. – Geological Journal Special Issue, **5:** 448 pp. Liverpool.

COSSMANN, M. (1916): Le Barrémien supérieur à faciès urgonien de Brouzet-les-Alais (Gard). – Mém. Soc. géol. France, Paléontologie, **31:** 10–43. Paris.

DHONDT, A. V. (1981): Répartition des Bivalves (sans Inocérames ou Rudistes) dans le Crétacé Moyen. – Cretaceous Research, **2:** 307–318. London.

Dhondt, A. V. (1983): Campanian and Maastrichtian Inoceramids: A Review. – Zitteliana, **10:** 689–701. München.
Dhondt, A. V. (1985): Late Cretaceous Bivalves from the A 10 Exposures in Northern Aquitaine. – Cretaceous Research, **6:** 33–74. London.
Dhondt, A. V. (1987): Bivalves from the Hochmoos Formation (Gosau-Group, Oberösterreich, Austria). – Ann. Naturhist. Mus. Wien, **88 A:** 41–101. Vienna.
Dhondt, A. V., & Dieni, I. (1988): Early Cretaceous Bivalves of Eastern Sardinia. – Memorie die Scienze geologiche, **40:** 1–97. Padova.
Dhondt, A. V., & Dieni, I. (1989): The Sardinian Early Cretaceous Bivalves and their Paleobiogeographic Affinities. in Wiedmann, J. (Ed.) Cretaceous of the Western Tethys. – pp. 281–297. Stuttgart.
Etayo-Serna, F. (1985). Trochoceramus del Campaniano – Maastrichtiano en la Formacion de la Cordillera Occidental de Colombia. – Geologia Norandina, **9:** 27–30. Bogota.
Hancock, J M., & Kauffman, E. G. (1979) The great transgressions of the Late Cretaceous. – Jl geol. Soc. Lond., **136:** 175–196. London.
Kakabadze, M. V. (1981): Ancyloceratids from the South of the USSR and their stratigraphic importance. – Geol Inst. Akad. Nauk Gruzh. SSR, Trudy NS, **71:** 197 pp. Tbilisi (in Russian).
Kakabadze, M. V. (1983): On the Hauterivian-Barremian correlation between the South of the USSR and certain Southern and Northern regions of Europe. – Zitteliana, **10:** 501–508 München.
Kauffman, E. G. (1973): Cretaceous Bivalvia. in Hallam, A.: Atlas of Palaegeography 353–383. Amsterdam.
Kauffman, E. G. (1975) Dispersal and Biostratigraphic Potential of Cretaceous Benthonic Bivalvia in the Western Interior. – in Caldwell, W. G. E., Ed.: The Cretaceous System in the Western Interior of North America. Geol. Ass. Canada Special Paper, **13:** 163–194. Waterloo, Canada.
Kelly, S. R. A., Dhondt, A. V., Zakharov, V. A. (1984): *Boreionectes* Zakharov 1965 (Bivalvia: Pectinidae) – a synonym of *Maclearnia* (Crickmay 1930). – J. Paleontology, **58:** 109–114. Kansas.
Kniker, H. T. (1919): Comanchean and Cretaceous Pectinidae of Texas. – Univ. Texas Bull., **1817:** 5–57. Texas.
Kollmann, H. A. (1978): Les Gastropodes cénomaniens de France et des régions voisines. – Géologie Méditerranéenne, **5:** 101–108. Marseille.
Kotetishvili, E. V. (1970): Stratigraphy and fauna of the *Colchidites* and adjoining horizons of western Georgia. – Geol. Inst. Akad. Nauk Gruzh. SSR, Trudy NS, **25:** 118 pp. Tbilisi (in Russian).
Kotetishvili, E. V. (1983): Sur la paléozoogéographie des bassins éocrétacées du Caucase. – Zitteliana, **10:** 375–386. München.
Lange, E. (1914): Die Brachiopoden, Lamellibranchiaten und Anneliden der Trigonia Schwarzi-Schicht, nebst vergleichender Übersicht der Trigonien der gesamten Tendaguruschichten. – Archiv Biontol., **3** (4): 187–289. Berlin.
Lefranc, J.-Ph. (1983). Etude de l'Huître du Cénomanien mésogéen *Exogyra (Costagyra) olisiponensis* Sharpe. – C. R., **108** Cong. nat. Soc. Sav. Grenoble 1983 Section Sciences I, Sciences Terre II: 59–70. Paris.
Malchus, N. (1990): Revision der Kreide-Austern (Bivalvia: Pteriomorphia) Ägyptens (Biostratigraphie, Systematik). – Berliner Geowissenschaftliche Abhandlungen, Reihe A, **125:** 1–231. Berlin.

MOORE, R. C. (Ed.) (1969): Treatise on Invertebrate Paleontology, Part N1 and 2, Mollusca 6 Bivalvia. 952 pp. – Boulder, Colorado (Geological Society of America).

MORONI, M. A., RICCO, S. (1968): Nuovi Studi sulla Fauna Cenomanian di Brancaleone (Calabria). – Lavori Ist. Geol. Univ. Palermo, **6:** 65 pp. Palermo.

MORTER, A. A., & WOOD, C. J. (1983): The Biostratigraphy of Upper Albian-Lower Cenomanian *Aucellina* in Europe. – Zitteliana, **10:** 515–529. München.

NAIDIN, D. P. (1986): Tethys, term and understanding. – Vestnik Mosk. Univ. Ser. 4 Geologija, **1986, 6:** 3–18. Moscow (in Russian).

PELLAT, E., & COSSMANN, M. (1907): Le Barrémien supérieur à faciès urgonien de Brouzet-les-Alais (Gard). – Mém. Soc. géol. France, Paléontologie, **37:** 5–42. Paris.

PERVINQUIERE, L. (1912): Etudes de Paléontologie tunisienne. II. Gastropodes et Lamellibranches des Terrains crétacés. – Carte géologique de la Tunisie. 352 pp. Paris.

REESIDE, J. B. (1929): Exogyra olisiponensis Sharpe and Exogyra costata Say in the Cretaceous of the Western Interior. – US Geol. Surv. Prof. Pap., **154 I:** 267–278. Washington DC.

SCOTT, R. W. (1970): Paleoecology and paleontology of the Lower Cretaceous Kiowa Formation, Kansas. – Univ. Kansas, Paleont. Publ., **52:** 94 pp. Lawrence, Kansas.

SOHL, N. F. (1987): Cretaceous gastropods: contrasts between Tethys and the Temperate provinces. – J. Paleontology, **61:** 1085–1111. Lawrence, Kansas.

STANTON, T. W. (1947): Studies of some Comanche Pelecypods and Gastropods. – U.S. Geol. Surv. Prof. Pap., **211:** 256 pp. Washington DC

STENZEL, H. B., in MOORE, R. C. (Ed.) (1971): Treatise on Invertebrate Paleontology, Part N3, Mollusca 6 Bivalvia. 953–1224. – Boulder, Colorado (Geological Society of America).

SUMMESBERGER, H. (1985): Ammonite Zonation of the Gosau Group (Upper Cretaceous, Austria). – Ann. Naturh. Mus. Wien, **87 A:** 145–166. Vienna.

THIEULOY, J. P. (1973): The occurrence and distribution of boreal ammonites from the Neocomian of southeast France (Tethyan Province). – in CASEY, R. & RAWSON, P. F. (Eds.): The Boreal Lower Cretaceous: 289–302. Liverpool.

THIEULOY, J. P. (1977): La zone à *callidiscus* du Valanginien supérieur vocontien (Sud-Est de la France). Lithostratigraphie, ammonitofaune, limite Valanginien-Hauterivien, corrélations. – Géologie Alpine, **53:** 83–143. Grenoble.

WIEDMANN, J. (1973): Ancyloceratina (Ammonoidea) at the Jurassic/Cretaceous Boundary. – in HALLAM, A.: Atlas of Palaeogeography: 309–316. Amsterdam.

YOUNG, K. (1972): Cretaceous Paleogeography: Implications of Endemic Ammonite Faunas. – Bur. Econ. Geol. Geol. Circ., **72, 2:** 13 pp. Austin, Texas.

ZAKHAROV, V. A. (1981): Buchiidae and the biostratigraphy of the boreal Upper Jurassic and Neocomian. – Akad. Sci. SSSP, Sibirskoe Otd. Inst. Geol. Geophysics. 269 pp. Moscow (in Russian).

Addendum

List of collections studied:

Austin: Palaeontological Collections of the Texas Memorial Museum, The University of Texas at Austin.

Basel: Naturhistorisches Museum Basel.

Berlin: Naturkunde Museum, Humboldt Universität.

Bonn: Paläontologisches Institut der Universität.

Brussels: Koninklijk Belgisch Instituut voor Natuurwetenschappen.
Budapest: Magyar Allami Földtani Intézet
Cambridge: Sedgwick Museum, University of Cambridge.
Denver: Paleontology and Stratigraphy Branch, US Geological Survey Coll.
Fukuoka: Kyushu University, Geology Dept.
Geneva: Muséum d'Histoire naturelle.
Hannover: Niedersächsische Landesanstalt.
Lausanne: Musée géologique.
Leningrad: Leningrad State University, Geology.
All Union Geological Survey (VSEGEI).
Lisbon: Portuguese Geological Survey Coll.
London· British Museum (Natural History).
Marseille: Musée d'Histoire naturelle.
Moscow: Moscow State University, Geological Museum.
Palaeontological Institute of the USSR Academy of Sciences (PIN).
Paris: Muséum national d'Histoire naturelle. Institut de Paléontologie.
Sendai: Tohoku University, Dept. Geology and Paleontology.
Sofia: University Kliment Ohridski, Geology.
Tbilisi: Tbilisi State University.
Tokyo: The University of Tokyo, University Museum.
Vienna: Naturhistorisches Museum Wien.
Washington DC: Paleobiology, US National Museum of Natural History, Smithsonian Institution.

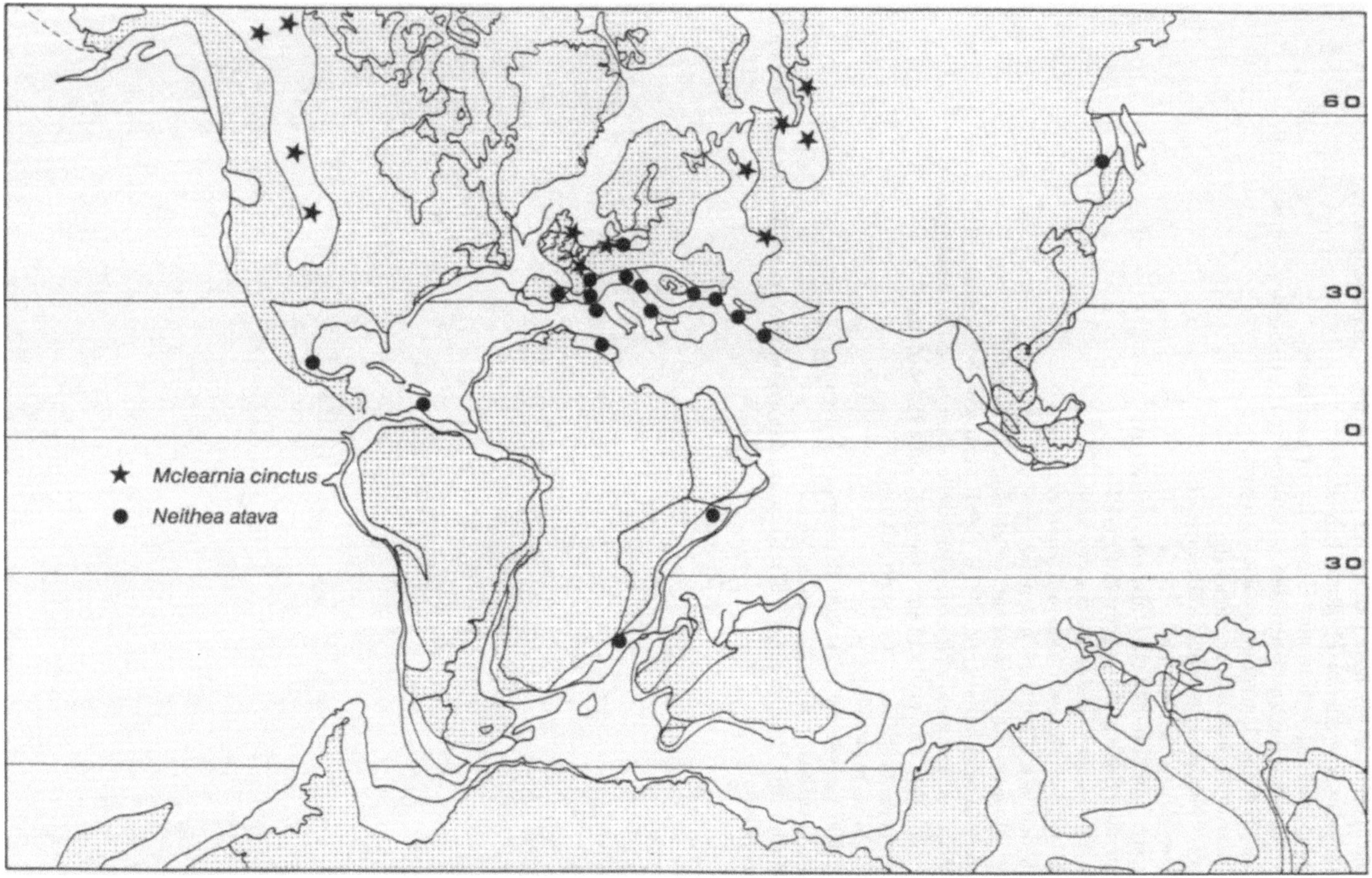

Fig. 1: Map at 120 million years (from BARRON et al., 1981) showing the distribution of *Neithea atava* (ROEMER) and *Mclearnia cinctus* (SOWERBY) in Valanginian-Hauterivian times. *N. atava* occurs in warm Temperate and in Tethys provinces, whereas *Mc. cinctus* is typical for Boreal (i. e. cold Temperate) provinces.

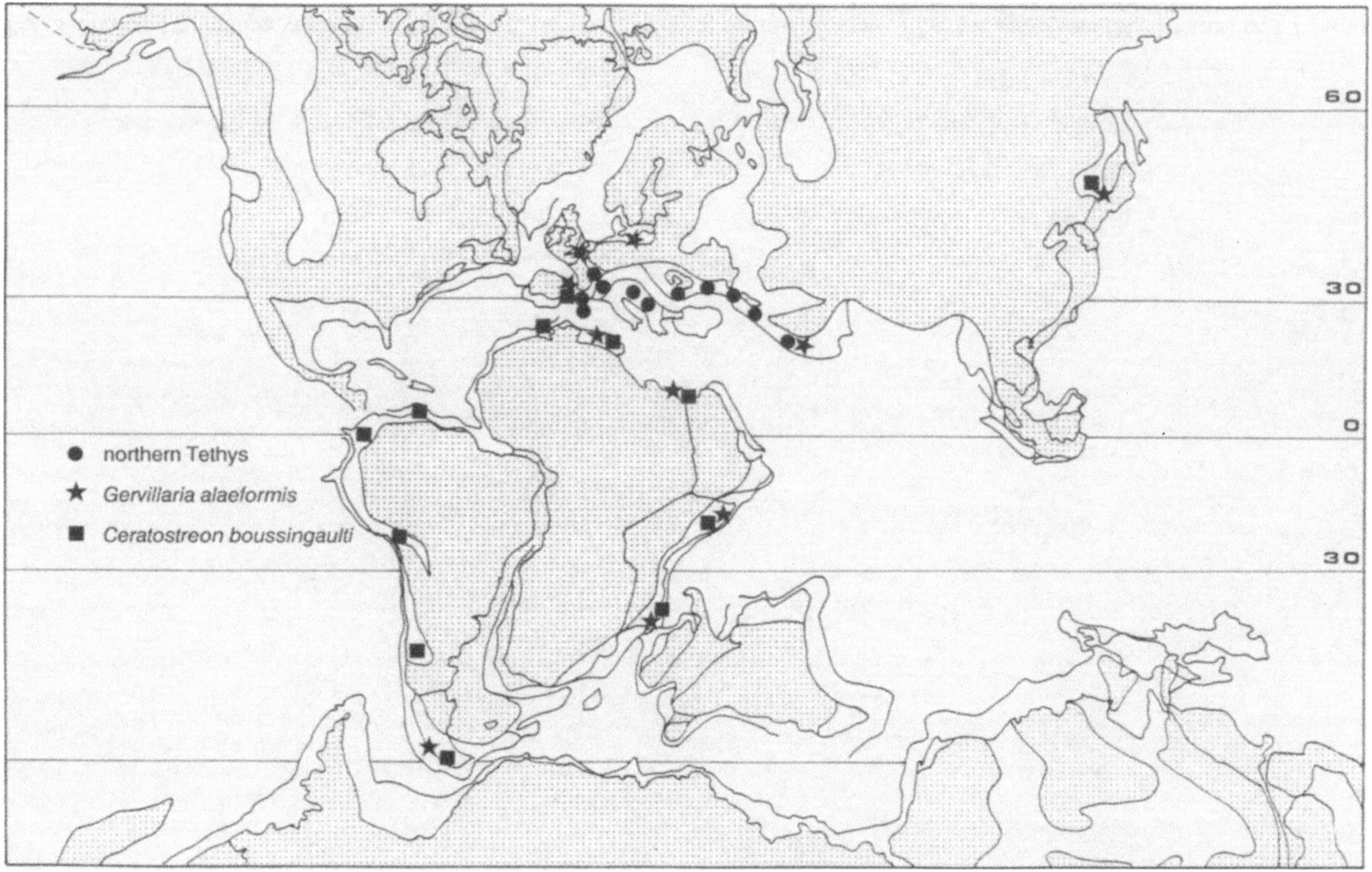

Fig. 2: Map at 120 million years (from Barron et al., 1981) showing the distribution of *Gervillaria alaeformis* (Sowerby) and *Ceratostreon boussingaulti* (d'Orbigny), and indicating the Valanginian-Hauterivian localities on the northern Tethys side which have very similar faunas (at least 80% of the bivalve species in common).

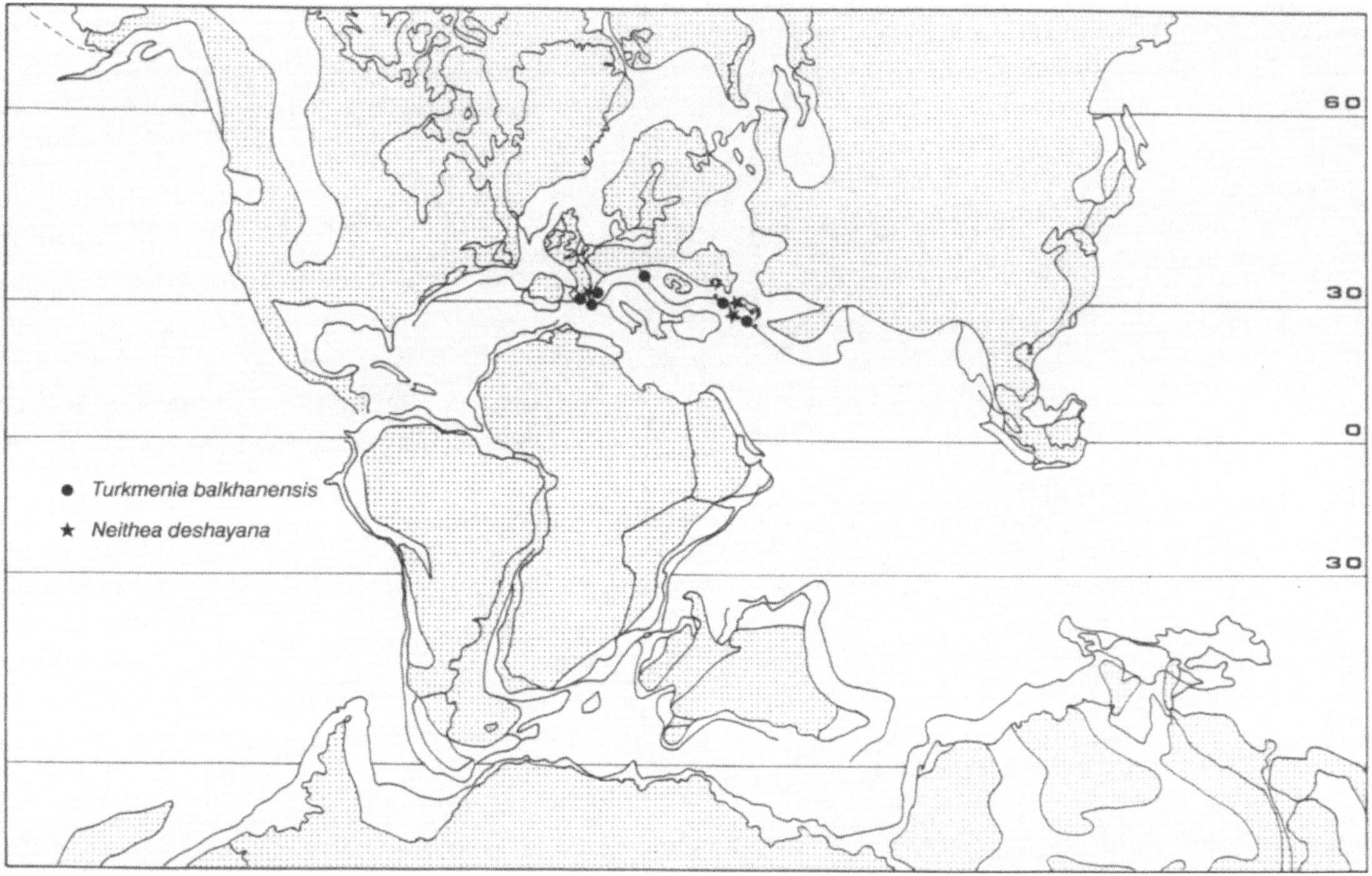

Fig. 3: Map at 120 million years (from BARRON et al., 1981) showing the distribution of *Neithea deshayana* (MATHERON) and *Turkmenia balkhanensis* KRIMHOLZ, species associated with rudist frameworks in Barremian times.

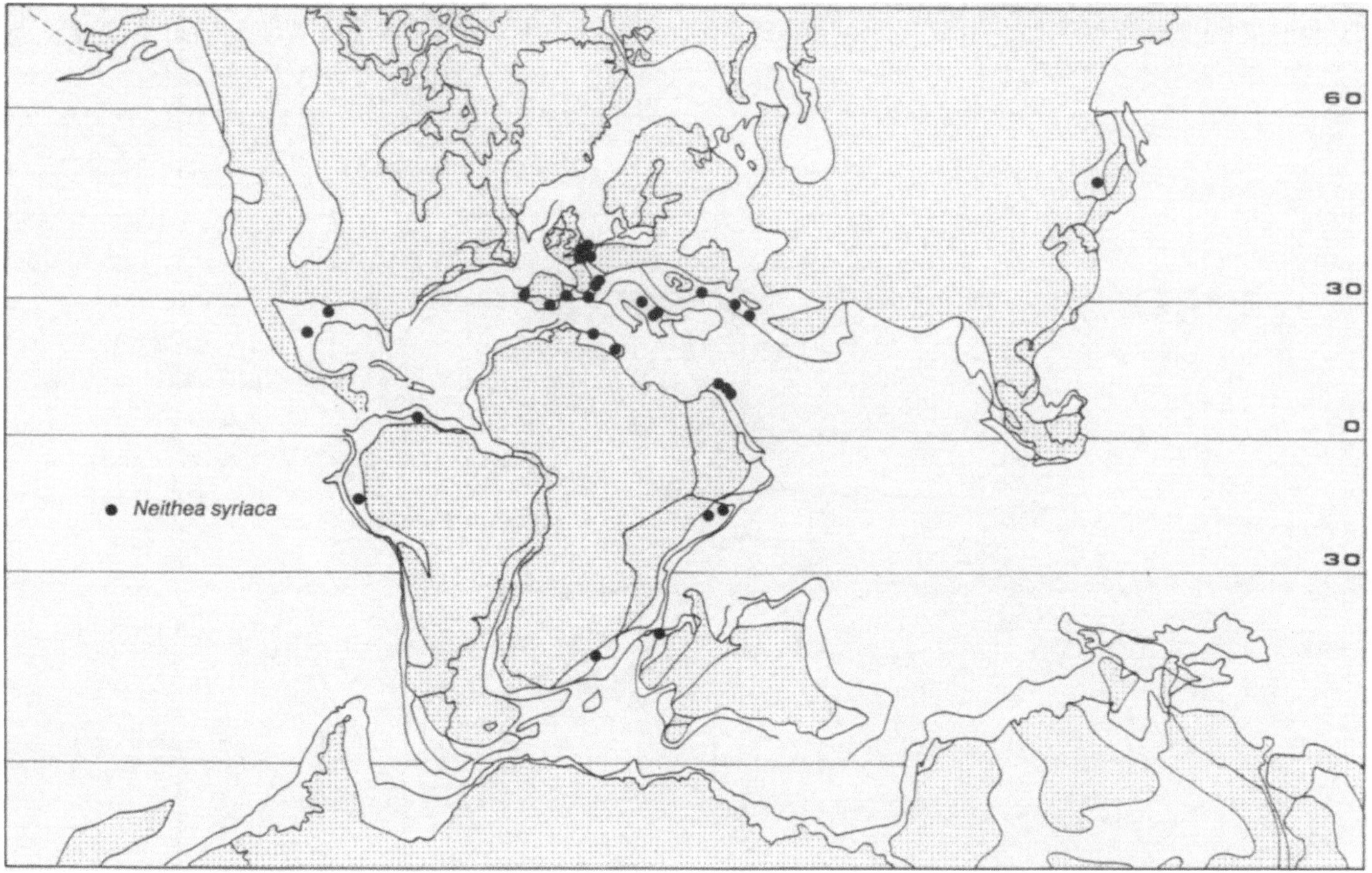

Fig. 4: Map at 120 million years (from BARRON et al., 1981) showing the distribution of *Neithea syriaca* (CONRAD) in the Aptian: from the West-coast of South-America to Japan, reaching the Isle of Wight in England to the North and Zululand in South Africa to the South.

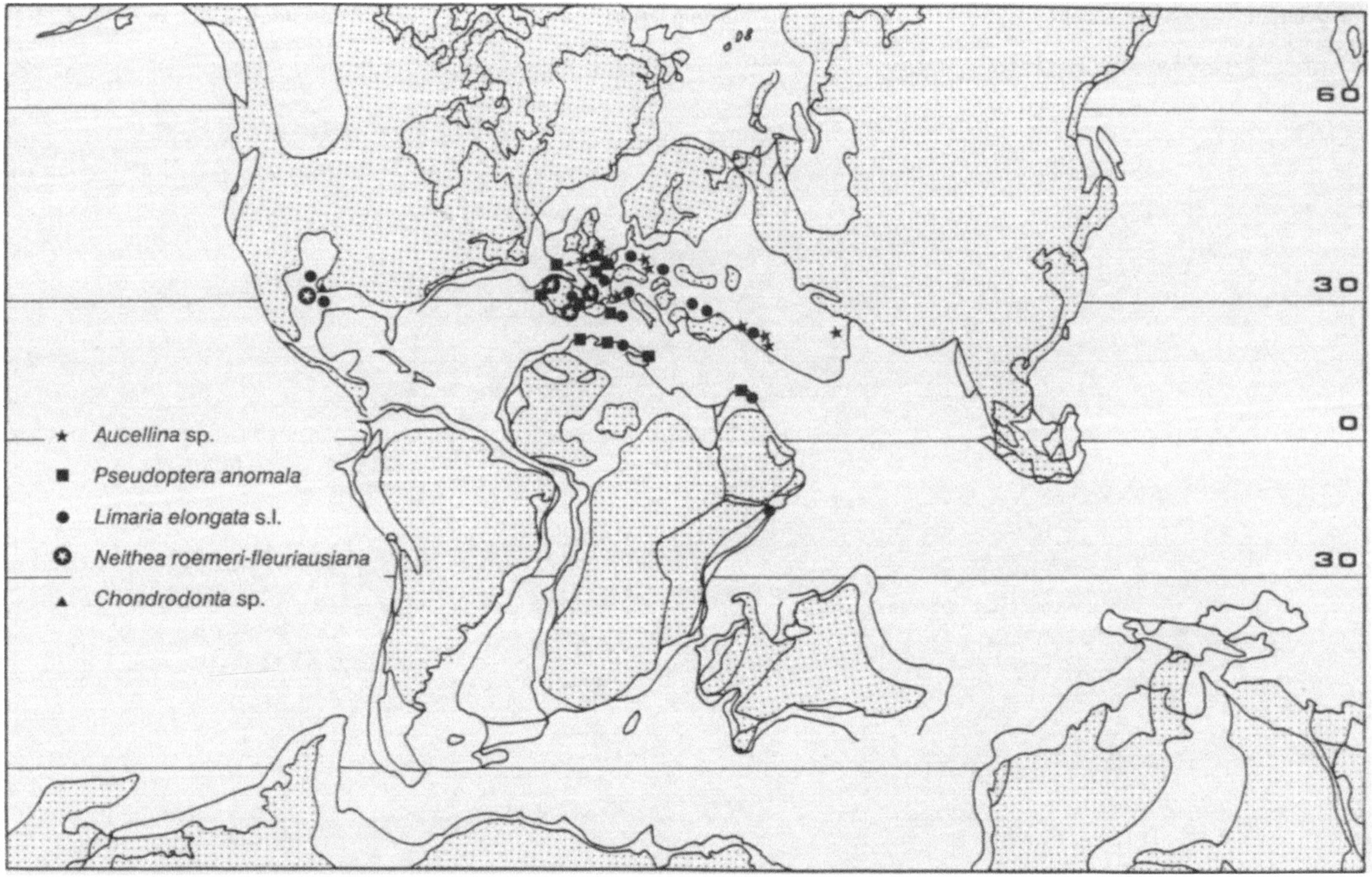

Fig. 5: Map at 100 million years (from Barron et al., 1981) showing the distribution of the cold water genus *Aucellina* in the Tethys (Upper Albian), of the bakevelliid *Pseudoptera anomala* (Sowerby), the limid plexus *Limaria elongata* (Sowerby) s. l. in the Upper Albian – Lower Cenomanian, of the rudist-framework associated species *Neithea roemeri – fleuriausiana*, and *Chondrodonta* spp. in the Upper Albian – Cenomanian.

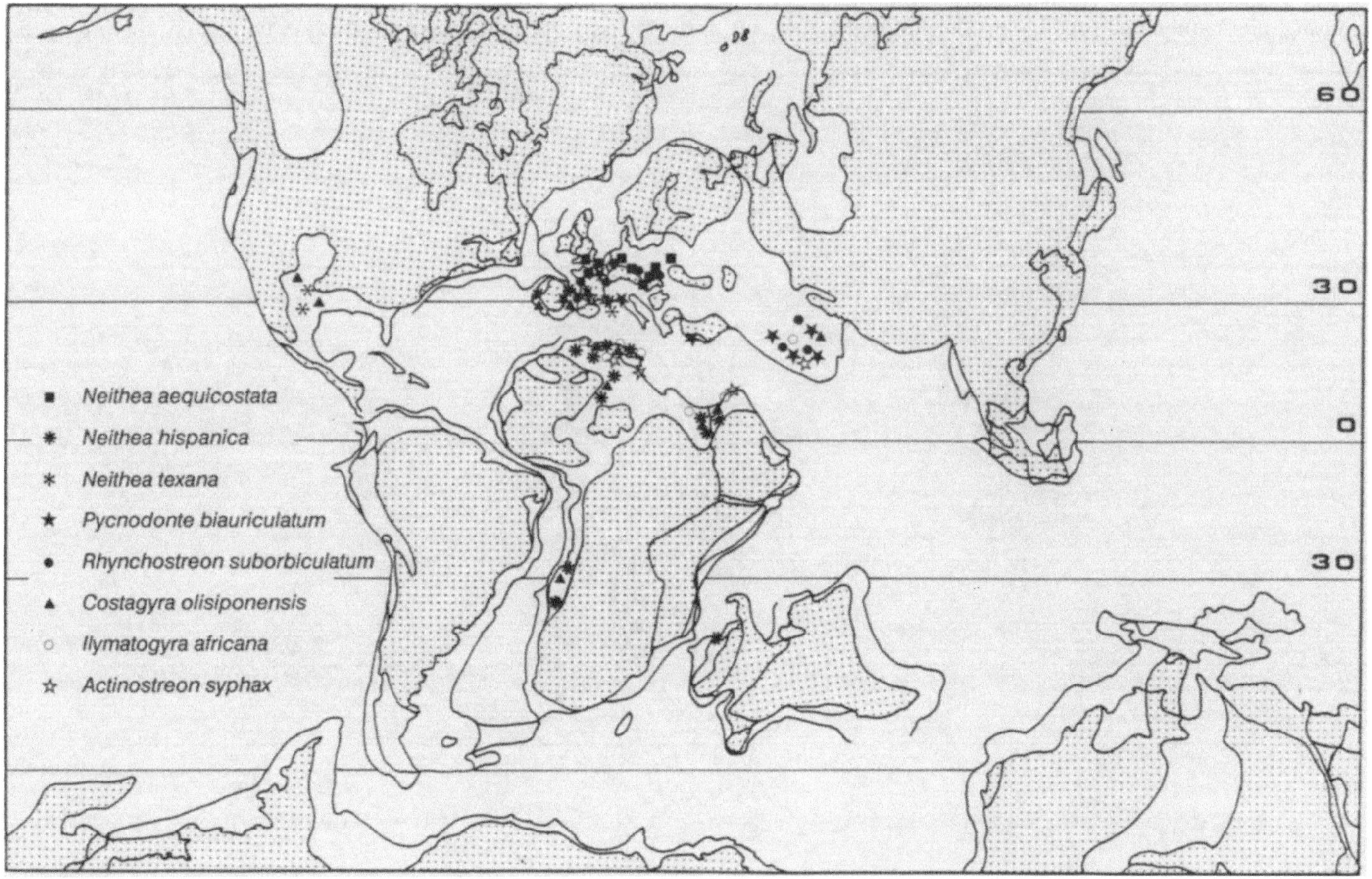

Fig. 6. Map at 100 million years (from Barron et al., 1981) showing the distribution during the Upper Cenomanian of the Temperate species *Neithea aequicostata* (Lamarck), Temperate and northern margin Tethys species *Pycnodonte biauriculatum* (Lamarck), *Rhynchostreon suborbiculatum* (Lamarck), and of Tethyan species *Neithea hispanica* (d'Orbigny), *N. texana* (Roemer), *Costagyra olisiponensis* (Sharpe), *Ilymatogyra africana* (Lamarck), *Actinostreon syphax* (Coquand).

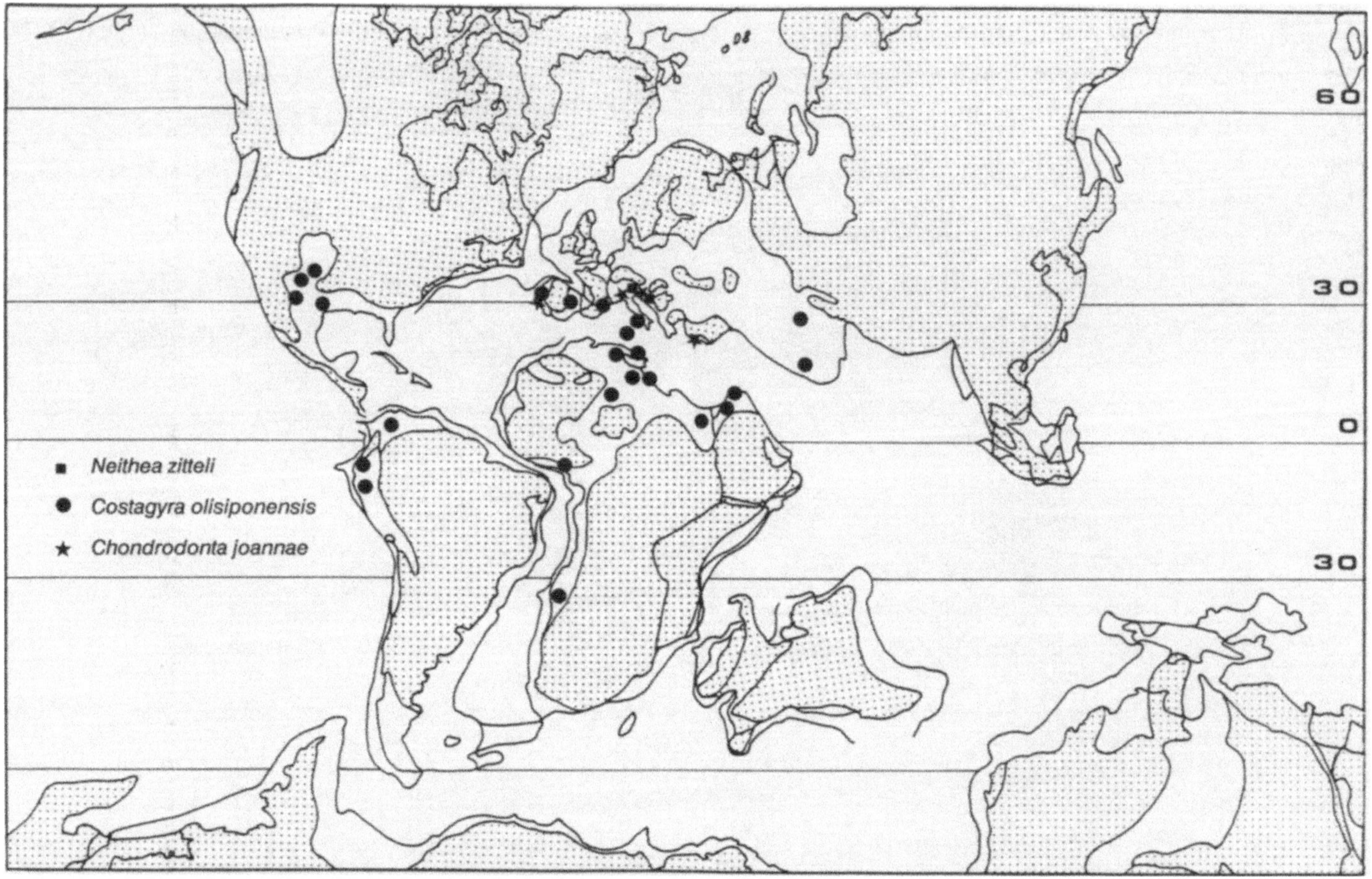

Fig. 7: Map at 100 million years (from BARRON et al., 1981) showing the distribution during the Upper Cenomanian and Turonian of the oyster *Costagyra olisiponensis* (SHARPE) (data partly out of REESIDE, 1929), and of the rudist-associated framework species *Neithea zitteli* (PIRONA) and *Chondrodonta joannae* (CHOFFAT).

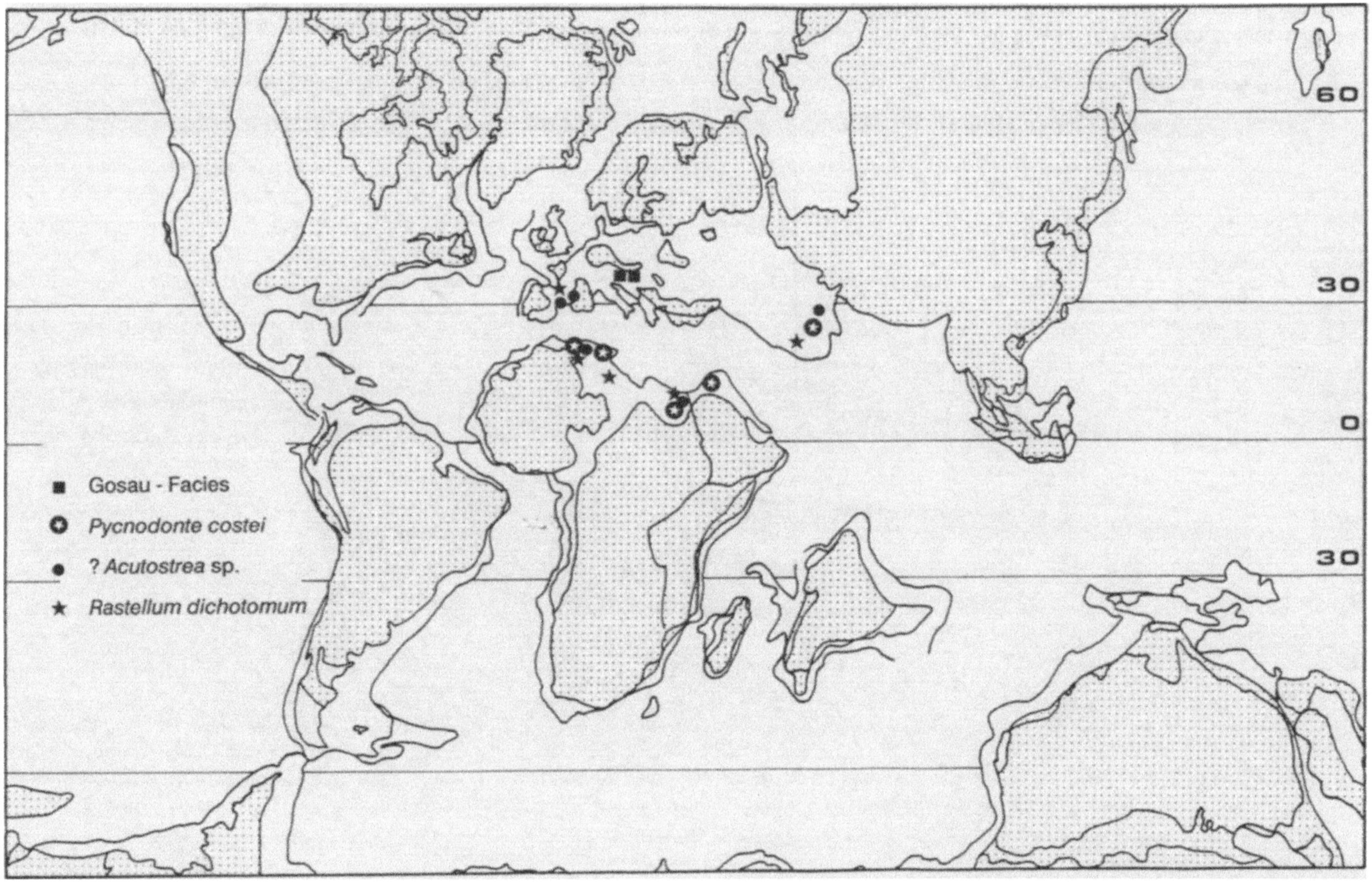

Fig. 8. Map at 80 million years (from BARRON et al., 1981) showing the distribution during the Coniacian-Santonian of oyster taxa typical for the north African oyster association: *Pycnodonte (Costeina) costei* (COQUAND), *Acutostrea acutirostris* (sensu COQUAND non NILSSON), *Rastellum dichotomum* (BAYLE); also indicated is the northern Tethyan Gosau region.

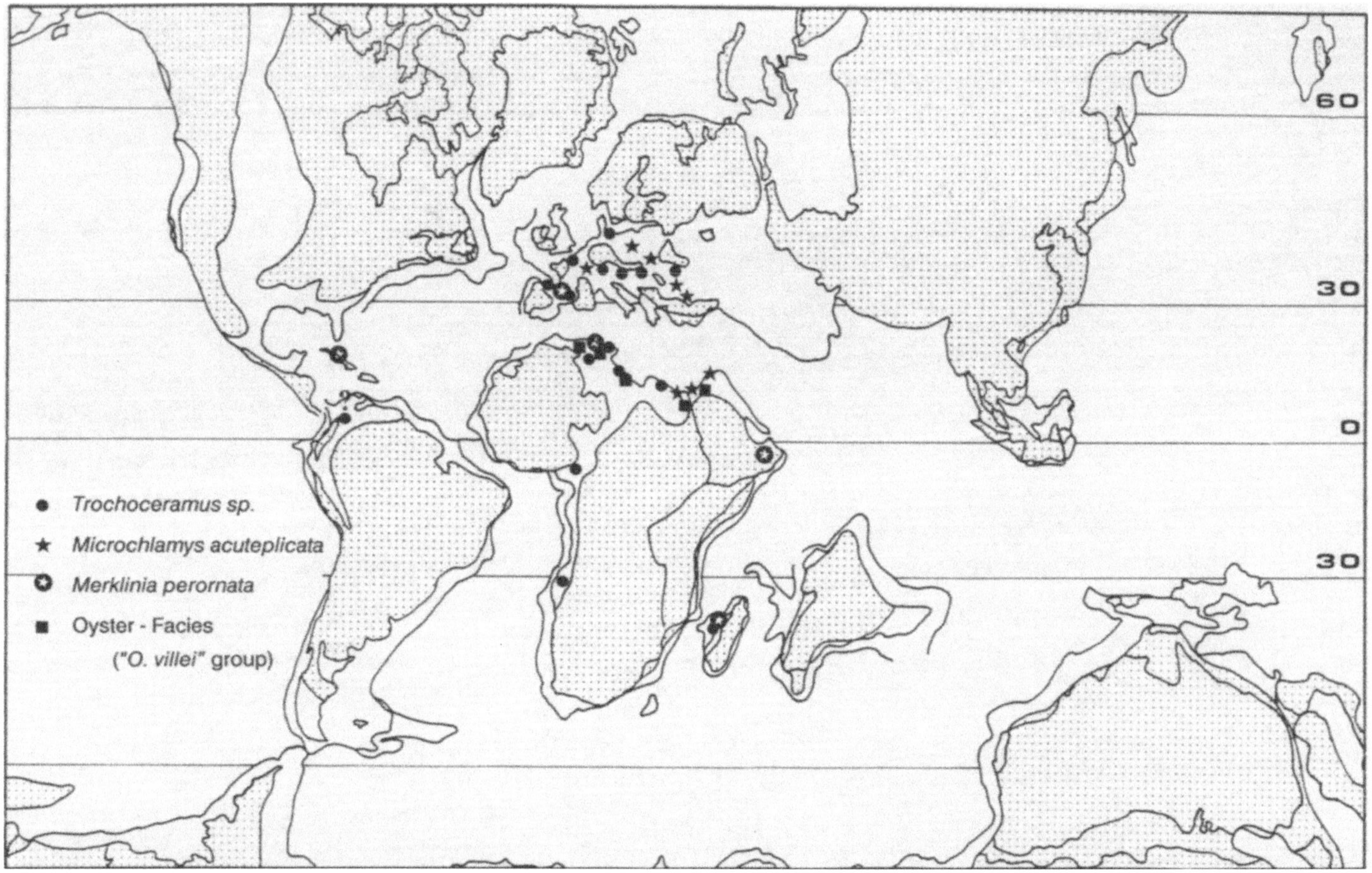

Fig. 9: Map at 80 million years (from BARRON et al., 1981) showing the distribution during the Upper Campanian-Maastrichtian of oyster taxa typical for the northern Africa oyster association: *Exogyra overwegi* (VON BUCH), *Actinostreon villei* (COQUAND), *Nicaisilopha nicaisei* (COQUAND); also indicated is distribution of the Tethyan pectinid species *Merklinia perornata* (COTTREAU), of the Temperate pectinid *Microchlamys acuteplicata* (ALTH) and of the inoceramid genus *Trochoceramus*.

New Aspects on Tethyan Cretaceous Fossil Assemblages.
Band 9 Schriftenreihe der Erdwissenschaftlichen Kommissionen der Österreichischen Akademie der Wissenschaften. 95–128.

2.5. Distribution of gastropods within the Cretaceous Tethyan realm

By Heinz A. Kollmann*)

Summary

Fossil assemblages of different geographical regions and within certain latitudes which are equivalent on the genus or family level indicate temperature-controlled realms in the geologic past. Paleobiogeographically, the Tethys is placed within the Tropical realm of the Mesozoic. In addition to the low-latitude Tropical realm, Warm Temperate, Temperate, and probably also Subpolar marine provinces can be distinguished by the distribution of fossils in Cretaceous rocks. In low latitudes, the horizontal distribution is to a varying degree vertically reproduced.

Like all benthic organisms the gastropods are restricted to specific environments. This distribution pattern is discussed for the nerineaceans (including the itieriids), the actaeonellids (*Actaeonella* and *Trochactaeon*), the cassiopids and the genera *Trajanella* and *Discotectus*. Valid biostratigraphic analyses can be obtained only by comparing stratigraphic ranges within the same environments.

Normally, marine gastropod assemblages of rocky surfaces or patches of hard substrates show high diversities of archaeogastropods. Level bottom communities are characterized by a high diversity of mesogastropods and cephalaspidean opisthobranchs. The evolution of neogastropods allows this group to become increasingly dominant during the Createaceous; the number of nerineacean genera is in the process of decreasing after the Cenomanian.

The origin of the neogastropods was within the Temperate marine zone. In the Tethyan realm, neogastropods were not recorded prior to the Upper Albian. It is assumed, that the first neogastropods were buccinids which have evolved from deposit-feeding groups.

The Tethyan gastropods were profoundly affected by the terminal Creta-

*) Naturhistorisches Museum, A-1014 Wien, Austria.

ceous events. The extinction of Tethyan gastropods is interpreted as having been primarily caused by a change of surface water temperature distribution in the marine biosphere.

1. Introduction

The composition of aquatic communities is controlled by the chemical and physical environment (Boucot, 1981) and by the interactions of their members. Chemical parameters controlling animal life are dependent on such factors as the mineral content of the water, Eh, pH, etc. Physical parameters include such primary factors as water temperature, light intensity and wave length, water pressure, water energy and substrate. In broad, pragmatic terms, only the substrate can be roughly identified by direct evidence through a petrological analysis of the rocks containing the fossil assemblage. The remaining parameters are deduced from observations of sediment structures and the composition and distribution of fossil assemblages by a critical actualistic analysis. However, due to the varying degrees of fossilisation potential within individual taxa large percentages of the ancient living communities are not represented in fossil assemblages.

Water temperature is probably the most critical factor among the physical parameters (Boucot, 1981). The distribution of animals is regulated through the physiological effect of water temperature on every single individuum of a species or even of taxonomically higher hierarchies to which species may be combined. The temperature dependence is not only due to the fact that the majority of marine invertebrates lives within the range of 0–35° C. Tolerance for maximum and minimum temperatures at which sufficient metabolic rates are maintained is specific for the representatives of a certain species. Temperatures above and below this range are, therefore, lethal for individuals of a taxon. However, it is much more important for the distribution of individual taxa that most organism require a critical temperature for reproduction (Gunter, 1957). Water temperature does not only have a primary regulative effect on the distribution of each taxon. The occurrence or non-occurrence of a taxon will influence an entire trophic system.

By far the majority of known megafossil assemblages are of shallow water origin; these assemblages reflect the local surface temperature. Consequently, shallow water assemblages of different geographical regions which contain taxa equivalent on the genus or even family level are indicative of temperature-controlled realms. It is assumed in all conclusions on the distribution of realms in geologic times that the temperature tolerance is a conservative element in the animal-environment relationship (Boucot, 1981). The deduction of relative temperature ranges based on the geographic distribution of fossil taxa together with sedimentological data is a rapid, pragmatic, and valid method for recognizing the distribution of temperature-related realms in the past.

2. Cretaceous realms

The various definitions of Tethys and the concept of Mésogée have been discussed by KOLLMANN and by MASSE in this volume. Tethys is utilized here in the sense of KAUFFMAN (1973, 1975) and SOHL (1971, 1987) as a paleobiogeographically defined realm which covers the area of distribution of algal-coral, coral-rudist, or rudist dominated organic framework build-ups. According to the present distribution of bioherms in the marine biosphere, the Tethys has to be considered as being within the Tropical realm during the Cretaceous. This definition does not agree with the original definition of Tethys by SUESS (1893), however, it is very convenient to work with. It has been emphasized by KAUFFMAN (1973, 1984) and SOHL (1971, 1987) that the Tethyan realm was not stable during the Cretaceous but has had varyable limits with time.

Gastropods do not provide any evidence for a Supertethys postulated by KAUFFMAN & JOHNSON (1988). It has been shown by BEAUVAIS & BEAUVAIS (1974) that corals are occurring in most parts of the belt considered as Supertethys. Although the distribution is sporadic it provides evidence against a Supertethys with hotter and more saline water than the rest of Tethys. ACCORDI, CARBONE & SIRNA (1982) have demonstrated that Upper Cretaceous coral-algal associations of south Italy are restricted to muddy sands. These sediments show a much lower diversity of rudists than adjoining rudist banks. In descriptions of lower Cretaceous occurrences of the Provence (TURNSEK & MASSE, 1973) and of Spain and Austria (REITNER, 1987) rudists generally occur in different beds than corals. All of these observations strongly support SCOTT, quoted by KAUFFMAN & JOHNSON (1988), that coral-algal dominated and rudist-dominated reef paleocommunities have preferred different habitats.

SOHL (1971, 1987) has developed the concept of a transitional marine belt north and south of the belt with Tropical organic build-ups. In suitable environments, this belt was inhabited by corals and rudists as well as Tethyan gastropods. Like the Tethyan realm this zone was varying in its geographical limits during Cretaceous times; these features can be explained also by comparison with modern marine realms:

According to WELLS (1957), modern coral reefs are best developed at mean annual water temperature between 23–25° C. No significant reefs occur where temperatures during the year fall below 18° C, except for a short period. The distribution of reefs therefore, is restricted to the Tropical belt. Hermatypic corals and calcareous algae, however, are not restricted to this belt but do also occur outside together with other Tropical elements. In fact, they have been reported into latitudes with mean annual temperatures as low as 15° C (WELLS, 1957), where they gradually diminish in number and diversity. These marine zones are called subtropical by VAUGHAN (1940) or Warm Temperate by STEPHENSON (1947). The extent of the zones has been summarized by HEDGPETH (1957).

It may therefore be assumed that the Warm Temperate zone of the Cretaceous marine biosphere is characterized by solitary rudists and corals. In the

Upper Cenomanian these fossil groups were spreading into Europe as far north as Saxony (see the faunas described by GEINITZ, 1872). In the Campanian rudists were found in abundance in southern Sweden (KÜHN, 1949) which is by far the northernmost distribution of this group. In the Maastrichtian corals and rudists were spreading as far north as Maastricht (Holland). It has been shown by SOHL (1971) that equivalent to modern seas, cypraeids are common gastropods of the Cretaceous Warm Temperate zone. This zone forms the southernmost extension of the archaeogastropod family Pleurotomariidae which gains importance further to the north. This gastropod family is therefore of great importance for distinguishing the Tethyan zone from the Temperate zones (SOHL, 1971; KOLLMANN, 1984).

The Warm Temperate marine zone is followed northward by a zone called Temperate in the biological literature. This zone is part of the Boreal zone of paleontologists. Tropical elements are missing; the gastropod faunas are not diverse and characterized by pleurotomariids. This zone seems to agree in its distribution with the Lower Cretaceous Hoplitinid province of OWEN (1971) and the Upper Cretaceous North Temperate realm as defined by the distribution of the cephalopod *Belemnitella* (KEGEL CHRISTENSEN, 1976).

There is not much evidence for a Cretaceous marine province equivalent to the Subpolar province of modern seas. Based on my own observations, a fauna from the island of Nugssuaq, west Greenland, dated as Maastrichtian with ammonites (ROSENKRANTZ, 1970) may represent this marine zone. It contains dominantly shells of infaunally living gastropods, mostly belonging to the opisthobranch group of Cephalaspidea. THORSON (1952) has shown that the percentage of infaunal species is highest in cold water environment of the Subpolar province as life within the sediment provides protection against strong temperature changes. Towards the Tropical zone, the percentage of infaunal species in the total fauna is diminishing while the number of taxa remains approximately the same.

The succession of temperature-controlled biogeographic zones in the northern hemisphere shows that there was, at least temporarily, a higher temperature gradient in the Cretaceous than assumed in the earlier literature (KAUFFMAN, 1973). This agrees with the observations of KEMPER (1987).

3. Vertical successions

Utilizing recent ostracodes, HAZEL (1970) has demonstrated that the temperature-controlled latitudinal marine realms of the Atlantic continental slope are to a certain extent vertically reproduced in lower latitudes. There also are some examples on various fossil groups that can be cited for the Cretaceous; they are:

WIEDMANN (1988) has recorded in the Vascogotic through of northern Spain a Lower Turonian association of mammitid ammonites of northern origin which had lived in water depths below 200 meters while shallow zones were inhabited by a typically Tethyan fauna with vascoceratid ammonites. Another, yet unpublished example, is the occurrence of a *Belemnitella* in the Upper Maastrichtian "couches de passage" of the Olonos-Pindos zone which was found during field work in association with I. MARIOLAKOS south of Tripolis (Peloponese, Greece). FLEURY (1980) considers the deposition depth of these sediments to be 2–3000 meters.

Examples involving gastropods include the Upper Aptian fauna of north Tunesia (MONGIN, 1971) which contains forms exclusively described from the Boreal Cretaceous of North Germany (WOLLEMANN, 1900, 1912), England (ABBASS, 1973), the Paris Basin (D'ORBIGNY, 1842–43) and Mangyslak in the Soviet Union (NACKIJ, 1916). A correlation between water depth and faunal composition has been observed in the Upper Cretaceous of the Charente in France by KOLLMANN (1985). Tethyan faunas are restricted here to the regressional phases of HANCOCK and KAUFFMAN (1979) while temperate ones with pleurotomariids and large archaeogastropods belonging to the genus *Coelobolma* are associated with transgressive pulses.

4. Systematic inventory of Cretaceous Tethyan and Warm Temperate gastropods

Scholars such as J. WALTHER (1908), DACQUÉ (1926) and DIENER (1925) have already emphasized the fact that an equatorial belt, named by authors southern sea-belt, mediterranean region or Mésogée was characterized not only by rudists, coral reefs and larger Foraminifera but also by *Actaeonella* and *Nerinea* among the gastropods. These genera, in the sense of the older authors, have undergone an extensive taxonomic subdivision. They have been revalorized taxonomically into family (Actaeonellidae) and superfamily (Nerineacea) rank but still can be used together with other forms for characterizing and delimiting the Tethyan realm (SOHL, 1971). In the Warm Temperate zone these gastropods are absent or very rare.

The most common gastropod genera of the Tethyan realm and of the Warm Temperate zone have been compiled by SOHL (1971). Additional observations have been published by SOHL (1987). With the exception of a few modifications, not much is to be added to the general inventory of Cretaceous Tropical and Warm Temperate gastropods (Table 1).

Family or genus	Tropical	Warm Temperate
Trochidae	diverse	few
Discotectus	common	absent
Fissurellidae	common	absent
Pleurotomariidae	absent	common
Neritidae	diverse, abundant	rare
Pileolus	widespread	rare
Cassiopeidae	rare	abundant
Turritellidae	few	abundant

Family or genus	Tropical	Warm Temperate
Aporrhaidae	common	common
Harpagodes	common	absent
Perissoptera	rare	common
Naticidae	common	abundant
Tylostoma	abundant	common
Lunatia?	rare	common
Nerineacea	diverse,abundant	absent or very rare
Actaeonellidae	abundant	absentor very rare

Table 1: Tropical and Warm Temperate gastropods. After SOHL (1971, 1987, slightly modified)

5. Occurrence of selected gastropod groups within the Tethyan realm

As stated by BOUCOT (1981), there is a high correlation between substrate type and benthic organisms within the realms. The use of benthic organisms for biostratigraphic correlation, therefore, is efficient only for sediments deposited under corresponding conditions. It should be the general rule never to extend correlation with benthic forms beyond the facies of their biocoenosis. It therefore seems to be of foremost interest to evaluate the life conditions of fossil gastropods. This has been done for some groups abundant in the Tethyan realm (SOHL, 1971). The state of knowledge is rather poor and all generalisations should be treated with caution until such time as more detailed studies become available.

5.1. Nerineacea

The nerineaceans were most diverse from the Upper Jurassic to the end of the Cenomanian. With the end of the Cenomanian the diversity becomes low.

Families and genera of nerineaceans are primarily distinguished by internal plications. The function of these plaits is still not fully understood. Obviously, some of them have served for subdividing the mantly cavity to separate the inhalant water current from the exhalant current. The exhalant current which carried out the waste material from the mantle cavity was expelled through a sutural notch. The position of the sutural notch allowed an expelling of the water well back of the head. It has been emphasized by SOHL & KOLLMANN (1985) that this is a feature characteristic for completely or partly infaunally-living gastropods.

The most striking feature of the nerineaceans is the mass occurrence in "Nerinea beds" which have been reported from virtually all areas of the Tethyan realm. It has been documented by DAUWALDER & REMANE (1979), WIECZOREK (1979) and KUOYOUMONTZAKIS (1987, 1989) that the heavy nerineacean shells are transported only over short distances and that their thanatocoenosis resembles the original biocoenosis to a high extent.

The nerineacean faunas of the earlier Cretaceous described among others by d'Orbigny (1842), Pictet & Campiche (1862), Pellat & Cossmann (1907), de Brun, Chatelet & Cossmann (1916), Delpey (1941) and Pchelintsev (1927, 1931, 1965) show a persistence of Upper Jurassic genera (or families in the not performable narrow taxonomic concept of Pchelintsev) into the Cretaceous. Most of these genera which have a short siphonal channel (Kollmann, 1976) and frequently a hollow columella become extinct during Aptian time. The process of the replacement of this Jurassic-early Cretaceous faunal stock is not yet documented sufficiently.

Observations on the environments of lower Lower Cretaceous nerineaceans are scarce. The analytical studies of Wieczorek (1979) on Upper Jurassic nerineacean faunas may be applied to lower Lower Cretaceous ones. According to this author, nerineacean assemblages are restricted to environments of the open shelf. Wieczorek (1979) concluded from sedimentological data that mass occurrences in nerineacean beds occurred in environments with low water energy, stable substrate and slow sedimentation rates. These would have been the optimal conditions for this gastropod group. From the Cretaceous, Rey (1979) has reported the occurrence of undertermined nerineaceans from wackestones to packstones of Estremadura (Portugal). These were deposited under relatively low energy conditions, too. Wieczorek (1979) reported the highest diversity of species from the "bahamite" or grapestone facies of the open shelf. The highest tolerance to environmental conditions has been observed by him in the genera *Nerinea*, *Nerinella* and *Ptygmatis* which extend into mud and pellet mud facies. This agrees with the observations of Shikama & Yui (1973), who have found Nerinea but no other nerineacean genera in sediments of various grain sizes.

There are only a few observations on the ecology of upper Lower and Upper Cretaceous nerineaceans. They do not provide a uniform picture of the environmental conditions. Accordi, Carbone and Sirna (1982) have described an Aptian fauna containing the nerineaceans *Neoptyxis* together with the Ceritiacean *Pseudonerinea* (generically determined after the figures in this publication). In Texas, these genera and *Multiptyxis* occur in low energy lagoonal environments (Kollmann, Decker & LeMone, unpublished). Occurrences in laminated calcarenites of probably high energy foreshore conditions are probably re-deposited.

From strata of probably Cenomanian age of the Rocca di Cave section, Carbone, Praturlon, and Sirna (1971) have recorded representatives of *Plesioplocus* and *Italoptygmatis* (generic determination changed after the figures of the above-mentioned work). According to these authors, the sediments have been deposited in a medium to high energy environment.

Nerineaceans which are herewith assigned to the genera *Nerinea*, *Neoptyxis*, *Haploptyxis* and *Italoptygmatis* have been recorded by Accordi, Carbone & Sirna (1982) from limestones of Cenomanian to Turonian age. These have been deposited in medium energy environments.

Rich faunas of nerineaceans belonging to the genera *Dalmatea*, *Nerinoides*,

Neoptyxis, Laevinerinea, Plesioptygmatis and *Haploptyxis* have been found by KOLLMANN (unpublished) in the massif of Akros, close to Epidavros (Greece). These sediments, dated as Cenomanian by DECROUEZ (1977) and MERMIGHIS (1989), consist of wackestones, floatstones, rudstones and grainstones deposited according to MERMIGHIS (1989, 1989a) in a low energy shallow environment of an internal platform.

KOUYOUMONTZAKIS (1985) recorded taxa of the genera *Haploptyxis* (revised, determined as *Plesioptygmatis* by this author) and *Neoptyxis* (the determination could not be verified as no significant specimen has been figured but it seems rather unlikely that it is correct). These occur in Upper Cenomanian to Turonian wackestones to grainstones, indicating a mobile substrate and low energy conditions. *Haploptyxis* is also occurring in the Upper Turonian sands of Uchaux as part of a diverse level bottom community in the sense of THORSON (1957). It has lived on the soft sandy bottom under low energy conditions.

The preferred environment of *Simploptyxis* which is the most common nerineacean of the Eastern hemisphere from Turonian upwards is different (HÖFLING, 1985). In the Gosau Group of the Lattengebirge (Bavaria) this genus has lived in an environment with relatively high water energy at a temporary low sedimentation rate. This agrees with observations of HERM (1977) who has recorded Simploptyxis in the Brandenberg area in Tyrol from coarse- to medium-grained sands deposited in unprotected environments of an external lagoon.

These scanty observations suggest the following environmental conditions for the Cretaceous nerineaceans:

1. Lower Lower Cretaceous nerineaceans were living exclusively under low energy conditions of the open shelf and may have had their optimal living conditions on consolidated surfaces.

2. Upper Lower Cretaceous and Upper Cretaceous nerineaceans were inhabiting various environments, but obviously always soft bottom substrates. *Neoptyxis* and *Haploptyxis* were inhabiting a low to medium energy environment. *Plesioplocus* and *Italoptygmatis* have lived under medium to high energy conditions. *Simploptyxis,* which is characteristic for the Upper Cretaceous from the Turonian upwards, was living under high energy conditions.

3. It has to be concluded that beginning with the Aptian nerineaceans were radiating into various environments.

5.1.1. Itieriidae

Itieriidae are well represented in the Cretaceous sediments. Like the Nerineacea mentioned in the previous chapter, the occurrence of lower Lower Cretaceous Itieriidae may be compared with that of the Upper Jurassic. WIECZOREK (1979) has described two genera, *Phaneroptyxis* and *Fibuloptygmatis,* from the Upper Jurassic of the Polish Holy Cross Mountains. These genera are restricted to the grapestone-facies of the carbonate bank which is interpreted to have been deposited in subtidal or intertidal shallow water with restricted water circulation. According to studies by CARBONE, PRATURLON, and SIRNA (1971), *Phaneroptyxis* is ranging stratigraphically upwards into the Upper Cenomanian (there are some doubts concerning this age as all other recordings of this genus are not younger than Barremian). Other occurrences in the platform limestone of the Aquilean Abbruzze Mountains have been described by SCHNARRENBERGER (1901) and PARONA (1909).

Beginning with the Aptian, another group of the Itieriidae developed. Its taxa have a short, beak-like siphonal channel but resemble in its general morphology the Actaeonellidae. Generally, the channel can be observed only in axial sections of shells. This group of itierids has primarily been distributed in siliciclastic sediments.

One itieriid genus belonging to this group is *Peruviella* which hitherto is known from West Africa (Angola), South America (Brasil, Peru), North America (Texas, British Columbia, unpublished) and the Soviet Union (POJARKOWA & DJALILOW, 1985). The age is Middle to Upper Albian, perhaps also Lower Cenomanian. For details on the distribution see: KOLLMANN & SOHL, 1979; CASSAB, 1982. Observations in West Texas (KOLLMANN, DECKER & LEMONE, unpublished) have shown that this genus occurred under different energy conditions and was exclusively controlled by the substrate (sand-sized sediments). Single specimens, sticking obliquely in the sediment as it has been described for the recent opisthobranch *Acteon* (FRETTER & GRAHAM, 1954) and, consequently, for the Actaeonellidae (SOHL & KOLLMANN, 1985), suggest an infaunal life for this gastropod group.

The genus *Sogdianella* DJALILOW (1972) is belonging to the same itieriid group as *Peruviella. (Omphalactaeonella* HACOBJAN (1972) is a younger synonym). This genus is widely distributed around the Mediterranean. *Sogdianella syriaca* (CONRAD) has been recorded from North Africa and Greece (KOLLMANN, 1985, 1987). Other species have been described by PCELINTSEV (1953), HACOBJAN (1976), DJALILOW (1972) and ALIEV & ALIEV (1986) from the southern Soviet Union and by KOLLMANN & SOHL (1979) from the Gosau Group of the Eastern Alps and from North America.

Sogdianella brandenbergensis (KOLLMANN), is the only species known from the Eastern Alps. It has lived in tidal flats (HERM, 1977). Sediments deposited here are coarse sands. In an occurrence in Greece described by KOLLMANN (1987), *Sogdianella* is observed in association with the cerithiids *Diozoptyxis* and *Itruvia*, several nerineaceans, the itieriid *Vernedia* and one *Trochactaeon* species. Similar assemblages have been observed in Armenia (HACOBJAN, 1978, personal observations). The sediments indicate a soft bottom. As in *Peruviella* the distribution was controlled by sediment grain size. Obviously, these genera were subject to infrequent strong water agitation as all gastropod specimens are eroded. Bivalves do not occur in these strata. Additionally, *Sogdianella* is rather common in the Tripolitsa limestone of Greece (KOLLMANN, 1985 and unpublished).

The genus *Eotrochactaeon* HACOBJAN, which is closely related to *Sogdianella*, was until recently only known from coarse-clastic, terrestrially-influenced sediments from the southern Soviet Union (HACAOBJAN, 1976), North Africa

(DELPEY, 1940) and Greece (KOLLMANN, 1982). The genus has recently been recorded by KOLLMANN, DECKER & LEMONE (unpublished) from shallow lagoon sediments of the Trinity sequence of West Texas.

The genus *Vernedia* MAZERAN (1912), which has been erroneously placed into the genus *Itruvia* by many authors (see: KOLLMANN & SOHL, 1979; KOLLMANN, 1987), is widely distributed from the upper Lower Cenomanian upwards. In Hölzelsau, Tyrol, *Vernedia* occurs as part of a diverse fauna described by RAHMAN (1967). The environment inhabited by this fauna indicates a sand bottom below wave base. In all other localities known to me, *Vernedia* occurs in large numbers in sands or other coarse clastic sediments and is part of a mollusc fauna of low diversity. As mentioned before under *Sogdianella*, the shells are generally highly eroded and, obviously, have been deposited under high energy conditions.

Summarizing the state of our knowledge on the Itieriidae the following statement can be made: A conservative stock of genera extends from the Jurassic into the Barremian, with single species perhaps also into the Cenomanian. Like the lower Lower Cretaceous nerineaceans, this group is restricted to parts of the carbonate platforms deposited under open shelf conditions. Like the nerineaceans the itieriids began to radiate into other environments in the Aptian.

5.2. Actaeonellidae

The ecology of Actaeonellidae has been treated in detail by KOLLMANN (1965, 1967) and SOHL & KOLLMANN (1985). Therefore, only a short summary of the results together with a discussion of recent literature is given here. Due to the lack of other data only the genera *Actaeonella* and *Trochactaeon* are discussed.

5.2.1. *Actaeonella*

It has been emphasized by KOLLMANN (1965) and by SOHL & KOLLMANN (1985) that *Actaeonella* is restricted to fully marine environments. HÖFLING (1985) has shown that *Actaeonella* occurred in lagoons under quiet water conditions. This agrees with occurrences of *Actaeonella* in the Upper Albian Meule de Bracquegnies of South Belgium (BRIART & CORNET, 1868; MARLIERE, 1939) and the Upper Turonian sand of Uchaux, France (ROMAN & MAZERAN, 1920). In both deposits *Actaeonella* is part of a diverse fauna characteristic for environments below wave base. The same environment has also been described by WAGREICH (1988) for the Santonian upper Streiteck Formation of the Gosau area where *Actaeonella laevis* (SOWERBY) is very abundant.

According to SOHL & KOLLMANN (1985) the sediments in which *Actaeonella* occurs are generally sands or sandy limestones indicative of an unstable bottom. There are only few occurrences known from carbonate platforms. In the area of Kent (Texas) the occurrence of *Actaeonella* is recorded in strongly bioturbated grainstones of the Buda Formation (KOLLMANN, LEMONE, DECKER, unpublished). *Actaeonella pecosensis* STANTON (1947) has been described from the Devils River Limestone of Texas.

Sedimentological data for these occurrences are not available. The occurrence of *Actaeonella* in the Marble of Almyropotamos (Euboea, Greece) described by KATSIKATSOS & KOLLMANN (1987) involves marbles which are completely recrystallized and do not reveal any prior sedimentological features.

The infaunal life of *Actaeonella* postulated by SOHL & KOLLMANN (1985) has been questioned by HÖFLING (1985) who believes this genus was part of the epifauna living as an active carnivore. It should be emphasized that the position of *Actaeonella* in the sediment, which has been described by SOHL & KOLLMANN (1985) agrees with observations on modern cephalaspidean gastropods which have a documented "bulldozing" mode of action (THAYER, 1983; VERMEJ, 1987). The elongated aperture is after LINSLEY's (1977) "law" nr. 5 characteristic for a water flow from anterior to posterior by passing a single gill. In *Actaeonella*, the water was leaving the mantle cavity by a funnel-shaped exhalant siphon at the posterior end of the shell. This feature is characteristic for the infaunally living cephalaspideans and occurs also in several prosobranch gastropods of the same life mode.

5.2.2. *Trochactaeon*

Similar to *Actaeonella*, a detailed interpretation of the environmental parameters of *Trochactaeon* has been given by KOLLMANN (1967) and by SOHL & KOLLMANN (1985). Otherwise than quoted by HÖFLING (1985), these papers emphasize the high salinity tolerance of *Trochacteon* compared with *Actaeonella*. The salinity range postulated by the former authors for *Trochactaeon* is extended by HÖFLING (1985). He interprets *Trochactaeon* of the Schneckenwand mass occurrence in the Gosau area (Austria) as inhabitants of a hypersaline environment.

Generally, *Trochactaeon* assemblages show a low diversity (HERM, 1977). The substrates inhabited by *Trochactaeon* species range from limestones and shales to coarse sands (HERM, 1977; HÖFLING, 1985; SOHL & KOLLMANN, 1985). As stated by the latter authors, *Trochactaeon* seems to have preferred generally more fine-grained sediments than *Actaeonella*.

According to HERM's (1977) studies on the Upper Cretaceous of Brandenberg (Tyrol), *Trochactaeon* inhabited tidal flats with high energy environments (The *Actaeonella* mentioned by HERM as co-occurring is the itieriied *Sogdianella*. (see SOHL & KOLLMANN, 1986). Sedimentological data of the Schneckenwand mass occurrence of *Trochactaeon* in the Gosau basin (HÖFLING, 1985) which was mentioned above suggest a calm, but temporarily agitated water (Plumley-Index II).

Because of high variability within the species of *Trochactaeon* and abundant effects of mechanical wear, and consequent shell repair by deposition of internal secondary shell layers SOHL & KOLLMANN (1985) concluded that *Trochactaeon* led a dominantly epifaunal life.

5.3. *Trajanella*

The occurrence of *Trajanella* in the "mediterranean" zone of the Cretaceous was first observed by DELPEY (1939). Localities where *Trajanella* occurs include the Turonian faunas of Uchaux (D'ORBIGNY, 1842–43; ROMAN & MAZERAN, 1920), the upper Lower Cenomanian fauna of Hölzelsau (RAHMAN, 1967), the Aptian-Albian Hiraiga Formation of Japan (KASE, 1984), and Cenomanian sandstones east of Horsetown, California (ANDERSON, 1958). The gastropods of these occurrences have been living in environments below the wave base in level bottom communities in the sense of THORSON (1957). In the Upper Cenomanian Pläner of Saxony (GEINITZ, 1871–75; DENINGER, 1905) and the equivalent Korycany Formation of Czechoslovakia (WEINZETTL, 1910), *Trajanella* has been found together with highly diverse faunas in conglomerates. KASE (1984) recorded *Trajanella* from sediments ranging from conglomerates to fine- to medium-grained sandstones together with patellid gastropods and several archaeogastropods. According to KASE (1984), the fauna was living on a rocky bottom. This suggests a wide range of habitats for *Trajanella*.

5.4. *Discotectus*

The Trochid *Discotectus* is recorded as having been distributed in a variety of environments within the Tethyan realm (SOHL, 1971, 1987). HÖFLING (1975) describes *Discotetcus* from the Gosau Group of the Lattengebirge in Bavaria. The genus occurs here in a nodular, grey, marly limestone containing irregularily distributed biomorphs and large bioclasts. According to HÖFLING, this originally mobile sediment has been deposited in the calm, temporarily agitated water of a back reef lagoon behind a hippuritid reef barrier. The numerous small specimens of this locality as well as of other alpine occurrences indicate that representatives of this genus may have lived on marine plants. *Discotectus* is also an abundant member of the diverse faunas of flats deposited below the wave base. In West-Texas *Discotectus* has been recorded exclusively on the fringing facies of a rudist reef (KOLLMANN, DECKER, LEMONE, unpublished). It occurs in Uchaux (ROMAN & MAZERAN, 1920), in Hölzelsau, Tirol (RAHMAN, 1967), in the Cenomanian of the Sarthe, France (GUERANGER, 1867; KOLLMANN, 1978), and in the Aptian Tanohata Formation of the Miyako area in Japan (KASE, 1984). There are many other occurrences such as the Plenus zone of Saxony (GEINITZ, 1871–75; DENINGER, 1905), the Korycany Formation of Czechoslovakia (WEINZETTL, 1910), the limestones of the Monti d'Ocre in Italy (PARONA, 1909) and the Edwards Formation (STANTON, 1947) for which no general environment interpretation can be given due to the diversity of facies.

5.5. *Tylostoma*

Tylostoma occurs most commonly in nodular limestones deposited in shelf basins (KOLLMANN, DECKER, LEMONE, unpublished) below storm wave base. In this facies, only internal molds are preserved. *Tylostoma* has been recorded by RAHMAN in Hölzelsau but otherwise representatives of this genus are rare within faunas living on siliciclastic substrates.

Systematically, *Tylostoma* is generally considered to be a naticid (WENZ, 1938–44). The shape of the molds is extremely variable and recordings of larger numbers of species from a single fauna as described by PERKINS (1960) have to be treated with caution.

5.6. *Cassiopidae*

The distribution of Cassiope (I do not follow the taxonomy of MENNESSIER, 1984, which seems to be too artificial), has been discussed in detail by CLEEVELY & MORRIS (1988). Earlier papers (HERM, 1977; KOLLMANN, 1975) have emphasized the occurrence of this genus in marginally marine to nonmarine environments utilizing the studies of MERTIN (1939). CLEEVELY & MORRIS (1988) have pointed out that the family lived in "essentially marine inshore/backshore environments".

6. Gastropod assemblages

This chapter gives an outline of the composition of those gastropod assemblages that do not consist exclusively of nerineaceans and actaeonellids. For following the evolution of communities it would be most advantageous to compare contemporaneous assemblages of each environment. This poses problems: On one hand the fossil record is not complete; on the other hand there are hardly any faunas known to me from siliciclastic sediments of the lower Lower Cretaceous because of the broad geographic distribution of carbonate platforms (SCHLAGER, 1981).

In the following Cretaceous gastropod assemblages of different statigraphic levels and substrates are split up into their archaeogastropod, mesogastropod, neogastropods and opisthobranchs taxa. The Mesogastropoda are further split into families (except for the Cerithiacea); the opisthobranchs are divided into Cephalaspidea and Entomotaeniata. The assemblages are discussed at the end of this chapter. To facilitate the reading of this paper the traditional taxonomic classification into orders is used.

6.1. Orgon and Brouzet-les-Alais. The fauna of Orgon has been described by COSSMANN (1916) while the fauna of Brouzet-les-Alais has been described by PELLAT and COSSMANN (1907) and DE BRUN, CHATELET, and COSSMANN (1916). The gastropod fauna is summarized in table 2.

26 Archaeogastropoda
29 Mesogastropoda belonging to the following families
    4 Naticidae
    5 Aporrhaidae
    1 Purpurinidae
    2 Colombellinidae
    12 Cerithiacea (not subdivided into families)
    1 Epitoniidae
    3 Rissoidae
    1 questionary Omalaxidae (*Pleurotomaria urgonensis* of Cossmann)
    5 Pseudomelaniidae
24 Opisthobranchia
    10 Cephalaspidea
    14 Entomotaeniata

Table 2. Composition of the gastropod fauna of Orgon (after PELLAT & COSSMANN, 1916)

The fauna of Brouzet-les-Alais is similar to the fauna of Orgon, as far as the diversity of the nerineaceans and archaeogastropods is concerned. The mesogastropods include aporrhaids and naticids, but the diversity is not as high as in Orgon. The same observation is made with the cephalaspidean opisthobranchs.

6.2. Aquilanian Abruzzos. Diverse faunas of Cenomanian age have been described from the Aquilanian Abbruzzos of Italy by SCHNARRENBERGER (1901) and PARONA (1909). The gastropod localities separated by Parona in his faunal list are compared in table 3:

Fossa Agnese	Northern margin of the Fossa Mezza Spada
27 Archaeogastropoda	14 Archaeogastropoda
13 Mesogastropoda	6 Mesogastropoda
4 Naticidae	2 Naticidae
1 Mathildididae	00
6 Cerithiacea	1 Cerithiacea
1 Aporrhaidae	2 Aporrhaidae
00	1 Cypraeidae
1 Colombellinidae	00
2 Neogastropoda	2 Neogastropoda
13 Opisthobranchia	5 Opisthobranchia
00	1 Cephalaspidea
13 Entomotaeniata	4 Entomotaeniata

Table 3. Comparison of Cenomanian gastropod faunas of the Aquilanian Abbruzzos (after PARONA, 1909)

6.3. Nodular limestones of the Washita-Edwards group of Texas. The faunas of this widely distributed facies have been investigated by CRAGIN (1893), STANTON (1947), PERKINS (1960) and KOLLMANN, DECKER, LEMONE (in preparation). PERKINS (1960) has shown that the Goodland Formation of the Fort Worth-Weatherford area contains a fauna dominanted by bivalves. It has not been explicitly mentioned by this author that the bivalves are preserved as internal molds of articulating specimens. The gastropods listed by PERKINS (1960) are mostly turritellids, aporrhaids and the naticid *Tylostoma*. Additionally, the trochid *Calliostoma*, a cerithiid belonging to the genus *Diozoptyxis* or a closely related genus (*Cerithium bosquensis* of STANTON, 1947) and an hitherto undescribed neogastropod genus (*?Turbo benbrookensis* STANTON) are recorded. Furthermore, this facies has yielded ammonites, many irregular echinoids and solitary corals.

The fauna of the Finlay Limestone Formation of West Texas (KOLLMANN, DECKER, LEMONE, in preparation) has virtually the same composition. Additionally, it contains abundant specimens of the neritid *Otostoma*, and the epitoniid *Confusiscala*. While the other molluscs are represented by internal molds the shell of Otostoma is always preserved.

The nodular limestones of the Washita-Edwards Group have been deposited under open shallow marine conditions at low water energy (ROSE, 1972; WILSON & JORDAN, 1983). They represent the facies zone 2 of WILSON (1975).

6.4. Northeastern Japan. The gastropod fauna of the Miyako Group in Japan (KASE, 1984) occurs in siliciclastic rocks of Upper Aptian-Albian age. It is obviously of marginally Tethyan origin. KASE has shown that patelliform species are restricted to conglomerates and to conglomeratic coquinal limestones while other archaeogastropods occur also in fine- to medium-grained sandstones but not in sandy shales. The mesogastropods are mostly restricted to the calcareous sandstones. Only *Trajanella* is abundant in conglomerates. *Cirsocerithium*, 2 genera of aporrhaids, *Gyrodes*, and 2 nerineaceans occur in sandy shales which otherwise show the lowest diversity of gastropods of all sediments of the Miyako Group.

6.5. Losenstein Formation. The Upper Albian fauna of the Losenstein Formation in Austria which was described by KOLLMANN (1976, 1978, 1979, 1982) contains 93 taxa which have undergone transportation. This fauna does not indicate a single level bottom community but represents a series of different environments (KOLLMANN, 1982).

6.6. Hölzelsau, Tyrol. A level bottom community is represented by the upper Lower Cenomanian gastropod fauna of Hölzelsau (Tyrol), described by RAHMAN (1967). See table 4:

9 Archaeogastropoda
27 Mesogastropoda
    3 Turritellidae

1 Architectonicidae
13 Cerithiacea (different families)
1 Epitoniidae
1 Pseudomelaniidae
1 Aporrhaidae
1 Cypraeaceae
5 Naticidae
5 Neogastropoda
28 Opisthobranchia
15 Entomotaeniata
13 Cephalaspidea

Table 4. Middle to Upper Cenomanian gastropods from Hölzelsau, Tyrol (RAHMAN, 1967)

6.7. Uchaux, France. The sand of Uchaux was deposited during the Upper Turonian in the "Golfe rhodanien" (SORNAY, 1950). Its rich and diverse fauna represents a level bottom community. The gastropod fauna described by D'ORBIGNY (1842–43) and ROMAN & MAZERAN (1920) has the following composition (Table 5)

9 Archaeogastropoda
30 Mesogastropoda
4 Pseudomelaniidae
1 Epitoniidae
4 Naticidae
5 Turritellidae
1 Xenophoridae
5 Cassiopidae
5 Cerithiacea (not subdivided into families)
1 Strombidae
4 Aporrhaidae
9 Neogastropoda
8 Opisthobranchiata
5 Cephalaspidea
3 Entomotaeniata

Table 5. Gastropods of the Upper Turonian of Uchaux, France (after ROMAN & MAZERAN, 1920)

6.8. Gosau Group. The Gosauschichten of the older literature do not represent a single facies. They occur in several geographically isolated areas of the Eastern Alps. These are called Gosau Basins after the village of Gosau in Upper Austria. The only comprehensive monograph on the Gosau gastropods (ZEKELI, 1852) is outdated in many respects. Critical revisions of this monograph by REUSS (1853) and STOLICZKA (1865) are outdated, too. More recent papers are those on the gastropods of the surroundings of Brandenberg

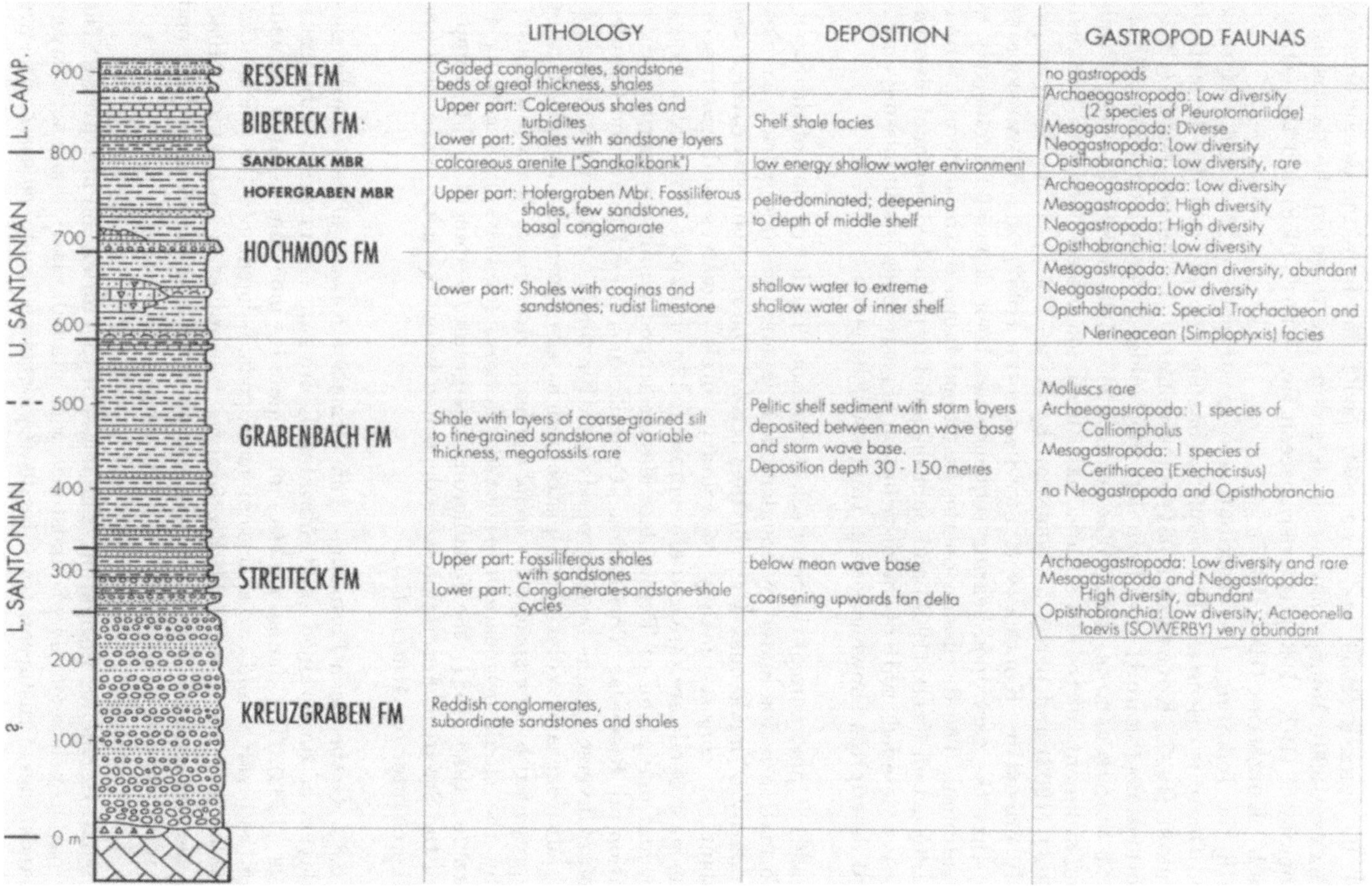

Fig. 1. The distribution of gastropods in the lower part of the Gosau sequence (lithostratigraphic column and partly environment interpretation after Wagreich, 1988).

(Tyrol) by SCHENK (1969) and on a gastropod fauna from the Gosau area by KOLLMANN (1980). Monographs on single groups are those on the melanopsids by KOLLMANN (1975, 1984), the nerineaceans by TIEDT (1958) and the actaeonellids by KOLLMANN (1965, 1967).

In the following, the distribution of gastropods in the rock sequence around Gosau is discussed. This sequence has been subdivided into formations by WEIGEL (1937), KOLLMANN (1981) and KOLLMANN & SUMMESBERGER (1982). The formations (figure 1) have been deposited under different conditions (KOLLMANN & SUMMESBERGER, 1982; WAGREICH, 1985, 1988, 1989). The megafossil-bearing sediments of the Gosau area have been considered partly as Turonian by FELIX (1908) and partly as Coniacian by KÜHN (1947). It has been subsequently proved by KOLLMANN & SUMMESBERGER (1982) and SUMMESBERGER (1985) that the gastropod-bearing sequence has been deposited exclusively in the Santonian. The discussion on the gastropod fauna of the different formations is based on FELIX (1908) and published and unpublished personal observations. As a detailed study of the distribution of gastropods is still missing only a general description showing certain trends will be given.

6.8.1. Upper part of the Streiteck Formation. The lower part of the Streiteck Formation is dominated by conglomeratic layers; the upper part is composed primarily of mudstones with siltstone layers. The megafauna is diverse and abundant. It consists mainly of solitary corals, bivalves and gastropods. Archaeogastropods are rare and are represented by a taxon of Trochidae (Angariinae) and one taxon of Turbinidae. Mesogastropod families occurring in this formation are Rissoidae, Turritellidae, Cerithiaceans, Aporrhaidae, and Cypraeidae. Most diverse are the neogastropods with Buccinidae, Galeodidae, Fasciolariidae, Mitridae and Volutidae. Opisthobranchs are especially abundant; they are represented by one species of cephalaspideans, *Actaeonella laevis* (SOWERBY), and by the small nerineacean *Aptyxiella flexuosa* (SOWERBY). According to WAGREICH (1985, 1987) the fossiliferous fine-grained sediments of the upper part of the Streiteck Formation were deposited below the mean wave base (see also the chapter on *Actaeonella*).

6.8.2. Grabenbach Formation. This formation consists of pelitic sediments with layers of fine-grained sand. Megafossils are rare (KOLLMANN & SUMMESBERGER, 1982). The fauna consists of echinoids, inoceramids and nuculids among the bivalves and two species of gastropods, belonging to the trochid *Calliomphalus* and the potamidid *Exechocirsus*. According to WAGREICH (1985, 1987) the Grabenbach Formation was deposited between mean wave base and storm wave base at a depth below 30 meters.

6.8.3. Hochmoos Formation. The facies of this formation are diverse. The lower part consists mainly of mudstones with graded, clastic biogenetic layers and sandstones. Crustacean trace fossils (*Ophiomorpha* and *Thalassinoides*) are abundant.

The gastropod faunas of the Nefgraben and the Wegscheidgraben of the Gosau basin listed by Felix (1908) are representative for the lower part of this formation. There are two taxa belonging to the archaeogastropod family Trochidae (2 species of *Angaria*). The mesogastropods are represented by the families Cerithiidae, Naticidae, Architectonicidae, Aporrhaidae, Potamididae, Xenophoridae and Rissoidae. The neogastropods are represented by Volutidae, the opisthobranchs by the nerineacean *Aptyxiella flexuosa*. The gastropod fauna is accompanied by diverse bivalve faunas and abundant solitary corals.

According to Wagreich (1988) the lower Hochmoos Formation has been deposited in shallow to extremely shallow water of the inner shelf under wave influence. A special facies of the Hochmoos Formation is the "Schneckenwand" which shows an accumulation of large *Trochactaeon*. These sediments have been interpreted by Höfling (1985) as deposits of an hypersaline lagoon (see also preceding chapter).

The upper part of the Hochmoos Formation is partly formed by the pelite-dominated Hofergraben Member (Kollmann, 1981). The gastropod fauna of the Hofergraben as recorded by Felix (1908) consists of the following: 3 taxa of the archaeogastropod family Trochidae; the mesogastropod families Rissoidae, Turritellidae, Potamididae, Aporrhaidae and Naticidae; the neogastropod families Fasciolariidae, Mitridae,Volutidae and Turridae, a single opisthobranch species which belongs to the Actaeonellidae. The bivalves and corals are very diverse.

The percentage of planktonic forms within the total number of foraminifera is higher in the Hofergraben Member than it is in the underlying part of the Hochmoos Formation (Wagreich, 1988). This indicates a deepening of the water.

The uppermost part of the Hochmoos Formation is formed by a fine-grained sandstone called Sandkalkbank (Weigel, 1937). The gastropods have been described by Kollmann (1980). The Archaeogastropoda are represented by 2 taxa of Pleurotomariidae. The Mesogastropoda are most diverse, consisting of: Rissoidae, Architectonicidae, Turritellidae, Potamididae, Aporrhaidae (3 taxa), Xenophoridae and Naticidae. The neogastropod families of the Sandkalkbank are Vexillinidae, Fasciolariidae, Fusinidae, Vasidae and Volutidae. The opisthobranchs comprise Acteonidae and Actaeonellidae. This gastropod fauna is accompanied by corals and a diverse fauna of bivalves (Dhondt, 1987 has recorded 50 taxa) as well as ammonites (Summesberger, 1979, 1980).

The sediment was deposited in a low energy shallow water environment (Kollmann, 1980; Wagreich, 1988).

7. Environment and composition of gastropod faunas

The preceding characterization shows the fundamental difference between gastropod faunas of carbonate platforms and of fine-grained siliciclastic sediments: The faunas of the carbonates of Orgon, Brouzet-les-Alais, and the Aquilanian Abbruzzos contain a high percentage of archaeogastropods. Modern representatives of this family are generally rasping off algae from rocky surfaces. Among the mesogastropods, cerithiaceans and aporrhaids which are deposit feeders are most diverse. Modern Cerithiaceans are living on hard substrates as well as on muddy bottom sediments (BANDEL & WERNER, 1987); aporrhaids are living buried in the surface layers of soft sediments (PURCHON, 1968). Soft substrates are also inhabited by the naticids and the large number of cephalaspidean opisthobranchs which are ploughing through the sediments searching for food.

This co-occurrence of gastropod groups preferring hard substrates and others living in and on soft substrates is characteristic for carbonate platforms. They provide the conditions for a diverse gastropod life. It also indicates the problem of the magnitude of ecological niches in this environment. This is equally the case in the clastic klippen facies of marginal warm temperate environments as in the Upper Cenomanian of Saxony and Bohemia where diverse faunas of gastropods with different environmental preference have been described by GEINITZ, 1874; DENINGER, 1905; WEINZETTL, 1910; and FRIC, 1911. A similar interpretation is made for the Maastrichtian of Maastricht (Holland) from which diverse gastropod faunas have been described by KAUNHOWEN (1898).

Deeper water environments of carbonate platforms are typically represented by nodular limestones of Texas and other areas of the Cretaceous Tethyan realm. The gastropod fauna consists mostly of large deposit feeders (turritellids, aporrhaids) and *Tylostoma* which systematically is placed within the carnivorous naticids. Nerineacea and actaeonellids, typical for the shallow water Tethyan sediments, are lacking in this facies. The number of archaeogastropoda is low. In West Texas the neritacean Otostoma is common in horizons with abundant oysters (KOLLMANN, LEMONE, and DECKER, unpublished) which may have formed the substratum for this archaeogastropod.

While the shallow water environments of the carbonate platforms provided sufficient hard surfaces for the growth of algae this was not the case on the level bottoms (Hölzelsau, Uchaux, Gosau). The communities are therefore characterized by low diversities of archaeogastropods. The number of taxa is decreasing with the grain size of the substrate. This has been shown by KASE (1984) for the Miyako Group. The Cenomanian fauna of Hölzelsau and the Turonian fauna of Uchaux are both found to occur in sandy-silty sediments and yield 9 archaeogastropod taxa each. With the exception of heavy-shelled neritopsids and a large fragmentary form determined as Astele by RAHMAN (1967), the archaeogastropods of Hölzelsau are small and may have lived as well on the sediment surface as on aquatic plants. This especially may be the case for small forms of *Discotectus* which are very abundant in fine-grained sediments of the Gosau Group (HÖFLING, 1985). In addition to Hölzelsau, the fauna of Uchaux shows large representatives of the trochid *Angaria* which also occurs in fine-grained sediments of the Gosau Group and was, because of its size, undoubtedly living on the bottom surface.

The main constituents of the faunas of Hölzelsau and Uchaux are the deposit-feeding cerithiaceans, among them the aporrhaids and the turritellids (for their ecology see ALLMON, 1980). This agrees with observations of JABLONSKI & BOTTJER (1983) that epifaunal deposit feeding molluscs prevail where the substrate becomes soft and muddy. In this respect, the gastropod faunas of the level bottoms resemble those of the nodular limestones but are much more diverse.

The same is the case with the carnivorous and ectoparasitic gastropod groups. The nodular limestones contain the naticid Tylostoma which occurs in large numbers and epitoniids. The level bottom assemblages of both, Hölzelsau and Uchaux, have yielded the carnivorous mesogastropod families Epitoniidae, Cypraeidae and Naticidae, further neogastropods and cephalaspidean opisthobranchs. The Entomotaeniata are diverse in Hölzelsau but are represented by only 2 species in Uchaux. This is consistent with the extinction of most nerineacean genera at the end of the Cenomanian.

As in the assemblages reviewed before the mesogastropods of the Streiteck Formation, the Hofergraben Member of the Hochmoos Formation and of the "Sandkalk-Bank" forming the uppermost part of the Hochmoos Formation are dominated by various cerithiacean families: among those abundantly represented are turritellids and aporrhaids. According to WAGREICH (1989), the sediments of these sequences have been deposited below the wave base (Streiteck Formation) of a "middle shelf" and in an environment with rare storm events. The fauna of the lower Hochmoos Formation which has been deposited in shallow to extremely shallow water is not as diverse, although the mesogastropod families represented are the same as in the other formations.

Neogastropods are dominating the Gosau gastropod faunas. This is due to the general development of this group in the Tethyan realm during the Cretaceous (see below). The diversity is low in the lower Hochmoos Formation which was deposited in shallow to extremely shallow water. In this part of the sequence the neogastropods are exclusively represented by Volutidae.

Opisthobranchs are occurring mainly in special facies, such as the lagoonal sediments of the Schneckenwand with its mass-occurrence of the actaeonellid Trochactaeon. Others contain abundant forms of the nerineacean Simploptyxis which is characteristic for the late Cretaceous after the Upper Turonian.

An unique feature is the occurrence of archaeogastropods belonging to the pleurotomariids in the "Sandkalk-Bank" which indicates a cold-water influence (KOLLMANN, 1980). This also is indicated by several bivalve taxa occurring mainly in the temperate marine zone (DHONDT, 1987).

A special facies is represented by the Grabenbach Formation of the Gosau area. Only 2 gastropod genera are represented here: the trochid *Calliomphalus* and the cerithiacean *Exechocirsus*. The low diversity of this fauna and especially the co-occurrence of the protobranch bivalve genus Nuculana indicates after KAUFFMAN (1976) a relatively poor benthonic environment. *Calliomphalus* is one of the few archaeogastropod genera which occur in the Upper Cretaceous

Ripley Formation of the Atlantic Coastal Plain (Sohl, 1960) and was probably feeding on organic material deposited on the sediment surface.

8. Evolution of Cretaceous gastropods

In the Permian and Triassic, the diverse stock of Palaeozoic archaeogastropods which formed a high percentage of the soft substrate communities was largely replaced by the deposit feeding and carnivorous mesogastropods. This gastropod group gained dominance at the end of the Triassic. It is especially diverse in fine-grained sediments where the potential for preservation is greatest. The turritellids, which are among the most common gastropods of soft substrates deposited from the Lower Cretaceous upwards, do not occur in shallow water sediments of the Tethyan realm before the Upper Albian (Kollmann, 1979; n. b., the Lower Albian fauna from the Corbières, France, described by Mongin, 1985 is not Tethyan in its composition).

Synchronously with the rise of the mesogastropods, the archaeogastropod families dominant in the Mesozoic and Cenozoic evolved. Modern representatives of these families are primarily grazers that are no longer restricted to soft substrates. Their diversity is highest on surfaces with sufficient algal growth, especially on rocks or patches of rock in shallow environments.

An important step in the evolution of gastropods is the first occurrence of the neogastropods which took place in the Cretaceous. The oldest fossil records of neogastropods are from the Hauterivien Hradiste Formation (Grodischter Schichten) of the Silesian nappe of the Western Carpathians (Ascher, 1906) and of Barremian sediments of North Germany (Wollemann, 1900; Kollmann, 1982). According to Joblonski & Potter (1990) these forms are problematic and may belong to mesogastropod groups as colombellinids. Actually, they have nothing in common with the colombellinids which have narrow dentate apertures. The forms discussed here have a short channel and are Buccinidae. This family is therefore occurring earlier in the geologic history than any of the Muricacea of which earliest representatives have been found at the Aptian-Albian boundary in North Germany (Wollemann, 1903). This fact is in opposition to Ponder's (1973) opinion who believes that the Muricacea are closest to the most primitive neogastropod type because of their unspecialized pallial cavity.

Ponder (1973) has indicated that the first evolutionary steps of neogastropods are within the Tropical marine zone. The fossil record disagrees with this view as all neogastropods earlier than Upper Albian have been found in the Temperate zone. The spreading into the Warm Temperate zone and into the Tethyan realm did not take place prior to this time (Stanton, 1947; Kollmann, 1976). The composition of assemblages presented in the preceding chapter clearly develops the gradual increase of neogastropod families in the Tethyan realm during the Upper Cretaceous. This agrees with Taylor and others (1980) who have stated that most predatory prosobranch families have evolved in higher latitudes and not in the tropics, although these groups show a strong latitudinal gradient increasing towards the equator in modern seas. This has been explained by MARGALEF (1968) who has called the tropical belt a refuge of types which settled there because of the stable climate but have originated in higher latitudes where the climatic fluctuation was more severe.

Opisthobranchs are diversifying considerably in the Mesozoic, too. The opisthobranch order Cephalaspidea of which earliest fossil representatives are known from the Lower Carboniferous (KOLLMANN & YOCHELSON, 1976) gives rise to a large number of genera and species (KITTL, 1894; HAAS, 1953) in the Triassic. The cephalaspideans have sturdy shells and consequently have developed an adequate fossil record. Modern cephalaspideans are primarily living infaunally and are generally carnivorous (e. g. ANDREWS, 1971; NORDSIECK, 1972). The evolution of the cephalaspideans is part of the evolution of bulldozing forms since the Permian (THAYER, 1983).

Actaeonellids and nerineaceans are the only gastropod groups of family and higher level in the taxonomic hierarchy which are restricted almost totally to the Tethyan realm. It may be puzzling that they are considered as opisthobranchs. Although they are extraordinarily large, compared with modern cephalaspideans the Actaeonellids are put into this opisthobranch group because of a sutural notch and a heterostrophic larval shell (KOLLMANN, 1967; SOHL & KOLLMANN, 1985). The earliest representatives of this group are species of *Cylindrites* of Middle Jurassic age. The major development took place in the higher Lower Cretaceous and in the Upper Cretaceous. The diversification of the actaeonellids as shown by SOHL & KOLLMANN (1985) is connected with spreading of the Atlantic and by the consequent interruption of the gene flow. This happened in the Coniacian for Trochactaeon. This is according to SOHL (1967) the time of a general change of American gastropod faunas.

Compared with modern opisthobranchs, many of the nerineaceans are equally extraordinary in size as are the actaeonellids. They are merged into the order Entomotaeniata COSSMANN together with the much smaller Pyramidellacea. The nerineaceans occur first in the Liassic (PCHELINTSEV, 1965). The major evolutionary pulses of the nerineaceans occurred between the Upper Jurassic and the Cenomanian on the open shelf. At our present stage of knowledge no provincialism of nerineaceans can be documented prior to the higher Upper Cretaceous. This indicates a high dispersal potential due to a long larval stage.

9. Instead of a foot-note: Remarks on the origin of neogastropods

A characteristic morphological feature of the neogastropod shell is the siphonal channel. This channel supports the siphon which serves to maintain an inhalant current of water into the pallial cavity. The water current on its way to the ctenidium which is a respiration organ passes the osphradium. This is particularly complex in carnivorous gastropods and may aid in the detection of the prey. As the inhalant siphon is directing the water current toward the osphradium, this development is clearly an advantageous morphogenetic feature.

The origin of the neogastropods certainly will remain always as a lively subject for discussion of their early history. Necessarily, this history is deduced from shell morphology of fossil forms which hardly can ever be reconciled with the anatomy based on soft part morphology of modern forms. A review of the different opinions has been assembled by TAYLOR and others (1980). These authors believe that the neogastropods originate from the Triassic to Upper Cretaceous family Purpurinidae ZITTEL. COSSMANN (1906) included in this family shells of diverse groups varying considerably in morphology. Many taxa belonging to this family do actually resemble neogastropods by their possession of an anterior notch, the shape of the aperture, and the sculpture.

Nevertheless, considerable doubt remains as all representatives of the Purpurinidae seem to have lived in warm water zones. As explained above the origin of neogastropods must be sought in the Temperate marine zone. The neogastropods must have evolved from a group that is represented in the Temperate zone, too.

One group from which the neogastropods may have evolved are the Amberleyacea which occur in the Temperate and the Warm Temperate zone. Representatives of this group which are closely related to each other occur in level bottom communities of the North German Lower Cretaceous. This seems to be unlikely for herbivorous forms. The nutrients and the physical conditions have been too uniform to generate a genetic pressures which may have led to the differenciation of different taxa. It therefore has been concluded by KOLLMANN (1982) that the amberleyaceans were carnivores and that the species may have evolved through different food preference. The amberleyaceans would seem to be a serious candidate for the origin of neogastropods, although, admittedly, the shell morphology is not nearly as convincing as that of the Purpurinidae.

The PONDER (1973) scenario has the neogastropods evolving from an archaeogastropod or a mesogastropod still possessing anatomical features of archaeogastropods (a "primitive" mesogastropod after PONDER). PONDER believes that the origin of neogastropods is related to a change from grazing to the feeding on a colonial animal. This is in accordance with his opinion that muricids are closest to the origin of neogastropods as this family is generally living on hard surfaces and therefore could have performed the step as described by changing its diet.

Actually, the fossil record is opposite to this opinion. Earliest known neogastropods have lived on soft bottoms which did not allow an algal growth. The Buccinids which are the first neogastropod group in the fossil record probably have evolved from deposit feeders by digesting small animals or carrion. The very general food of Buccinidae (TAYLOR and others, 1980) confirms this opinion. In this evolutionary concept, the neogastropods developed methods and devices which led from browsing towards active predatory hunting (Fig. 2). There is no absolute necessity for neogastropods to evolve from forms which already had a developed siphonal channel. It was only with the adoption of

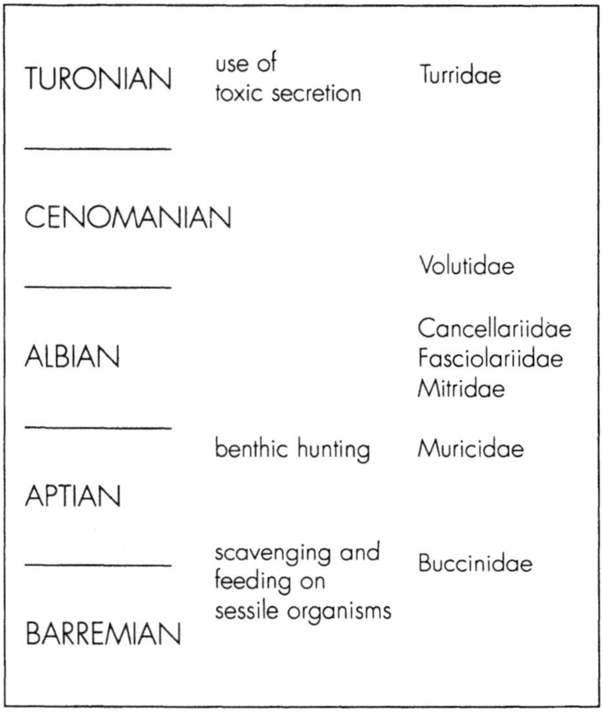

Fig. 2: Evolution of neogastropod feeding in the Cretaceous.

carnivorism and the subsequent increasing metabolism that a device was developed for obtaining sufficient food. This device was the siphonal channel which together with the osphradium allowed the efficient location of prey.

10. Tethyan gastropods at the beginning and at the end of the Cretaceous

The Tethyan gastropod faunas do not show any dramatic changes at the Jurassic-Cretaceous boundary. This may be due to the fact that Tethyan gastropods are known only from carbonate platforms which obviously were not affected by environmental changes at this time.

With the exception of *Tylostoma* which occurs in abundance in the Danian sediments of Denmark and Greenland (Kollmann & Peel, 1983) typically Tethyan groups did not survive the Cretaceous-Tertiary boundary. They shared this fate with other major Tethyan groups such as the rudists, most hermatypic coral genera (Beauvais & Beauvais, 1974) and significant numbers of the planktonic foraminifera. As these fossils have lived in shallow environments they are generally not associated with groups which allow an exact dating of the sediments. For this reason it is impossible to document the fact of whether the extinction occurred synchronously at a single event for all above-mentioned Tethyan groups or over a longer time-span.

If one examines the record of the extinction and decimation of selected fossil taxa at the end of the Cretaceous the following occurs: The majority of the affected taxa are within the Tethyan realm. The warm temperate and the cold temperate zone seem to be much less affected by the terminal Cretaceous event(s). It has been pointed out by KOLLMANN (1979a) that Lower Danian gastropod faunas do not show any differentiation related to water temperatures. A warm temperate fauna was extended from Central Asia and North Africa into Greenland. The faunas on both sides of the Atlantic are different which probably is related to the breakdown of the equatorial current of the Atlantic. This clear separation of faunal provinces ends with the Thanetian where the temperature-controlled marine zones are re-established. It seems therefore that the extinction of Tethyan gastropods was primarily caused by a change of surface water temperatures. A bolide impact which should have effected the entire gastropod fauna, is definitely not proved by the paleontological record.

11. Gastropods and the "Mesozoic Revolution"

VERMEJ (1977, 1978) has drawn attention to the fact that predation was spreading in the Mesozoic among different animal classes. He has called this evolutionary important process "Mesozoic Revolution". The gastropods are affected in two ways by this process: On the one hand they developed a sturdiness of shells to resist predation (VERMEJ, 1977, 1978); on the other hand carnivorous gastropod groups gained steadily in importance. This does not only concern the opisthobranchs and the neogastropods among the prosobranchs, but also the mesogastropods. TAYLOR and others (1980) have pointed out that the carnivorous mesogastropod groups Naticacea, Cymatiidae, Bursidae, Cassidiidae, Tonnidae and Ficidae had their first occurrence in the Mesozoic or at the Cretaceous-Tertiary boundary. The Naticidae which after KITTL (1892) and HAAS (1953) were already diverse in the Triassic obviously have changed their feeding methods during Mesozoic. Naticid drillholes do not occur before the Upper Albian where they have been documented with some frequency in the Blackdown Greensand (TAYLOR and others, 1980). SOHL (1969) correlated the occurrence of the drillholes with the first occurrence of the naticid subfamily Polynicinae.

The reason for the spreading of predation has been explained by VERMEJ (1977) as a response to the breakup of continents which according to VALENTINE (1969) and VALENTINE and MOORES (1970) created numerous biogeographic provinces as a consequence of geographic isolation. This certainly may explain the burst of biotic diversification of benthic groups following the interruption of the gene flow between members of population. Genetic isolation will lead to the development of different non interbreedable species which consequently may be broken up taxonomically into higher systematic hierarchies.

This taxonomic procedure is totally artificial showing the grade of similarity between species which form the base of the hierarchy. It does not represent the rise of predation itself because in mature ecosystems the amount of energy transferred through the various trophic levels is about the same at equal physical conditions (TAPPAN, 1968, 1971 a, b), independent of the number of taxa.

TAYLOR and others (1980) have pointed out that "the diversification and adaptive radiation of predatory families during the late Cretaceous may have arisen as a consequence of competive interaction between taxa, giving rise to broad habitat separations and specializations upon particular food type". The increase in predation requires an increase in prey availability in the same environment which, consequently, can be only the result of a higher primary (phytoplankton) production. TAPPAN (1968, 1971 a, b) has shown that the fossil record indicates as well an increase of the number of phytoplankton taxa as an increase of phytoplankton production in the Jurassic. This is probably controlled by changes in ocean physiography. This increase in phytoplankton productivity which is deduced from an increase of the oxygen level is probably the actual reason for the "Mesozoic Revolution" which has affected virtually all marine environments.

Acknowledgments

Funding by the Austrian Science Foundation and by the Austrian IGCP Committee is gratefully acknowledged. I am grateful to Prof. David V. LEMONE of the University of Texas at El Paso for his help in improving the style of the manuscript.

12. Literature

ABBASS, H. L. (1973): Some British Cretaceous Gastropods belonging to the Families Procerithiidae, Certhiidae and Cerithiopsidae (Cerithiacea). – Bull. Brit. Mus. (Nat. Hist.), Geology, **23/2:** 105–175.

ACCORDI, G., F. CARBONE, F., & SIRNA, G. (1982): Relationships among tectonic settings, substratum and benthonic communities in the Upper Cretaceous of northeastern Matese (Molise, Italy). – Geologica Rom., **21:** 755–793.

ALIEV, O. B., & ALIEV, G. A. (1986): Novye vidy gastropod iz santonskich otlozenij. – Pal. Jurn., **1986/2:** 108–110.

ALLMON, W. D. (1988): Ecology of recent turritelline gastropods (Prosobranchia, Turritellidae): Current knowledge. – Palaios, **1988/3:** 259–284.

ANDERSON, F. M. (1958): Upper Cretaceous of the Pacific coast. – Geol. Soc. Am. Mem., **71:** 1–378.

ANDREWS, J. (1977): Shells and shores of Texas. – pp. I–XX, 1–365.

ASCHER, E. (1906): Die Gastropoden, Bivalven und Brachiopoden der Grodischter Schichten. – Beitr. Pal. Geol. Österr.-Ungarn, **19:** 135–172.

BANDEL, K., & WERNER, E. (1987): Hydroid, Amphineuran and Gastropod zonation in the Litoral of the Caribbean sea, Columbia. – Senckenbergiana marit., **19/1–2:** 1–129.
BRIART, A., & CORNET, F.-L. (1865): Description mineralogique, géologique et paléontologique de la Meule de Bracquegnies. –Mém. Cour. Sav. étrang., **34:** 1–92.
BRUN, P. DE, CHATELET, C., & COSSMANN, M: (1916): Le Barrémien supérieur a facies urgonien de Brouzet-Les-Alais, part 2. – Mem. Soc. Geol. Fr., Paléont., **21/4:** 1–56.
CARBONE, F., PRATURLON, A., SIRNA, G. (1971): The Cenomanian shelf-edge facies of Rocca di Cave (Prenestini Mts, Latium). – Geol. Rom., **10:** 131–198.
CLEEVELY, R. J., & MORRIS, N. J. (1988): Taxonomy and ecology of Cretaceous Cassiopidae (Mesogastropoda). – Bull. Brit. Mus. (Nat. Hist.) Geol., **44/4:** 233–291.
COSSMANN, M. (1895, 1896, 1899, 1901, 1903, 1904, 1906, 1909, 1912, 1915, 1918, 1921, 1925): Essais de Paléoconchologie comparée, vol. 1–14.
COSSMANN, M. (1916): Les coquilles des calcaires d'Orgon (B.- du- R.). – Bull. Soc. Geol. Fr., (4) **16:** 336–431.
CRAGIN, F. W. (1893): Contributions to the invertebrate paleontology of the Texas Cretaceous. – Geol. Surv. Texas, 4th Ann. Rep. for 1882: 139–294.
DACQUÉ, E. (1926): Paläogeographie. pp. 1–196.
DAUWALDER, P., & REMANE, J. (1979). Etude du Banc á Nérinés á la limite «Kimmeridgien-Portlandien» dans le Jura Neuchatelois méridional. – Paläontologische Zeitschrift, **53:** 163–181.
DECROUEZ, D. (1977): Etude stratigraphique au Crétacé d'Argolide (Peloponnèse septentrional, Grèce). 1-Introduction générale et la Formation de l'Akros (domaine ophiolithique externe). Note Lab. Pal. Univ. Genève, **3,** Nov. 1977: 8.
DELPEY, G. (1939): Trajanella Popovici-Hatzeg, genre méditerranéen du Crétacé supérieur. – Bull. Soc. Geol. Fr., (5) **7:** 297–303.
DELPEY, G. (1940): Paléogéographie du sud-ouest de la France au Crétacé supérieur. – Bull. Soc. Hist. Nat. Toulouse, **73:** 250–266.
DENINGER, K. (1905): Die Gastropoden der sächsischen Kreideformation. – Beitr. Pal. Österr.-Ung., **18/1:** 1–35.
DHONDT, A. V. (1987): Bivalves from the Hochmoos Formation (Gosau Group, Oberösterreich, Austria). – Ann. Naturhistor. Mus. Wien, **88A:** 41–101.
DIENER, C. (1925): Grundzüge der Biostratigraphie. – pp. I–IV, 1–304.
DJALILOV, M. R. (1972): K sistematike Akteonellia (Gastropod). – Pal. Jurn., **1971/1:** 16–23.
DOUVILLE, H. (1900): Sur la distribution géographique des rudists des orbitolines et des orbitoides. – Bull. Soc. Geol. Fr., (3) **28:** 222–235.
FELIX, J. (1908): Studien über die Schichten der oberen Kreideformation in den Alpen und den Mediterrangebieten. Die Kreideschichten bei Gosau. – Palaeontographica, **54:** 251–339.
FITTON, W. H. (1836): siehe Sowerby, J. de C.
FLEURY, J. (1980): Les zones de Gavrovo-Tripolitza et du Pinde-Olonos (Grèce orientale et Péloponnese du Nord). Soc. Geol. Nord, Publ., **4:** 1–651.
FRETTER, V., & GRAHAM, A. (1949): The structure and mode of life the Pyramidellidae, parasitic opisthobranchs. – Journ. Marine Biol. Ass., **28:** 493–532.
FRIC, A. (1911): Studien im Gebiete der böhmischen Kreideformation. Ergänzung zu Band I. Illustriertes Verzeichnis der Petrefacten der cenomanen Korycaner Schichten. – Arch. f. d. naturw. Landesdurchforschung von Böhmen, **15/1:** 1–101.
GEINITZ, H. B. (1871–1875): Das Elbthalgebirge in Sachsen, 1. Teil. – Palaeontographica, **20:** 1–319.

GUERANGER, E. (1867): Album paléontologique du département de la Sarthe. pp. 1–20, 25 pl.
GUNTER, G. (1957): Temperature. – Geol. Soc. Am., Mem., **67/1:** 159–184.
HAAS, O. (1953): Mesozoic invertebrate faunas of Perm. Part 2. Late Triassic gastropods from Central Peru. – Bull. Am. Mus. Nat. Hist., **101:** 7–328.
HACOBJAN, V. T. (1972): Novyj rod pozdnemelovych akteonellid. – Dokl. Akad. Nauk. Arm. SSR, **55:** 230–234.
HACOBJAN, V. T. (1976): Pozdnemelovye gastropody Armenskoj SSR. – pp. 1–440.
HACOBJAN, V. T. (1978): Biostratigrafija verchemelovych otlozenij armenskoj SSR. – pp. 1–287.
HANCOCK, J. M., & KAUFFMAN, E. G. (1979). The great transgressions of the Late Cretaceous. – Journ. Geol. Soc., **136/2:** 175–186.
HAZEL, J. E. (1970): Atlantic Continental Shelf and Slope of the United States – Ostracode Zoogeography in the Southern Nova Scotian and Northern Virginian Faunal Provinces. – Geol. Surv. Prof. Pap., **529-E:** I–V, E1–E21.
HEDGPETH, J. W. (1957): Marine Biogeography. – Geol. Soc. Am. Mem., **67/1:** 359–382.
HERM, D. (1977): Zyklische Regressions-Sedimentation und Fossil-Vergesellschaftungen in der Gosau (Santonium) von Brandenberg/Tirol. – Mitt. Bayer. Staatsslg. Paläont. hist. Geol., **17:** 257–277.
HÖFLING, R. (1985): Faziesverteilung und Fossilvergesellschaftungen im karbonatischen Flachwasser-Milieu der alpinen Oberkreide (Gosau-Formation). – Münchner Geowiss. Abh., Series A, **3:** 1–240.
JABLONSKI, D., & BOTTJER, D. J. (1983): Soft-bottom epifaunal suspension-feeding assemblages in the Late Cretaceous. – Biotic interactions in recent and fossil benthic communities (ed. M. J. S. Tevesz & P. L. McCall), pp. 747–812.
KASE, T. (1984): Early Cretaceous marine and brackish-water gastropods from Japan. – pp. 1–262.
KATSIKATSOS, G. CH., & KOLLMANN, H. A. (1987): An Upper Cretaceous Mollusc Fauna from the Marbles of Almyropotamos (Euboea, Greece). – Ann. Naturhistor. Mus. Wien, **88A:** 103–116.
KAUFFMAN, E. G. (1973): Cretaceous Bivalvia. In: Atlas of Palaeobiogeography (A. Hallam, Ed.), pp. 353–383.
KAUFFMAN, E. G. (1975): Dispersal and biostratigraphic potential of Cretaceous benthonic Bivalvia in the Western Interior. – Geol. Assoc. Can., Spec. Pap., **13:** 163–194.
KAUFFMAN, E. G. (1976): Deep-sea Cretaceous macrofossils: Hole 317 A, Manihiki Plateau. – Init. Rep. Deep Sea Drilling Proj., **33:** 503–528.
KAUFFMAN, E. G. & CL. JOHNSON (1988): The morphological and ecological evolution of the Middle and Upper Cretaceous reef-building rudistids. – Palaios, **5/3:** 194–216.
KAUNHOWEN, F. (1998): Die Gastropoden der Maastrichter Kreide. – Pal. Abh. N.S., **4/1:** 1–132.
KEGEL CHRISTENSEN, W. (1976): Biogeography of late Cretaceous belemnites of Europe. – Paläont. Z, **50/3–4:** 113–129.
KEMPER, E. (1987): Das Klima der Kreidezeit. – Geol. Jb., **A 96:** 5–185.
KENNEDY, W. J., & COBBAN, W. (1976): Aspects of ammonite biology, biogeography and biostratigraphy. – Spec. Pap. Palaeont., **17:** 1–94.
KENNEDY, W. J., & KOLLMANN, H. A. (1979): Lower Albian ammonites from the Tannheim Formation near Losenstein, Upper Austria. – Beitr. Paläont. Österreich, **6:** 1–25.
KITTL, E. (1891–94): Die Gastropoden der Schichten von St. Cassian der südalpinen Trias. – Ann. Naturhistor. Mus., **VI, VII, IX:** 1–277.

KOLLMANN, H. A. (1965): Actaeonellen (Gastropoden) aus der ostalpinen Oberkreide. Ann. Naturhistor. Mus. Wien, **68:** 243–262.
KOLLMANN, H. A. (1967): Die Gattung Trochactaeon in der ostalpinen Oberkreide. Zur Pylogenie der Actaeonellidae. – Ann. Naturhistor. Mus. Wien, **71:** 199–261.
KOLLMANN, H. A. (1975): Zur Systematik und Paläobiologie der Gastropodengattung Tanaliopsis COSSMANN. – Ann. Naturhistor. Mus. Wien, **79:** 37–46.
KOLLMANN, H. A. (1976): Gastropoden aus den Losensteiner Schichten der Umgebung von Losenstein (Oberösterreich). Part 1: Euthyneura und Prosobranchia 1. – Ann. Naturhistor. Mus. Wien, **80:** 162–206.
KOLLMANN, H. A. (1978): Gastropoden aus den Losensteiner Schichten der Umgebung von Losenstein (Oberösterreich). Part 2: Naticidae, Colombellinidae (Mesogastropoda). – Ann. Naturhistor. Mus. Wien, **81:** 173–201.
KOLLMANN, H. A. (1979): Gastropoden aus den Losensteiner Schichten der Umgebung von Losenstein (Oberösterreich). Part 3: Cerithiacea (Mesogastropoda). – Ann. Naturhistor. Mus. Wien, **82:** 11–51.
KOLLMANN, H. A. (1979a): Distribution patterns and evolution of gastropods around the Cretaceous-Tertiary boundary. – Proc. Symp. Cretaceous-Tertiary Boundary Events, pp. 83–87.
KOLLMANN, H. A. (1980): Gastropoden aus der Sandkalkbank (Hochmoosschichten, Obersanton) des Beckens von Gosau (OÖ.). – Ann. Naturhistor. Mus. Wien, **83:** 197–213.
KOLLMANN, H. A. (1981): Gosauablagerungen im Becken von Gosau. In: Erläuterungen zu Blatt 95 – St. Wolfgang. – Geol. K. Republ. Österr. 1 : 50.000 (B. PLÖCHINGER, Ed.) pp. 30–34.
KOLLMANN, H. A. (1982): Gastropoden aus den Losensteiner Schichten der Umgebung von Losenstein (Oberösterreich). Part 4: Archaeogastropoda und allgemeine Bemerkungen zur Fauna. – Ann. Naturhistor. Mus. Wien, **84/A:** 13–56.
KOLLMANN, H. A. (1982a): Cenomane Gastropodenfaunen aus den Ophiolith-Konglomeraten Böotiens (Griechenland). – Ann. Geol. Pays Helléniques, **31:** 333–358.
KOLLMANN, H. A. (1985): Gastropoden aus dem Tripolitza-Kalk von Krioneri, Zentrales Griechenland. – Praktika Akad. Athinon, **60:** 485–499.
KOLLMANN, H. A. (1985a): Upper Cretaceous gastropods from excavations for the Highway A 10 (Charente, France). – Cretaceous Research, **6:** 85–111.
KOLLMANN, H. A. (1987): Eine cenomane Gastropodenfauna aus Nea Nikopolis bei Kozani (Mazedonien, Griechenland). – Ann. Naturhistor. Mus. Wien, **89:** 37–56.
KOLLMANN, H. A. (1991): Tethys – the evolution of on idea (in press).
KOLLMANN, H. A., & PEEL, J. S. (1983): Paleocene gastropods from Nugssuaq, West Greenland. – Greenlands Geol. Unders., Bull., **146:** 1–115.
KOLLMANN, H. A., & SOHL, N. F. (1979): Western hemisphere Cretaceous Itieriidae gastropods. – Geol. Surv. Prof. Pap., **1125-A:** 1–15.
KOLLMANN, H. A., & SUMMESBERGER, H. (1982): Excursions to Coniacian-Maastrichtian in the Austrian Alps. pp. 1–105.
KOLLMANN, H. A., & YOCHELSON, E. (1976): A survey of Paleozoic gastropods possibly belonging to the subclass Opisthobranchia. – Ann. Naturhistor. Mus. Wien, **80:** 207–220.
KOYOUMONTZAKIS, G. (1987): Paleobiogéographie et paléoecologie des Nerineidae (Gastropoda) du Crétacé supérieur de Provence, Bassin du Beausset (Var, France, S. E.) – Géol. Méditerranéenne, **12–13:** 13–27.
LINSLEY, R. M. (1977): Some "laws" of gastropod shell form. – Paleobiology, **3:** 196–206.
MARGALEF, R. (1968): Perspectives in ecological theory. – pp. 1–111.

Marlière. R. (1939): La transgression albienne et cénomanienne dans le Hainaut. – Mem. Mus. Roy. Hist. Nat. Belg., **89:** 1–440.

Masse. J. P. (1991): The Lower Cretaceous Mesogee: The state of the art (in Press).

Mazeran, P. (1912): Sur un genre nouveau de gastéropodes du Crétacé supérieur. – Ann. soc. Linn. Lyon. **59:** 163–172.

Mennesier. G. (1984): Révision des gastéropodes appartenant à la famille des Cassiopidae Kollmann (= Glauconiidae Ptchelintsev). – Trav. Dép. Géol. Univ. Picardie, **1:** 1–190.

Mermighis. A. (1988): Plates-formes carbonatées et recifs a rudists du Crétacé supérieur de l'Argolide septentrionale (Peloponnese NE, Grèce). Thèse Univ. de Provence. pp. 1–171.

Mertin. H. (1939): Über Brackwasserbildungen in der oberen Kreide des nördlichen Harzvorlandes. – Nova Acta Leopoldina, N. F.. **7:** 141–262.

Mongin. D. (1971). Quelques gastéropodes nouveaux dans l'Aptien supérieur de Tunesie. – Notes Services Géol. Tun., **34:** 21–44.

Mongin. D. (1985). Les Mollusques de l'Albien des Corbières (Sud de la France). Modes de vie et paléoécologie (with an introduction on the stratigraphy by B. Peybernes). – Privately printed by D. Mongin. pp. 1–129.

Nackij. A. D. (1916). Gastropodi septarievych glin Mangyslaka. – Trv. Mus. Geol. Min. Petrograd. **2:** 23–48.

Nordsieck. F. (1972): Die europäischen Meeresschnecken. Opisthobranchia mit Pyramidellidae. Rissoacea. pp. I–XIII. 1–327.

d'Orbigny. A. (1842–43): Paléontologie Francaise. Terr. Crétacé, vol. 2 Gastéropodes. – pp. 1–456.

Owen. H. G. (1971): Middle Albien Stratigraphy in the Anglo-Paris Basin – Bull. Brit. Mus. (Nat. Hist.). Geology. Suppl., **8:** 1–164.

Parona. C. F. (1909): La fauna coralligena dei Cretaceo dei Monti d'Ocre. – Mem. serv. descriz. carta geol. Ital.. **7:** 1–242.

Pcelintsev. V. F. (1927): The Jurassic and Lower Cretaceous Fauna of the Crimea and the Caucasus. – Mém. Com. Géol., N. S., **172:** I–XVI, 1–320.

Pcelintsev. V. F. (1931)· Brjuchonogie verchnej Jury i niznego mela Kryma. – Glav. Geol. Razv. Doc. Upra. v.. pp. I–XIX. 1–252.

Pcelintsev. V. F. (1953): Fauna brjuchonogich verchnemelovych otlozenij Zakavkazyja i srednei Azii. – Geologiceskij Muzej A. P. Karpinskogo, Izdatellstvo Akad. Nauk SSR, pp. 1–391.

Pcelintsev. V. F. (1965): Murchisoniata mesozoja gornogo Kryma. – Akad. Nauk SSR. pp. 1–215.

Pellat. R.. & Cossmann. M. (1907)· Le Barrémien supérieur a facies urgonien de Brouzet-Lez-Alais (GARD). Déscription des gastropodes & pélécypodes. – Mém. Soc. Geol. France. **37:** 6–42.

Perkins. B. F. (1960). Biostratigraphic studies in the Comanche (Cretaceous) series of northern Mexico and Texas. – Geol. Soc. Am. Mem., **83:** 1–138.

Pictet. F. J., & Campiche. G. (1864): Déscription des fossiles du terrain Crétacé des environs de Sainte-Croix. – Mat. Pal. Suisse, ser. 3 part 2. pp. 1–752.

Ponder, W. F. (1973): The origin and the evolution of the neogastropoda. – Malacologia. **12/2:** 295–338.

Purchon, R. D. (1968): The biology of the Mollusca. pp. I–XXV + 1–560.

Rahman. A. (1967). Die Gastropoden der Oberkreide (Ober-Cenoman) von Hölzelsau bei Niederndorf in Tirol. – Mitt. Bayer. Staatssamml. Paläont. hist. Geol.. **7:** 23–134.

Rey, J. (1979): Les formations bioconstructes du Crétacé inférieur d'Estremadura (Portugal). – Geobios, Mem. spec., **3:** 89–99.
Roman, F., & Mazeran, P. (1920): Monographie Paléontologique de la Faune du Turonien du Bassin d'Uchaux et de ses dépendances. – Arch. Mus. d'Hist. Nat. Lyon, **12:** 1–137.
Rose, P. R. (1972): Edwards Group, Surface and Subsurface, Central Texas. – Bureau of Economic Geology, Univ. Texas at Austin, Rep. of Investigations, **74:** 1–198.
Rosenkrantz, A. (1970): Marine Upper Cretaceous and Lowermost Tertiary deposits in west Greenland. – Medd. Dansk Geol. Foren. Kobenhavn, **19:** 406–453.
Schlager, W. (1981): The paradox of drowned reefs and carbonate platforms. – Geol. Soc. Am. Bull., **92:** 197–211.
Schnarrenberger, K. L. (1901): Über die Kreideformation der Monte d'Ocre-Kette in den Aquilaner Abruzzen. – Thesis, pp. 1–39.
Shikama, T., & Yui, S. (1973): On some Nerineid gastropoda in Japan (Preliminary report). – Science Rep. Yokohama Nat. Univ., Sect. II, **20:** 9–58.
Sohl, N. F. (1960): Archeogastropoda, Mesogastropoda and stratigraphy of the Ripley, Owl Creek, and Prairie Bluff Formations. – Geol. Surv., Prof. Pap., **331 A:** 1–151.
Sohl, N. F. (1967): Upper Cretaceous gastropod assemblages of the Western Interior of the United States. – Paleoenvironments of the Cretaceous Seaway-a Symposium. pp. 1–37.
Sohl, N. F. (1969): The fossil record of shell boring by snails. – Am. Zool., **9:** 725–734.
Sohl, N. F. (1971): North American Cretaceous biotic provinces delineated by gastropods. – Proc. North Am. Paleont. Conv. L: 1610–1638.
Sohl, N. F. (1987): Cretaceous gastropods, Contrasts between Tethys and the temperate provinces. – Journ. Paleont., **61/6:** 1085–1111.
Sohl, N. F., & Kollmann, H. A. (1985): Cretaceous actaeonellid gastropods from the western hemisphere. – Geol. Surv. Prof. Pap., **1304:** I–IV, 1–104.
Sornay, J. (1950): Etude stratigraphique du Crétacé supérieur de la vallée du Rhone entre Valence et Avignon. – Trav. Lab. Geol. Univ. Grenoble, **27:** 35–278.
Stanton, T. W. (1947): Studies of some Comanche pelecypods and gastropods. – Geol. Surv. Prof. Pap., **211:** 1–253.
Stephenson, T. A. (1947): The constitution of the intertidal fauna and flora of South Africa. Part III. – Ann. Natal Mus., **11/2:** 207–324.
Suess, E. (1893): Are great oceans permanent? – Nat. Sci, **2:** 180–187.
Summesberger, H. (1979): Eine obersantone Ammonitenfauna aus dem Becken von Gosau. – Ann. Naturhistor. Mus. Wien, **82:** 109–176.
Summesberger H. (1980): Neue Ammoniten aus der Sandkalkbank der Hochmoosschichten (Obersanton; Gosau, Österreich). Ann. Naturhist. Mus. Wien, **83:** 275–283.
Summesberger, H. (1983): Ammonite zonation of the Gosau Group (Upper Cretaceous, Austria). – Ann. Naturhistor. Mus. Wien, **87:** 145–166.
Tappan, H. (1968): Primary production, isotopes, extinctions and the atmosphere. – Palaeogeogr. Palaeoclimatol., Palaeoecol., **4:** 187–210.
Tappan, H. (1971a): Faunal extinction. – Yearbook of Science and Technology 1971, pp. 214–217.
Tappan, H. (1971b): Microplancton, ecological succession and evolution. – North Am. Pal. Convention, Part H: 1058–1103.
Taylor, J. D., Morris, N. J., & Taylor, C. N. (1980): Food spezialization and the evolution of predatory prosobranch gastropods. – Palaeontology, **23/2:** 375–409.
Taylor, J. D. W., & Sohl, N. F. (1962): An outline of gastropod classification. – Malacologia, **1/1:** 7–32.

THAYER, CH. W. (1983): Sediment-mediated biological disturbance and the evolution of marine benthos. – Biotic interactions in recent and fossil benthic communities (ed. M. J. S. Tevesz & P. L. McCall), pp. 480–625.
THORSON, G. (1952): Zur jetzigen Lage der Bodentier-Ökologie. – Zool. Anz., Suppl., **16:** 276–327.
THORSON, G. (1957): Bottom Communities (Sublittoral or Shallow Shelf). – Geol. Soc. Am. Mem., **67/1:** 461–534.
VALENTINE, J. W. (1973): Evolutionary paleoecology of the marine biosphere. – pp. I–XV, 1–511.
VALENTINE, J. W., & MOORES, E. M. (1970): Plate-tectonic regulation of faunal diversity and sea level: A model. – Nature, **228:** 657–659.
VAUGHAN, T. W. (1940) Ecology of modern marine organisms with reference to paleogeography. – Geol. Soc. Am. Bull., **51:** 433–468.
VERMEJ, G. J. (1977) The Mesozoic marine revolution: Evidence from snails, predators and grazers. – Paleobiology, **3:** 245–258.
VERMEJ, G. J. (1978) Biogeography and adaption. Patterns of marine life. – pp. I–XI, 1–332.
VERMEJ, G. J. (1987): Evolution and Escalation. An ecological history. – pp. I–X, 1–527.
VILANOVA D J. (1863): Ensayo de description geognostica de la provincia de Teruel. – pp. 1–312.
WAGREICH, M. (1985). Zur Lithofazies der Gosauschichten im Gebiet von Gosau und Rußbach (Oberösterreich und Salzburg). – Anz. Österr. Akad. Wiss., math.-naturw. Kl., **121:** 13–17.
WAGREICH, M. (1988). Sedimentologie und Beckenentwicklung des tieferen Abschnittes (Santon-Untercampan) der Gosauschichtgruppe von Gosau und Rußbach (Oberösterreich-Salzburg). – Jb. Geol. B. A., **131/A:** 663–686.
WAGREICH, M. (1989): Coarsening-upward fan-delta sequences in the Lower Streiteck Formation (Santonian) of the Gosau Group near Gosau (Upper Austria). – N. Jb. Geol. Paläont. Mh., **1989/1:** 47–64.
WALTHER, J. (1908): Geschichte der Erde und des Lebens. – pp. 1–570.
WANDERER, K. (1909): Tierversteinerungen aus der Kreide Sachsens. – pp. I–XXII, 1–80.
WEIGEL, O. (1937): Stratigraphie und Tektonik des Beckens von Gosau. – Jb. Geol. B. A., **87:** 11–40.
WEINZETTL, V. (1910). Gastropoda Ceskeho Kridoveho útvaru. – Palaeontographica Bohemiae, **8:** 1–56.
WELLS, J. W. (1957): Coral Reefs. – Geol. Soc. Am. Mem., **67/1:** 609–631.
WIECZOREK, J. (1979): Upper Jurassic nerineacean gastropods from the Holy Cross Mts (Poland). – Acta Palaeont. Polon., **24:** 299–350.
WIEDMANN, J. (1988): Plate tectonics, sealevel changes, climate and the relationship to ammonite evolution, provincialism, and mode of life. – WIEDMANN, J., & KULLMANN, J. (Eds): Cephalopods – Present and Past, pp. 737–765.
WILSON, J. L. (1975). Carbonate Facies in Geologic History. – pp. 1–471.
WILSON, J. L., & JORDAN, C. (1983). Middle shelf environment. – AAPG Mem., **33:** 297–343.
WOLLEMANN, A. (1900): Die Bivalven und Gastropoden des deutschen und holländischen Neocoms. – Abh. Preuß. Geol. L. A., NF., **31:** 1–180.
WOLLEMANN, A. (1903): Die Fauna des mittleren Gaults von Algermissen. – Jahrb. Preuß. Geol. L. A., **24:** 22–42, pl. 4–5.

Wollemann, A. (1912): Die Bivalven und Gastropoden des norddeutschen Gaults (Aptiens und Albiens). – Jahrb. Kgl. Preuß, Geol. L. A., **27:** 259–300.
Zekeli, F. (1852): Die Gastropoden der Gosaugebilde. – Abh. Geol. R. A., **1/2:** 1–124.

3. Paleogeographic implications of Tethyan Cretaceous faunas and floras

New Aspects on Tethyan Cretaceous Fossil Assemblages.
Band 9 Schriftenreihe der Erdwissenschaftlichen Kommissionen der Österreichischen Akademie der Wissenschaften, 131–154.

3.1. Les assemblages des pollens normapolles du Cretacé supérieur des rivages ouest européens de la Tethys

Par J. MÉDUS*)

Upper Cretaceous normapolles pollen assemblages from the West European Tethys margin

Key Words: Normapolles, Paléogéographie, Crétacé supérieur, Tethys

"Yet the coastline is one of the most ephemeral features of paleogeography ..." (SMITH et al., 1981, p. 1).

Résumé

La trace d'un rivage occidental de la Téthys est dégagée a) de la comparaison des palynoflores de gisements du Turonien, Santonien et Campanien de Sardaigne, Provence, vallée du Rhône, Aquitaine, Nord de l'Espagne et Portugal, puis b) de l'analyse des affinités morphologiques de formes de *Pseudoromeinipollenites*.

Abstract

Evidence for a western sea-shore of the Tethys are obtained from:

a) the comparison of palynoflores of Turonian, Santonian and Campanian localities in Sardinia, the Provence, the Rhone valley, the Aquitaine, northern Spain and Portugal

b) an analysis of the morphological affinities of forms of *Pseudoromeinipollenites*.

La quantité importante de documents paléobotaniques obtenus par la

*) Faculté Sciences et Techniques de Saint Jérôme, Lab. Bot. Historique & Palynol., C. 451, 13397 Marseille Cédex 13.

palynologie sur les ensembles floristiques du Crétacé supérieur permet certes de bien délimiter les grandes zones de répartition de la flore terrestre de cette époque et d'étudier leurs rapports avec le climat (Batten, 1984; Góczán et al., 1967; Herngreen & Chlonova, 1981; Samoilovitch, 1967; Srivastava, 1978, 1981; Zaklinskaja, 1967). Sur la base de déterminations plus fines ou/et de la localisation géographique de certains taxons, une sub-division de ces empires floraux crétacés peut même parfois être proposée (Batten & Li, 1987; Frederiksen, 1987; Médus, 1973). Mais curieusement, en dépit de ces résultats, peu de travaux de paléopalynologie portent précisémment sur les zones continentales, leur position géographique ou leurs limites (Jekhowsky, 1963). Aussi, à l'aide d'une comparaison floristique d'abord et de l'analyse de la diversification d'un taxon ensuite, le but de la présente note est de montrer a) que les mêmes plantes prospéraient sur les rivages ouest-européens de la Tethys d'est en ouest durant le Sénonien et b) que les vicissitudes de cet environnement très mobile sont intervenues directement sur l'évolution des flores de la même façon qu'elles ont participées à celle des faunes (Adams & Ager, 1967).

W. Krutzsch avait dès 1966 reconnu l'existence d'une «région méditerranéenne» à l'intérieur de la Province des Normapolles sur la base d'un ensemble de taxons (e. g. *Capipollis oculis* Goczan *Oculopollis maximus* Krutzsch, *O. semimaximus* Krutzsch, *Pseudoromeinipollenites* Kedves) caractéristiques du domaine tethyan de Bohème méridionale, Hongrie et France du sud (Góczán

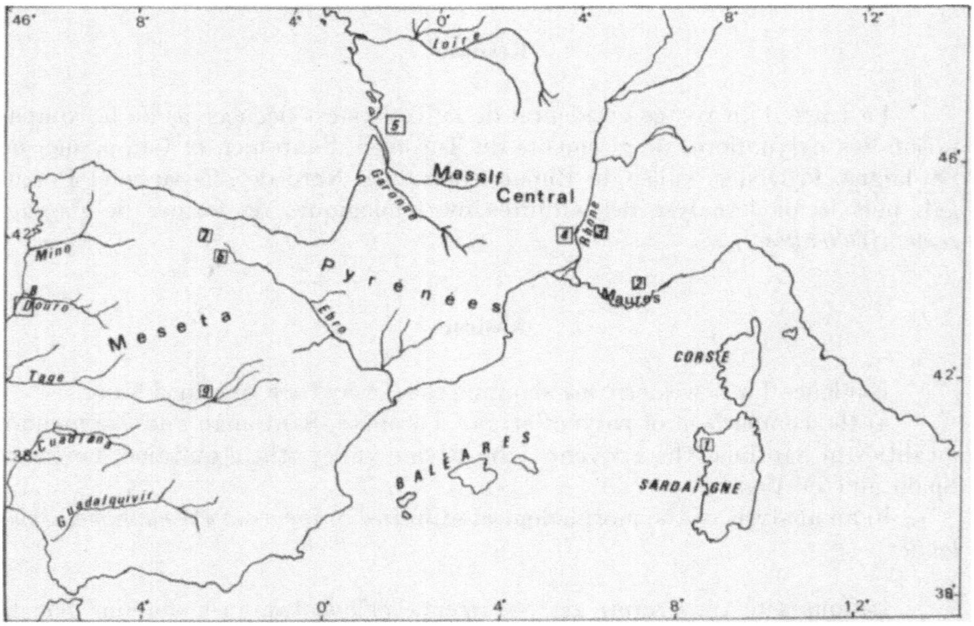

Fig. 1: Localisation géographique des gisements. 1, Brunestica; 2, Brenon; 3, Drôme, Vaucluse; 4, Gard; 5, Charentes; 6, Sedano; 7, Formation Voznuevo; 8, Portugal; 9, Madrid.

et al., 1967). Sur la situation et l'extension de la (des) masse(s) continentale(s) soumise(nt) à l'influence de la Téthys à ce moment et à son environnement et qui en constituai(en)t son rivage nord, l'analyse géographique de quelques assemblages du Crétacé supérieur de Sardaigne, du Sud de la France, de l'Espagne et du Portugal fournit déjà des indications.

Pour simplifier l'exposé, les gisements seront examinées de l'Est vers l'Ouest (see. fig. 1) et il faut préciser que seuls les résultats obtenus sur celui de Brenon en Provence sont inédits.

Les Sites

Le gisement de Brunestica se trouve au NE d'Olmedo en Sardaigne (Filigheddu & Oggiano, 1984): il correspond à une lentille d'argile ligniteuse située au toit de la bauxite de la Nura, directement surmontée par des calcaires du Coniacien (Cherchi, 1987). L'assemblage palynologique trouvé dans un échantillon comporte principalement des grains normapolles (tab. 1 et planches 1 et 2) avec, en faible pourcentage des grains tricolpés, tricolporés, monocolpés ornementés, gemmulés ou clavulés réticulés et des spores (e. g. *Patellasporites distaverrucosus* (Brenner) Kemp, *Vadaszisporites sacali* Deák et Combaz).

Le gisement de Brenon au Sud de Castellane (fig. 2) est constitué par une alternance d'argiles ligniteuses, grès et sables visible le long de la route D52 avant le village de Brenon. Cette formation repose sur des calcaires micritiques

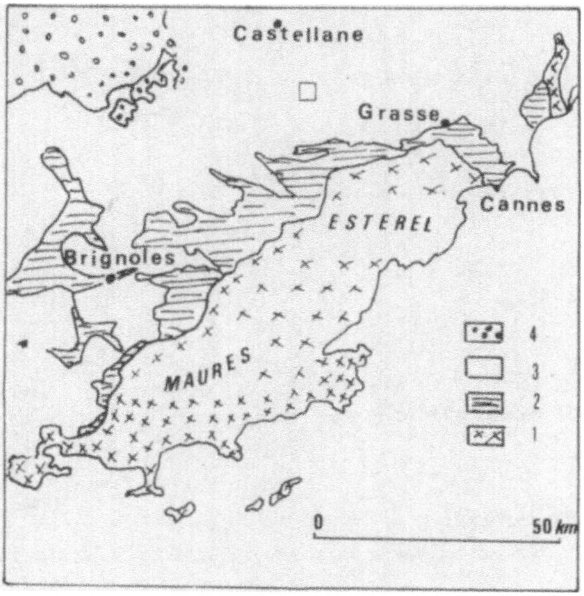

Fig. 2 Situation géologique simplifiée du gisement de Brenon. 1. Paléozoique: 2. Trias-Jurassique: 3. Crétacé: 4. Tertiaire.

Fig. 3: Situation géologique simplifiée des gisements de la Drôme, du Vaucluse et du Gard.

et marnes interstratifiées ayant livrées un assemblage de Globotruncanidae du Turonien supérieur-Coniacien basal. Les principaux éléments de la composition palynofloristique des niveaux favorables sont répertoriés dans le tableau 1. Il s'agit de grains normapolles pour la majorité (tab. 1 et pl. 1 et 2) et comme précédemment, le pourcentage du spectre laissé aux grains prolés tricolpés, tricolporés et aux monocolpés est faible: quelque spores sont remarquables telles: Spore trilète lisse in Médus, 1970b: *Leiotriletes (foveaperturites* typ.) in Azéma et al., 1972: *Cicatricosisporites* cf. *matesovae* in Médus et al., 1980 ou le genre *Camarozonosporites*.

Plus à l'ouest et encore en France se trouvent les gisements du massif de l'Etoile, entre Aix-en-Provence et Marseille (cf. Médus, 1972), des Martigues, des massifs plus ou moins proches de la vallée du Rhone (fig. 3): Saou, Nyons, Dieulefit Piolenc, de la Cèze . . . etc. Enfin, ceux d'Aquitaine et des Charentes (fig. 1). En Espagne, il y a les gisements de Sedano (Pce de Burgos) et de la Formation Voznuevo (fig. 4) au nord de Leon (Amerom, 1965: Cramer et al., 1979: Gaillard, 1982: Ipert, 1975, 1976: Médus, 1970 b, c, 1981, 1987a: Médus et al., l.c.: Samuel & Gaillard, 1984: Vazquez Reyero, 1983: cf. Herngreen & Chlonova, ibid. pour d'autres references). Vers le sud, il faut citer les gisements des environs de Madrid (Alvarez Ramis & Doubinger, 1980: Médus & Alvarez Ramis, 1989). Enfin au Portugal, existent plusieurs gisements du Crétacé supérieur sur le bord occidental du Massif hespérique (fig. 5) dans la région de Aveiro, Coimbra, Leira (Batten & Morrison, 1987: Médus et al., ibid: Médus, 1981). La liste taxonomique de chacune de ces palynoflores figure dans le tableau 1. Pour ettayer notre propos, les gisements de la bauxite des Coms prés de Artesa de Segre, Pce de Lerida, en Espagne et celui de la bauxite de Villeveyrac en Languedoc près de Montpellier seront cités aussi (fig. 6).

Paléogéographie floristique

On peut remarquer tout de suite que le nombre de taxons en commun entre les diférentes listes dépend d'une façon évidente plus de la position stratigraphique que géographique des gisements; ce qui permet d'ailleurs d'en inférer que l'âge probable des assemblages de Brunestica est Turonien supérieur et celui de Brenon Santonien. Il est clair d'autre part que ce tableau fait apparaître une forte continuité floristique entre les diverses zones géographiques. Or, si tel un floriste l'on admet 1) que les taxons utilisés correspondent à des espèces linéennes et 2) qu'il est légitime de rapporter à la même espèce des grains tout à fait semblables entre eux ou au type mais trouvés dans des sites différents, ce phénomène ne peut résulter que de la colonisation par une même flore d'un territoire unique ou peu morcelé pour que les populations spécifiques forment un continuum. Ainsi, le bloc Sarde devait nécessairement être proche des blocs paléozoiques de Provence et de Bohème: car de plus, la présence des plantes

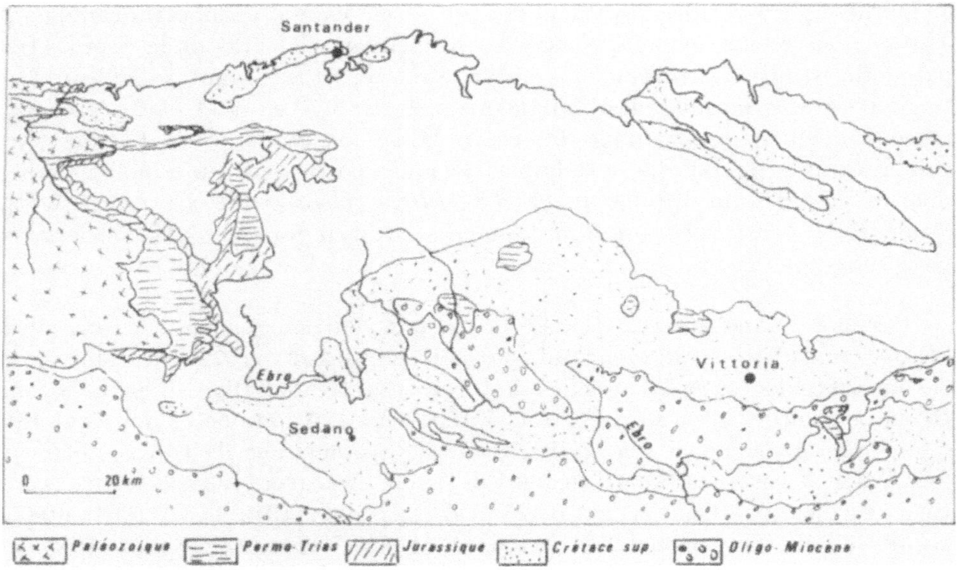

Fig. 4: Situation géologique simplifiée du gisement de Sedano.

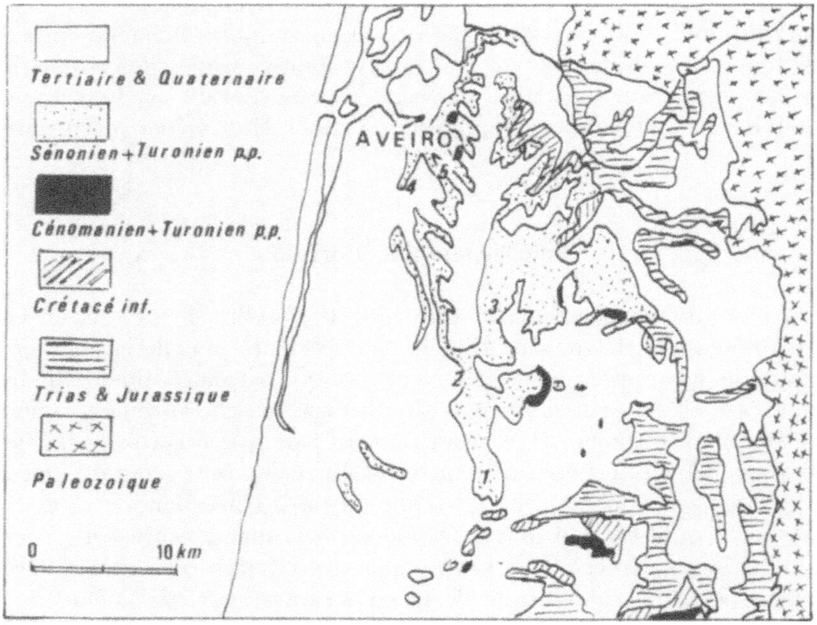

Fig. 5: Situation géologique simplifiée des gisements du Portugal. 1, Siadouro; 2, Vale; 3, Nariz; 4, Ilhavo; 5, Presa; 6, Chousa do Fidalgo.

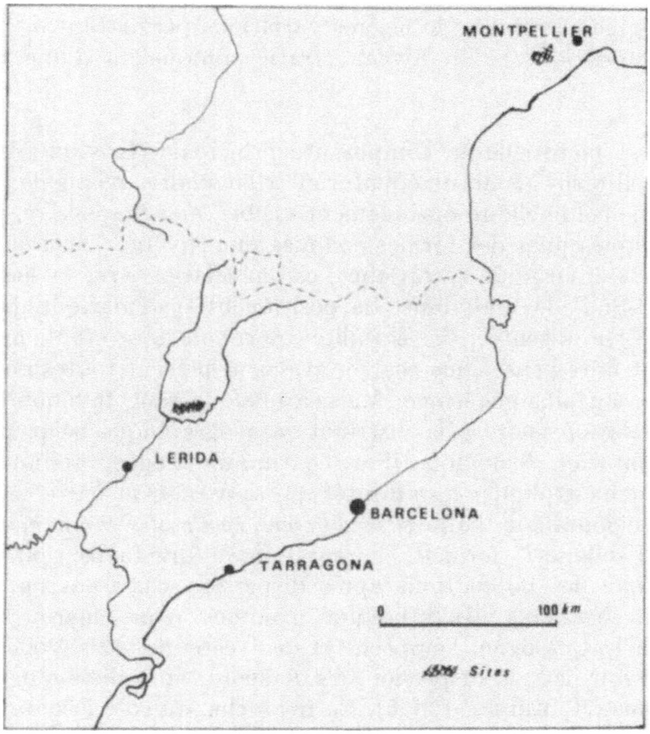

Fig. 6. Localisation géographique des bauxites des Coms près Lérida et de Villeveyrac près Montpellier.

productrices de *Capipollis oculis* en Sardaigne au Turonien puis au environs de Budapest au Santonien ne peut guère avoir plus simple explication.

Dans certains cas, comme les assemblages étudiés ne comportent pas, ou presque pas, d'élément marin ils caractérisent des végétations de marais littoraux et de ce fait la disposition géographique des gisements indique, approximativement, la ligne de rivage, par exemple le long des bords occidental et septentrional du Massif Hespéro-asturien ou des rives Est et Ouest du Massif Central français. Cependant, s'il est possible de visualiser les continents susceptibles de relier les gisements d'Espagne puis ceux de France, on voit encore dificilement comment ceux-là peuvent être reliés entr'eux, malgré certaines tentatives (Combes, 1969).

Paléogéographie biosystématique

A ce sujet cependant l'analyse morphologique de variations de *Pseudoromeinipollenites* fourni des indications (Médus, 1987b).

Ce taxon est connu en différents gisements d'Espagne, de France et en Hollande. Il présente une morphologie assez remarquable en microscopie optique et, bien qu'assez stable de légères variations permettent de distinguer des espèces caractéristiques d'un niveau stratigraphique ou d'une région ou des deux a la fois.

L'analyse factorielle en composante principale (KOVACH, 1987) de neufs caractères qualitatifs (contour équatorial triangulaire, triangulaire convexe ou circulaire; zone columellaire optiquement visible, marque polaire, pore grand ou petit) de quelques unes des formes connues montre (fig. 7) qu'une partie de la variation (35% environ de la variance) est en relation avec la situation géographique (Nord-Sud) des individus. La position biosystématique des populations peut certes être discutée: la stabilité morphologique tout au long de la séquence peut faire penser que ces variations se placent toutes au niveau spécifique, ou tout au plus générique. Mais, a quelque rang taxonomique qu'on les situe, cette relation claire à la distribution géographique ne peut résulter que de la fragmentation écologique d'un continuum biogéographique. Il est bien évident que pour expliquer les affinités qui apparaissent entre certaines formes de la zone occidentale de l'aire (*Pseudoromeinipollenites campanicus-P. irregularis-P.* forme I sedano-P. forme C charente), une individualisation écologique ou/et géographique des populations après dispersion pas-à-pas ne peut pas être évoquée. Des barrières géographiques connues (zone marine profonde sur l'Aquitaine et la Gascogne) l'empèche et des vecteurs actifs (zoochorie) doivent parfois intervenir dans la dispersion des diaspores et la dissémination de certaines populations (cf. MÉDUS, 1987b). En revanche, du coté oriental de l'aire, une succession de sites ayant fourni des assemblages continentaux (des mines de bauxite de Villeveyrac et des Coms près de Lérida, fig. 6) permet de relier les points extrèmes entre l'Espagne et la Provence. L'interprétation proposée (fig. 8) résume ces faits et questions.

Quelques rivages occidentaux de la Tethys

En d'autres termes, entre les blocs hespériques, pyrénéens et provencaux (Maures, Esterel . . .), puis entre ces blocs et le Massif Central français des basses plaines marécageuses ont sans doute facilité les échanges de flore (MÉDUS & IPERT, 1977), mais leur largeur a évidemment subi de multiples avatars avec les changements paléogéographiques du Sénonien (El Crétacico de Espana, 1982). Il est sur qu'au Maastrichtien non seulement cette zone et d'autres vers le Nord étaient largement émergées; ce qui d'ailleurs explique bien les affinités apparues entre les *Pseudoromeinipollenites* de Provence et *P. laevigatus* (KEDVES et HERGREEN) LEGOUX de Hollande. Vers le Sud au contraire, les conditions de gisement des flores du Nord de Madrid, recouvertes par des calcaires marins, montrent un ennoyage de cette zone après leur dépot au Campanien. Vers l'Est, comme les affinités floristiques durant le Sénonien l'indiquent (GOCZAN, 1964; GOCZAN et al., 1967 et illustration pl. 1–4), les relations continentales entre la Provence et les Monts Bakony étaient étroites. Une espèce telle que *Pseudoculo-*

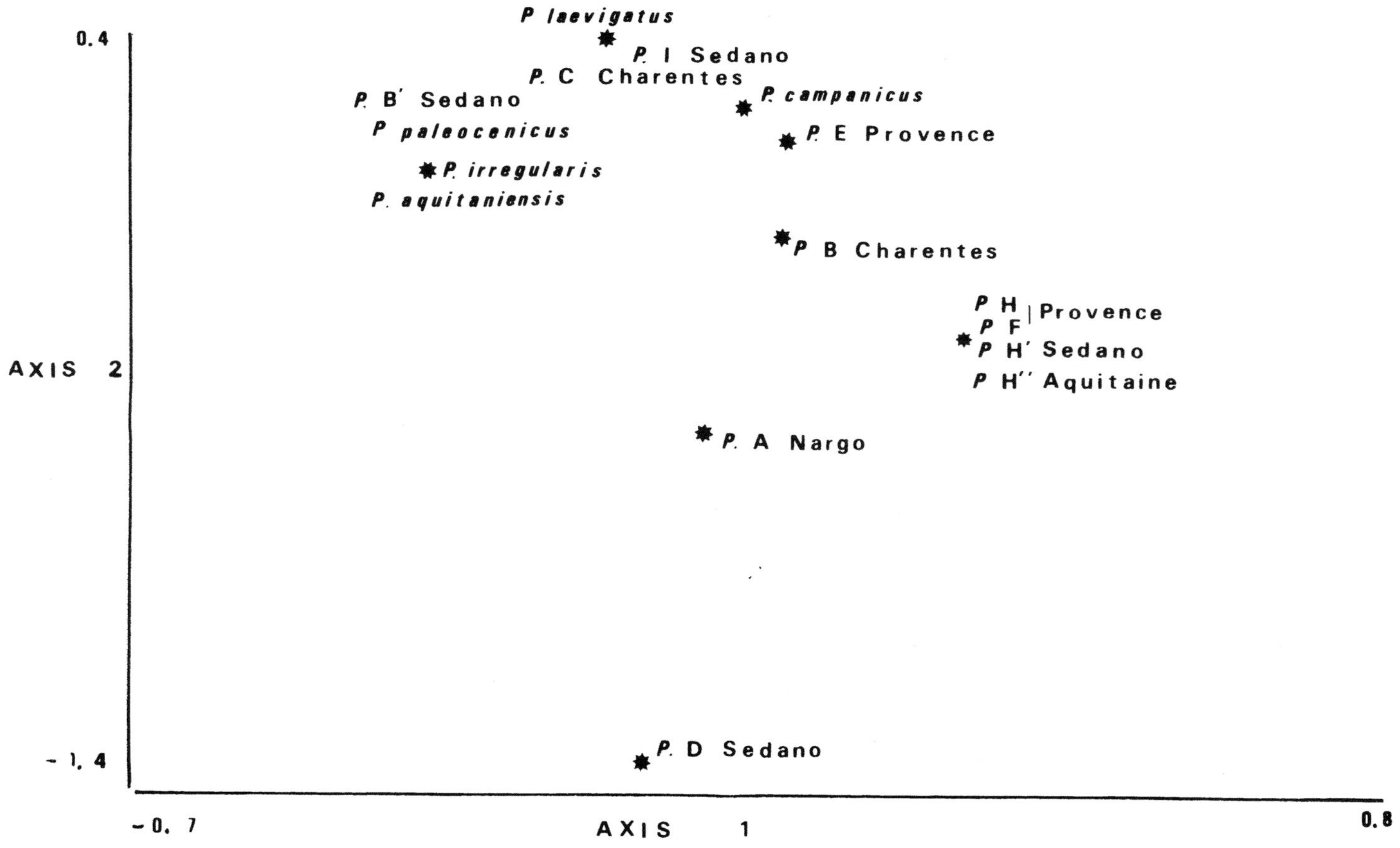

Fig 7· Répartition des individus dans le plan 1–2 pour l'analyse factorielle de neuf caractères de *Pseudoromeinipollenites*

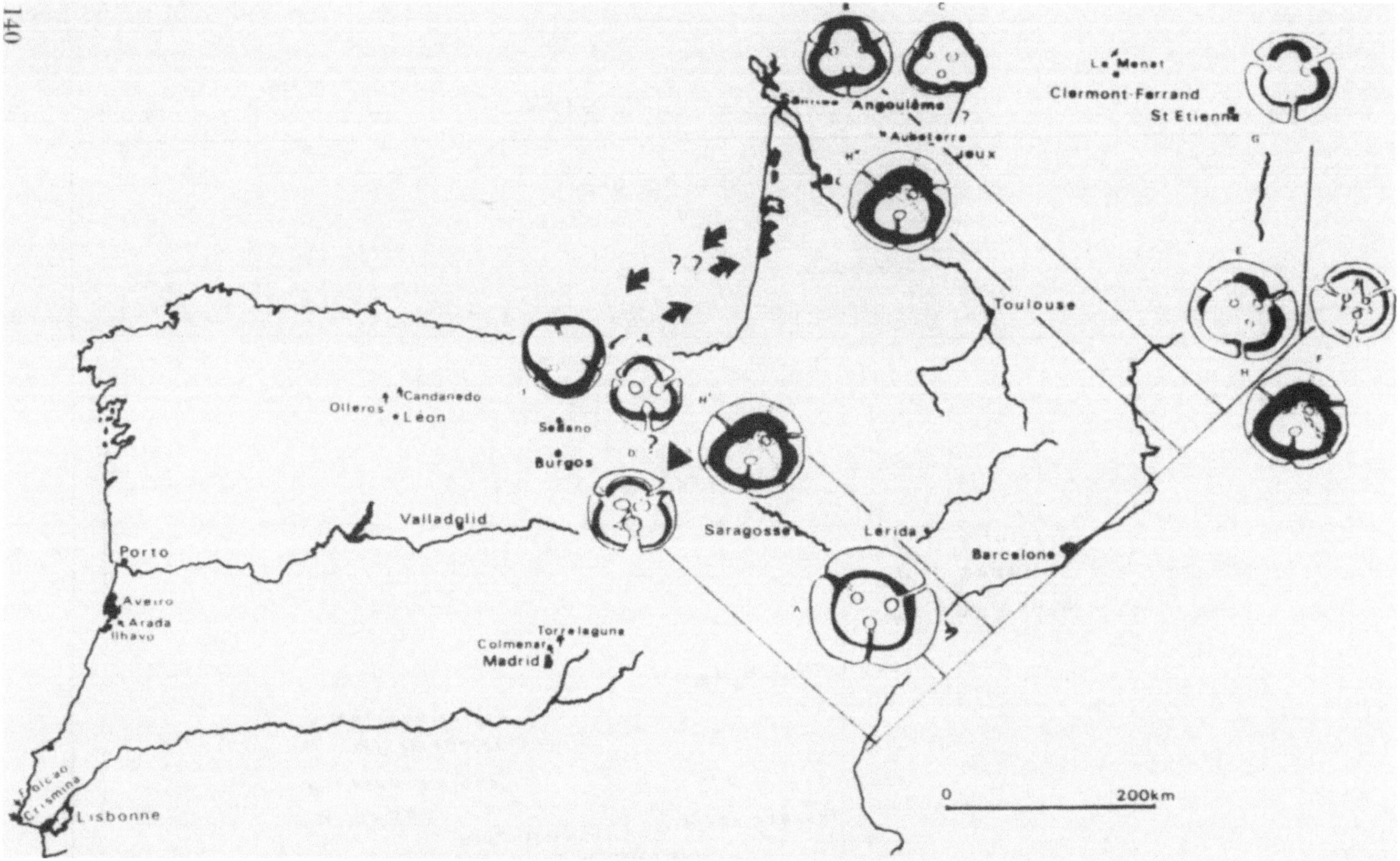

Fig. 8: Schéma des relations possibles entre les espèces de *Pseudoromeinipollenites*. Les traits et le triangle indiquent des relations continentales; les flèches un transport aérien ou marin.

pollis sp de Piolenc (Médus, 1970a) par exemple, pour exister aussi pendant le même étage – Santonien – aux abords du Massif bohémien (*Bohemiapollis* type A in Pacltova 1981, pl. 10, fig. 1) devait corespondre à des plantes productrices dont l'environment (palustres ou marécageux comme les milieux qui sont à la limite des domaines marins et continentaux) est un facteur important de leur diffusion.

Conclusion

Cette analyse géographique des paléoflores à grains normapolles, qui emprunte sa méthode a la phytogéographie ouvre des perspectives interessantes sur les différents types de végétation et leur répartition au Crétacé supérieur. L'étude paléobiologique d'un taxon comme on l'a constaté d'ailleurs repose sur les conditions géographiques. En outre, connaître la distribution des angiospermes sur les continents est également le moyen de savoir comment se produisent les changements floristiques au lieu de seulement les observer dans les successions stratigraphiques certes, celles-ci nous montrent de façon évidente les périodes (brutales ou progressives) de bouleversement floristique soit locales soit générales: c'est-à-dire d'ordre géomorphologique ou climatique-.

Dans le présent essai, les résultats acquis montrent que, sous certaines précautions concernant les taxons évidemment, l'étude phytogéographique des paléoflores conduirait à préciser les contours paléomagnétiques ouest européens de la Téthys au Campanien (barron et al., 1981). La mobilité de ces rivages a constituée manifestement un facteur déterminant dans la diversification des taxons: toutefois l'on devra essayer e savoir qu'elle part reviend aux modifications climatiques dans la diversification spécifique.

References

Adams, C. G., & Ager, D. V. (1967): Aspects of Tethyan biogeography. – The Systematics Association, Publication, **7:** 336 pp.

Alvarez-Ramis, C., & Doubinger, J. (1980): Los microfosiles vegetales del Cretacico superior de Torrelaguna (Madrid). – Bol. R. Soc. Espanola Hist. Nat. (Geol.), **78:** 205–215.

Amerom, H. W. J. (1965): Upper-Cretaceous pollen and spores assemblages from the so-called "Wealden" of the Province of Leon (Northern Spain). – Pollen et Spores, **7:** 93–133.

Azéma, C., Durand, S., & Médus, J. (1972): Des miospores du Cénomanien moyen. – Paléobiologie continentale, Montpellier, **3:** 1–54.

Barron, E. J., Harrison, C. G. A., Sloan II, J. L., & Hay, W. W. (1981): Paleogeography, 180 million years ago to the present. – Eclogae geol. Helv., **74: 443–470.**

Batten, D. J. (1984): Palynology, climate and the development of Late Cretaceous floral provinces in the Northern Hemisphere; a review. – Fossils and Climate, P. Brenchley ed., John Wiley & Sons Ltd., 127–164.

Batten, D. J., & Li W. (1987): Aspects of palynomorph distribution, Floral Provinces and climate during the Cretaceous. – Geol. Jb., **96:** 219–237.

Batten, D. J., & Morrison, L. (1987): Morphology and occurence of the Normapolles pollen genus Papillopollis in the Cretaceous of Portugal. – Palynology, **11:** 133–154.

Cherchi, A. (ed.) (1987): Groupe Français du Crétacé. Excursion en Sardaigne 24–29 mai 1987. Livret-Guide, Univ. Gagliari, 195 pp.

Combes, P. J. (1969): Recherches sur la genèse des bauxites dans le Nord-Est de l'Espagne, le Languedoc et l'Ariège (France). – Mém. Centre d'Etudes et de Rech. Géol. et Hydrogéol., **3–4:** 342 pp., Fondation C.E.R.G.A., Montpellier.

Cramer, F. H., Diez, M. d. C. R., & Vazquez, J. J. (1979): Nota sobre las palinofloras cretacicas del norte de Espana. Paleobiologie continentale, Montpellier, **10:** 7–12.

Diniz, F., Kedves, M., & Simoncsics, P. (1974): Les sporomorphes principaux de sédiments crétacés de Vila Flor et de Carrajao, Portugal. – Comm. Serv. Geol. Portugal, Lisboa, **58:** 161–178.

El Cretacico de Espana, 1982. Universidad Complutense, Madrid, pp. 680.

Filigheddu, R., & Oggiano, G. (1984): Contributo alla stratigrafia delle bauxiti e del Cretaceo della Nura Mediante lo studio di un livello pollinico. – Atti Soc. Tosc. Sci. Nat. Mem., Serie A, **91:** 111–118.

Frederiksen, N. O. (1987): Tectonic and Paleogeographic setting of a new latest Cretaceous floristic Province in North America. – The Society of Economic Paleontologists and Mineralogists, Research Reports, pp. 533–542.

Gaillard, M. G. (1982): Etude palynologique de gisements à lignite du Crétacé supérieur du département du Gard. Stratigraphie – Paléogéographie. – Thèse Dr. Troisième Cycle, Univ. Claude Bernard, Lyon, 141 pp.

Góczán, F. (1964): Stratigraphic palynology of the Hungarian Upper Cretaceous. – Acta Geologica, **8:** 229–264.

Góczán, F., Groot, J. J., Krutzsch, W., & Palctova, B. (1967): Die Gattungen des «Stemma Normapolles Pflug 1953b» (Angiospermae). Neubeschreibungen und Revision europäischer Formen (Oberkreide bis Eozan). – Palaont. Abh. B., **2:** 427–633.

Herngreen, G. F. W., & Chlonova A. F. (1981): Cretaceous microfloral Provinces. – Pollen et Spores, **23:** 441–555.

Ipert, C. (1975): Palynologie des faciès sablo-ligniteux du Sénonien de Dieulefit, Nyons et Saou (Drôme). – Thèse Dr. spécialité, Fac. Sc. St. Jérôme, Marseille, 61 pp.

Ipert, C. (1976): Données palynostratigraphiques sur les formations des sables à lignites de Dieulefit, Nyons et Saou (Drôme). – Géobios, **9:** 97–100.

Jekhowsky, B. de (1963): Variations latérales en palynologie quantitative et passage du continental au marin: le Dogger supérieur du Sud-Ouest de Madagascar. – Rev. Institut Français du Pétrole, **18:** 977–995.

Kedves, M., & Diniz, F. (1981): Contribution à la connaissance des pollens d'angiospermes du Crétacé supérieur du Portugal. – Sociedade Geologica de Portugal, **22** (1980–1981): 19–31.

Kedves, M., & Diniz, F. (1983): Les Normapolles du Crétacé supérieur en Europe: implications paléobiogéographiques. – Geobios, **16:** 329–345.

Kedves, M., & Pittau, P. (1979): Contribution à la connaissance des pollens des Normapolles du «groupe papilloide» du Crétacé supérieur du Portugal. – Pollen et Spores, **21:** 169–209.

Kovach, W. L. (1987): MVSP. A multivariate statistical package for the IBM PC and Compatibles. ver. 1.31. Dep. Biol. Indiana Univ. Bloomington, Ind.

Krutzsch, W. (1966) Die Sporenstratigraphische Gliederung der Oberkreide im nördlichen Mitteleuropa. Methodische Grundlagen und gegenwärtiger Stand der Untersuchungen. – Abh. zentr. geol. Inst., **8:** 79–111.

Médus, J. (1970a): Une analyse morphologique de quelques populations polliniques du Santonien de Piolenc (France). Palaeontographica, B. **130:** 1–11.

Médus, J. (1970b) Première contribution à la palynologie stratigraphique du Crétacé supérieur du sud-est de la France et du nord-est de l'Espagne. – Rev. Paléobiologie continentale, Montpellier, **1:** 3–14.

Médus, J. (1970c): Contribution à la connaissance des associations polliniques du Crétacé terminal dans le SE de la France et le NE de l'Espagne. – Rev. Micropal., **13:** 45–50.

Médus, J (1972) Palynological zonation of the Upper Cretaceous in southern France and northeastern Spain. – Rev. Palaeobot. Palynol., **14:** 287–295.

Médus, J. (1973). Contribution palynologique à la géobotanique de l'Eurasie au Crétacé supérieur. – 96e Congrès national des sociétés savantes, Toulouse, 1971, sciences **5:** 223–232.

Médus, J. (1981) Pollens Normapolles de coupes stratotypiques du Crétacé supérieur des Charentes et du Sénonien du Portugal. – Com. Serv. Geol. Portugal, **67:** 19–28.

Médus, J. (1982): Morphologie et ultra-structure de pollens du Turonien du genre *Pflugipollis* W. Kr. – Rev. Micropal., **24:** 238–243.

Médus, J. (1987a): Analyse quantitative des palynoflores du Campanien de Sedano, Espagne. – Rev. Palaeobot. Palynol., **51:** 309–326.

Médus, J (1987b). Paléogéographie et systématique du genre Pseudoromeinipollenites au Crétacé supérieur. – J Civis Llovera, & M. F. Valle Hernandez (eds). Acta de Palinologia (Actas del VI Simp. de Palinol., A.P.L.E.). Salamanca, sept. 1986, pp. 321–326.

Médus, J., & Ipert, C. (1977): Introduction de l'analyse multifactorielle dans le traitement de données palynofloristiques santoniennes (Crétacé supérieur). – Rev. Palaeobot. Palynol., **24:** 141–154.

Médus, J., & Alvarez Ramis, C. (1989): Des pollens normapolles d'un niveau du Crétacé supérieur de la région de Guadalix, Prov. de Madrid, Espagne. – Rev. Esp. Micropal **21:** 139–144. Madrid.

Médus, J., Boch, A., Parron, C., Lauverjat, J., & Triat, J. M. (1980): Turonian Normapolles From Portugal and Southern France; correlations. – Rev. Palaeobot. Palynol., **31:** 105–153.

Onoratini, G., & Azéma, C. (1973): Une palynoflore du Santonien inférieur dans le Gard. Considérations sur la structure de grains de pollen de Normapolles. Rev. Micropal., **16:** 214–222.

Palctova, B. (1981): The evolution and distribution of Normapolles pollen during the Cenophytic. Rev. Palaeobot. Palynol., **35:** 175–208.

Pacltova, B., & Krutzsch, W. (1970). Neue Pollen- und Sporenarten aus der mittleren Oberkreide, insbesondere Mitteleuropas. – Palaeobotanik, B, **3:** 573–598.

Pflug, H. (1953): Zur Entstehung und Entwicklung des Angiospermiden pollens in der Erdgeschichte. Palaeontographica, B, **95:** 60–171.

Samoilovich, S. R. (1967): Tentative botanico-geographical subdivision of Northern Asia in Late Cretaceous time. – Rev. Palaeobot. Palynol., **2:** 127–139.

Samuel, E., & Gaillard, M. G. (1984): Les gisements à flore fossile d'âge crétacé supérieur en France: localisation, stratigraphie et essai de corrélations des données macro et microfloristiques. – Bull. Soc. Linn. Lyon, **53:** 213–223.

Skarby, A. (1968): Extratriporopollenites (Pflug) emend. from the Upper Cretaceous of Scania, Sweden. – Stockholm Contributions in Geology, **16:** pp. 60.
Srivastava, S. K. (1978): Cretaceous spore-pollen Floras: A global evaluation. – Biological Memoirs, **3:** 2–130.
Srivastava, S. K. (1981): Evolution of Upper Cretaceous phytogeoprovinces and their pollen flora. – Rev. Palaeobot. Palynol., **35:** 155–173.
Vazquez Reyero, J. J. (1983): Aportaciones para la identificacion de los generos Papillopollis e Interporopollenites pertenecientes al Cretacico superior de la Formacion Voznuevo, Provincia de Leon, noroeste de Espana. – Actas del IV Simposio de Palinologia, N. Solé de Porta, & M. Suarez Cervera, eds. Barcelona, pp. 387–396.
Zaklinskaia, E. D. (1967): Palynological studies on Late Cretaceous-Paleogene floral history and stratigraphy. – Rev. Palaeobot. Palynol., **2:** 141–146.

Taxons/Gisements	Sa	Tu	Br	RP	Aq	PS
Complexiopollis turonis in Diniz et al., 1974 (*Complexiopollis* sp. 3 in Médus et al., 1980)	+	+				
Complexiopollis (?) sp. 6 in Médus et al., 1980	+	+				
Extratriporopollenites sp. in Médus et al., 1980	+	+				
E. nonperfectus in Médus et al., 1980	+	+				
E. perlucidus Skarby	+	+				
E. vestibulus in Médus et al., 1980	+	+				
Extrapollis hastaclarus (Weyl. et Krg.) Krutzsch	+	+				
Hungaropollis sp. in Médus, 1981	+	+				
Oculopollis sp. 2 in Médus et al., 1980	+	+				
Pflugipollis sp. in Médus, 1982 = *Pseudoculopollis* sp. in Médus et al., 1980	+	+				
Portaepollenites aff. *P. ornatus* Kedves et Diniz	+	+				
Triangulaepollis turonicus in Gaillard, 1982	+	+				
Trudopollis sp.	+	+				
Trudopollis sp. 2	+	+				
Turonipollis sp. 3 in Médus et al., 1980	+	+				
Vacuopollis microconcavus Pacltova et Krutzsch	+	+	+			
V. percentus in Gaillard, 1982	+	+				
V. pyramis in Gaillard, 1982	+	+				
E. vestibulus (Ross) Skarby	+	+	+			
Trudopollis sp. 3	+		+			
Piolencipollis piolencensis Groot et Krutzsch	+		+	+		
Basopollis sp. 2 in Médus et al., 1980		+	+			
Complexiopollis (?) sp. 6 in Médus et al., 1980		+	+			

Taxons/Gisements	Sa	Tu	Br	RP	Aq	PS
E. desidens in Médus et al., 1980		+	+			
Cf. *E. hemiperfectus* in Médus et al., 1980		+	+			
Latipollis sp. 2 in Médus et al., 1980		+	+			
Neotriangulipollis sp. 13 in Médus et al., 1980		+	+			
Oculopollis sp. 1 in Médus et al., 1980		+	+			
Papillopollis sp. 3 in Médus et al., 1980		+	+			
Cf. *Pompeckjoidaepollenites* sp. in Médus et al., 1980		+	+			
T. cf. *T. pertrudens* in Médus et al., 1980		+	+			
Turonipollis christae in Médus et al., 1980		+	+			
Vacuopollis sp. 2 in Médus et al., 1980		+	+			
Vacuopollis sp. 5 in Médus et al., 1980		+	+			
Vacuopollis sp. 8 in Médus et al., 1980		+	+			
Vacuopollis sp. 9 in Médus et al., 1980		+	+			
Complexiopollis vancampoe in Médus et al., 1980		+	+	+		
Semioculopollis sp. 2 in Médus et al., 1980		+	+	+		
Extrapollis hastaclarus (Weyl. et Krg.) Krutzsch		+		+		
Bohemiapollis nemejci in Médus, 1981			+	+		
Longanulipollis bajtayi in Onoratini & Azéma, 1973			+	+		
Magnoporopollenites praemagnoporatus Krutzsch			+	+		

Taxons/Gisements	Sa	Tu	Br	RP	Aq	PS
M. germicrassicus KRUTZSCH			+	+		
Megatriopollenites santonius GROOT et KRUTZSCH			+	+		
Oculopollis semimaximus KRUTZSCH			+	+		
Proteacidites cf. *annularis* in MÉDUS, 1970a			+	+		
Spore trilète lisse in MÉDUS, 1970b			+	+		
Suemegipollis triangularis GÓCZÁN			+	+	+	
Oculopollis maximus KRUTZSCH			+	+	+	
Extratriporopollenites cf. *longianulus* in MÉDUS, 1970b				+	+	
Heidelbergipollis tilioides KRUTZSCH				+	+	
Lusatipollis gallicus GROOT et KRUTZSCH				+	+	
Trudopollis sp. 4 in MÉDUS, 1981				+	+	
Convexipollis convexigerminalis KRUTZSCH				+	+	+
Pseudopapillopollis praesubherzynicus				+	+	+
Trudopollis bulborus in MÉDUS, 1981				+	+	+
Papillopollis aradaensis KEDVES et PITTAU				+		+
Cf. *Intratriporopol magnoporatus* in MÉDUS, 1981				+	+	+
Conclavipollis purgatus PFLUG					+	+
Interporopol. f. stephanocolpus in MÉDUS, 1987					+	+
Papillopollis barbosae BATTEN et MORRISON					+	+
Platycarya					+	+
Prenudopollis prezensis KEDVES et DINIZ					+	+
Trudopollis endanulens in MÉDUS, 1981					+	+
T. verruco-échinulé in MÉDUS, 1981					+	+

Tableau 1: Distribution de quelques taxons des palynoflores de Brunestica (Sa), du Turonien du Gard et Vaucluse (Tu), de Brenon (Br), du Santonien-Campanien de la vallée du Rhône et Provence (RP), du Santonien-Campanien des Charentes-Aquitaine (Aq) et du Portugal et Sedano (PS).

Planche 1

(Toutes les figures X 1000)

Fig. 1 *Trudopollis* sp. 2
Fig. 2. *Pecakipollis* sp.
Fig. 3· cf. *Triangulaepollis* sp. in ALVAREZ RAMIS & DOUBINGER. 1980
Fig. 4 *Emscheripollis* typ.
Fig. 5. 6· *Gothanipollis* sp.
Fig. 7: *Trudopollis* sp. 3
Fig. 8. 9: *Vacuopollis* typ. recyclés
Fig. 10. Ticolporé *Siberiapollis* typ
Fig. 11 *Longanulipollis* sp.
Fig. 12 *Trudopollis fossulotrudens* (PFLUG) PFLUG
Fig. 13 *Trudopollis* sp.

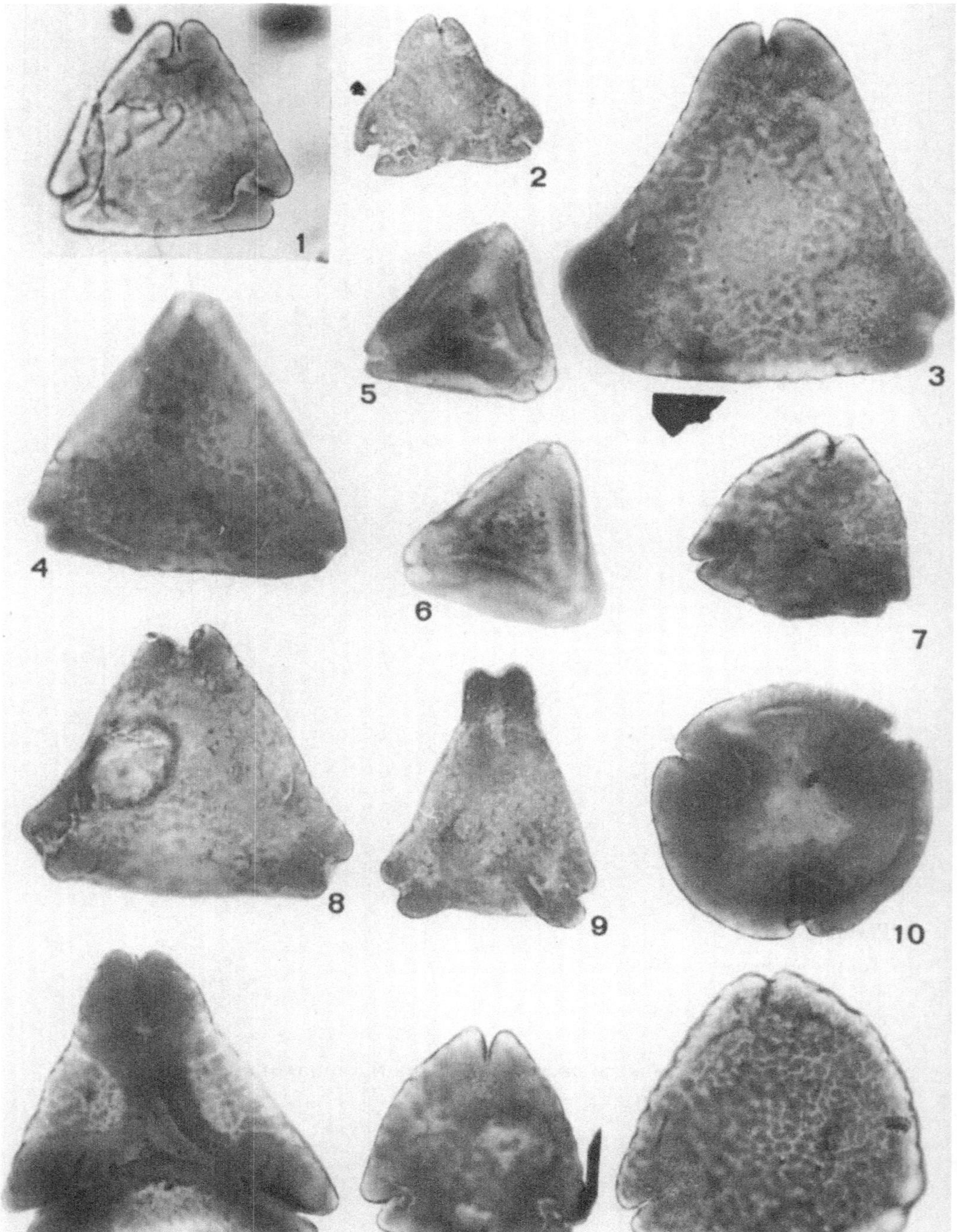
1
2
3
4
5
6
7
8
9
10
11
12
13

Planche 2

(Toutes les figures X 1000)

Fig. 1: *Papillopollis* sp. in ANTONESCU, 1973, 21, 7
Fig. 2, 5, 6: *Capipollis oculis* GOCZAN
Fig. 3: Pentacolporé
Fig. 4: Tricolporé recyclé *Borealipollis* typ.
Fig. 7 et 10: *Pseudoculopollis* typ.
Fig. 8: aff. *Basopollis orthobasalis* (PFLUG) PFLUG
Fig. 9: *Trudopollis* sp.
Fig. 11: aff. *Plicapollis* fsp. (KUNERT & LENK, 1964)
Fig. 12: *Vacuopollis percentus* in GAILLARD.

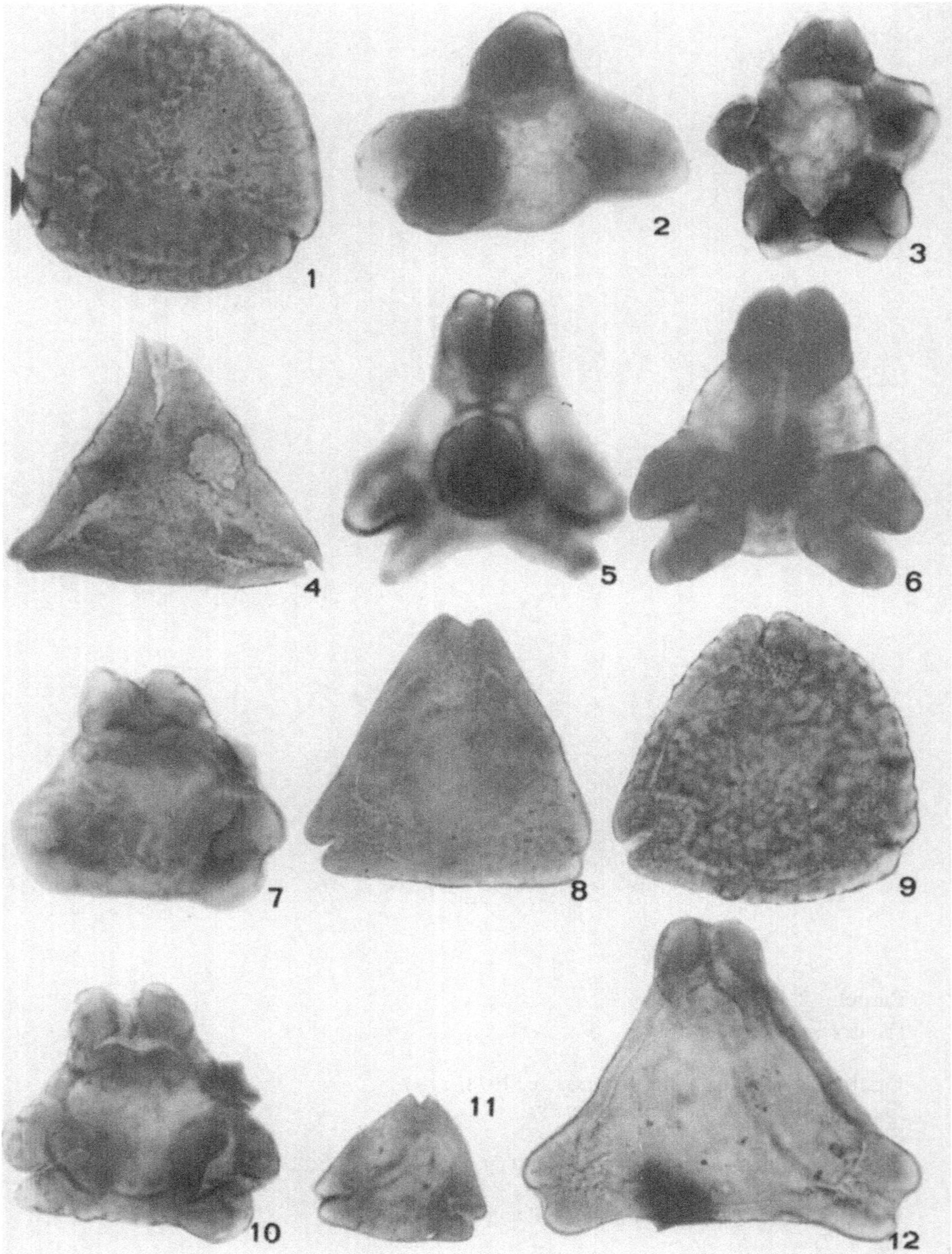
1
2
3
4
5
6
7
8
9
10
11
12

Planche 3

(Toutes les figures X 1000)

Fig. 1. 2: *Krutzschipollis spatiosus* GOCZAN
Fig. 3. 4: *Extrapollis* typ.
Fig. 5. 6: *Portaepollenites* aff. *P. ornatus* KEDVES et DINIZ. (Gard 515/21)
Fig. 7: *Pseudoplicapollis paleacaenicus* KRUTZSCH (cf. *Trevisaepollenites* sp. KEDVES et DINIZ)
Fig. 8: *Extratriporopollenites vestibulus* (ROSS) SKARBY
Fig. 9 *Vacuopollis* typ.
Fig. 10 *Extratriporopollenites triceps* SKARBY

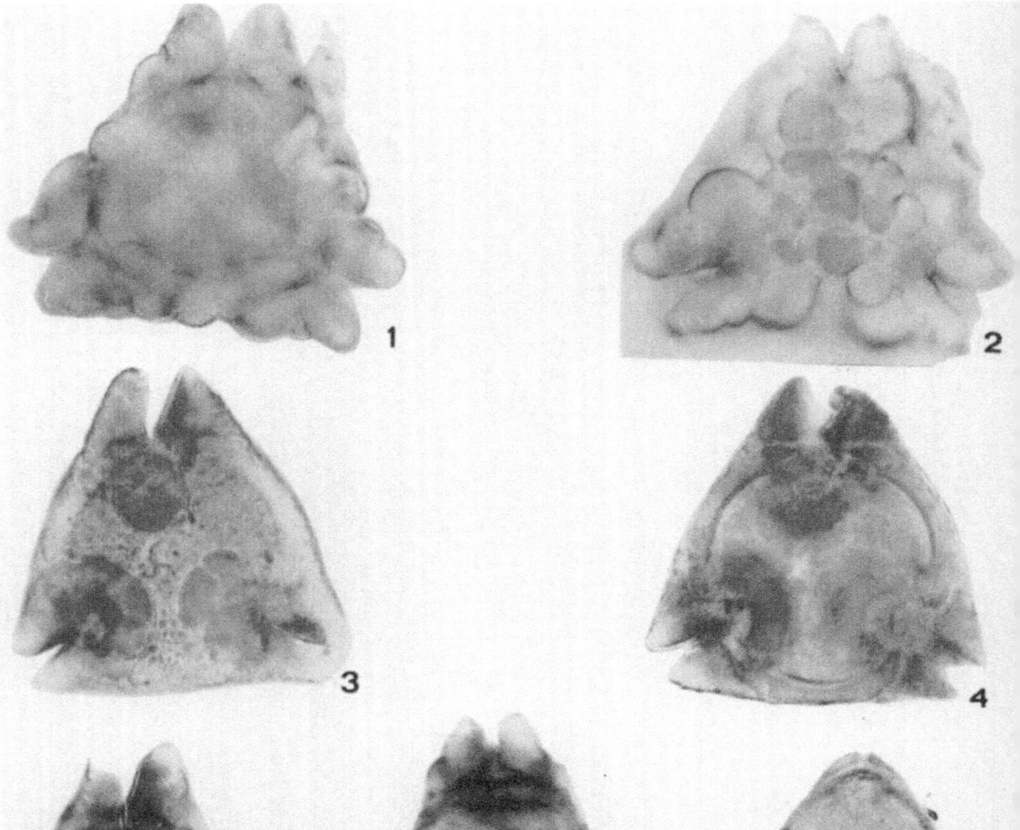
1
2
3
4
5
6
7
8
9
10

Planche 4

(Toutes les figures X 1000)

Fig. 1: *Longanulipollis longianulus* GOCZAN
Fig. 2, 3. 4: *Oculopollis zaklinskaiae* GOCZAN (cf. *Verruoculopollis tschudyi* KEDVES et DINIZ)
Fig. 5, 6: *Triporopollenites robustus* PFLUG in GOCZAN, 1964
Fig. 7: *Extratriporopollenites crassus* GOCZAN
Fig. 8: *Vacuopollis microconcavus* in MÉDUS et al., 1980
Fig. 9: *Basopollis orthobasalis* (PFLUG) PFLUG
Fig. 10. *Pseudovacuopollis intraconvacus* KRUTZSCH
Fig. 11, 12: *Krutzschipollis magnoporus* GOCZAN

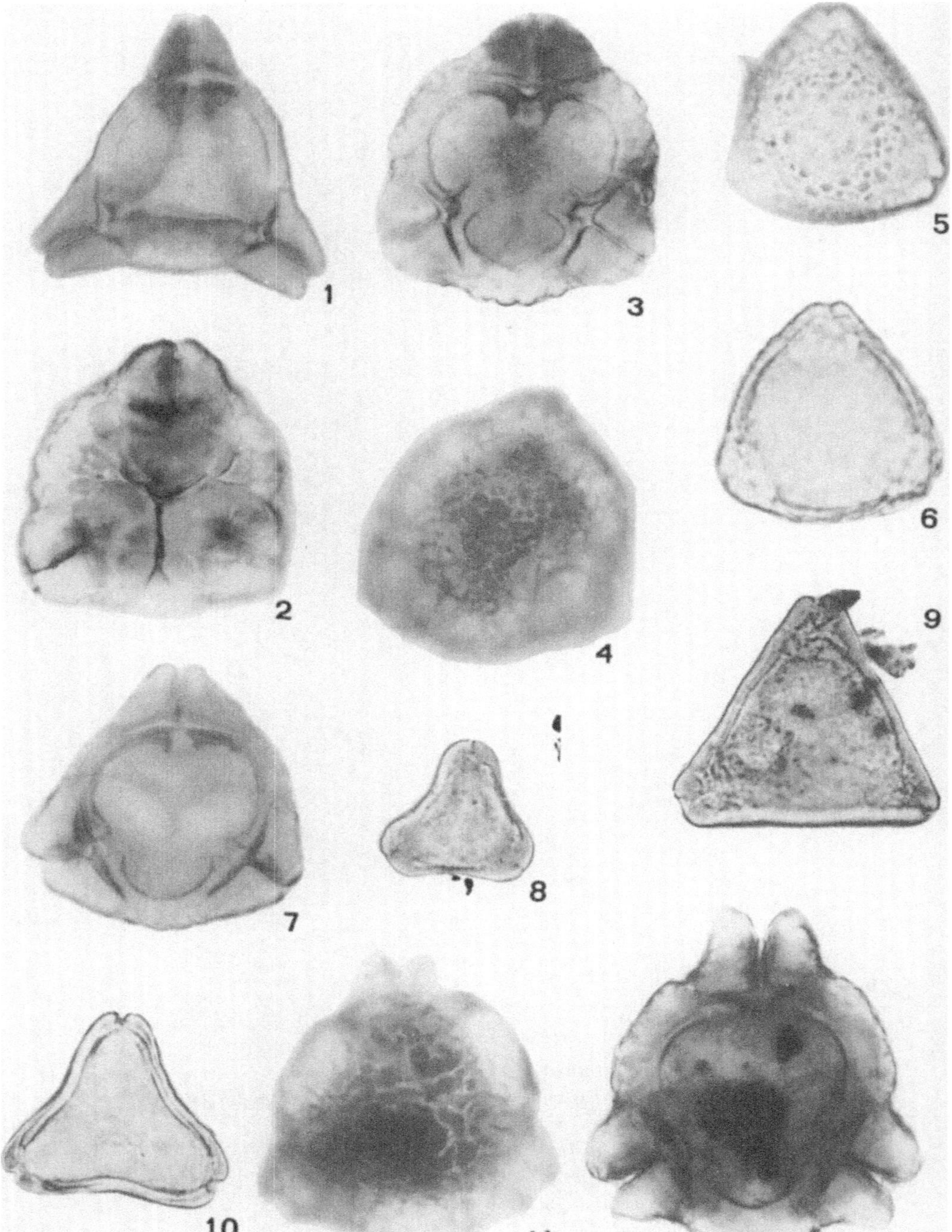
1
3
5
6
2
4
9
7
8
10
11
12

New Aspects on Tethyan Cretaceous Fossil Assemblages.
Band 9 Schriftenreihe der Erdwissenschaftlichen Kommissionen der Österreichischen Akademie der Wissenschaften. 155–170.

3.2. Tethyan Cretaceous corals in Yugoslavia

By Dragica TURNŠEK*)

Abstract

The distribution of corals within the paleo-geotectonic units of Yugoslavia is varying during the Cretaceous period. Lower Cretaceous corals are restricted to the Dinarid Platform and to the Carpathian-Balcanid unit. Cenomanian-Turonian-Coniacian corals occur exclusively in the Drina-Ivanjica unit. The Santonian-Campanian and Maastrichtian corals are restricted to the northern part of the Inner Dinarids. The prosperity and obliteration of corals on the paleo-geotectonic units of Yugoslavia was caused by environmental changes due to tectonic activities.

1. Short review of Cretaceous corals in the Tethyan region

Corals played an important role in the formation of the Cretaceous sediments throughout the shallow Tethys (Fig. 1). Fossil remains of these organisms may be found in almost all Cretaceous intervals. From some stages recordings are rare; in others, however, corals are extraordinarily abundant. The total number of known coral genera from the Cretaceous is 362 (WELLS, 1986). The number of species exceeds 1200.

While corals are hardly known from Berriasian to Valanginian sediments the number of species recorded from the Hauterivian (and "Neocomian") is almost 300. In Barremian and Aptian times this group developed equally successfully in all shallow marine regions of the Tethys. A major crisis occurred at the end of the Aptian, when the number of species decreased to some tens. The largest number of species is known from the Cenomanian although coral faunas are only known from a small number of localities. In the Turonian, the number of species decreases again and reaches a minimum of 12 in the Coniacian. In the

*) Centre of Scientific Research of the Slovenian Academy of Sciences and Arts. Novi trg 5. 61000 Ljubljana. Slovenia.

Santonian, the corals flourished again. They are spread all over the world, the number of species is more than 300. Due to an unsufficient stratigraphic resolution many of the Santonian-Campanian corals are ascribed the Senonian in general. In the Maastrichtian, the number of genera and species decreases and only a few of them persists into the Paleocene.

A number of researchers have attempted to delineate the world-wide development of corals during the Cretaceous. The most accomplished presentation of species is given in the catalogue "Anthozoa Cretacea" by FELIX (1914). It reviews all species known up to this time. More recently, important revisions of corals were published by VAUGHAN & WELLS (1943), ALLOITEAU (1952), and WELLS (1956). The distribution of Upper Cretaceous corals in the entire Tethys was studied by M. and L. BEAUVAIS (1974), who presented all the localities and quoted some of the species important for individual horizons. TCHECHMEDJIEVA (1986) provided a review of the Upper Cretaceous corals of Bulgaria, KUZMICHEVA (1986) for the Upper Cretaceous of the Soviet Union, SIHARULIDZE (1979) for the Urgonian corals of Georgia, and TURNŠEK (1989) for the Mesozoic corals of NW Yugoslavia. Among others, monographic works on corals from the following localities are important: Upper Cretaceous localities of the Gosau Group, Austria (M. BEAUVAIS, 1982), Lower Cretaceous of the Polish and Romanian Carpathians (MORYCOWA 1964, 1971), Eastern Serbia (TURNŠEK & MIHAJLOVIČ 1981), Catalonia in Spain (BATTALER 1937), Texas, Mexico and other regions in North America (WELLS 1932, 1933), Madagascar (ALLOITEAU 1958).

The present paper deals with the stratigraphic distribution of corals in the territory of Yugoslavia and its relationship to paleogeography. A complete data base of Cretaceous coral species will be presented in the near future under the auspicies of IGCP Project 262.

2. Geographic and stratigraphic distribution of Cretaceous corals in Yugoslavia

No corals have yet been found in the Berriasian.

Three species of Valanginian corals have been found in Banjska planota in western Slovenia; all three are known from Jurassic strata. They are accompanied by large tintinnids and the stromatopore *Milleporidium variocellatum* (TURNŠEK & BUSER, 1974).

In Hauterivan deposits of Ozren and Devica in eastern Serbia four coral species have been found together with various fossils such als *Cidaris, Rhynchonella, Terebratula,* and others (SUČIĆ, 1953). A locality with *Cladocoropsis cretacica* in Montenegro is also ascribed to this stage (TURNŠEK, 1968).

In the Barremian-Aptian an extraordinary increase of coral species has been registered. 97 species have been found. Their localities are on Banjška planota in western Slovenia, where the stromatopore *Dehornella* occurs besides corals. They are accompanied by other fossils as *Chaetetopsis,* nerineacean gastropods as *Phanerophtyxis,* the foraminiferan genera *Orbitolina, Chofatella, Cosci-*

nolina, the algae *Macroporella, Triploporella,* and *Salpingoporella dinarica* (TURNŠEK & BUSER, 1974). Some corals of the same stratigraphic interval are mentioned from Montenegro (RADOIČIĆ, 1960). Corals are very numerous in eastern Serbia in the territory between the South Morava river and Stara Planina (KOCHANSKY, 1951; MARKOVIĆ, 1951; TURNŠEK & MIHAJLOVIĆ, 1981). A review of the Urgonian facies of this territory was provided by JANKIČEVIĆ (1978), who cited in addition to corals 15 species of algae, 15 species of foraminifera, 29 brachiopod species, 83 bivalve species, 45 sea-urchin species, and 6 cephalopod species.

No corals have yet been found in the Albian of Yugoslavia.

Three coral species from the locality Duboki potok in western Serbia were ascribed to the Cenomanian (MAKSIMOVIĆ & MARKOVIĆ, 1953). Besides of corals this locality contains gastropods, sea-urchins, brachiopods, bivalves and other fossils.

Only two coral species have been found in Turonian strata. They have been discovered in Kosjerić in western Serbia, where the section continues into the Coniacian with 6 species of corals (PAŠIĆ, 1957). They are accompanied by the rudists *Requienia, Chondrophora, Sauvagesia,* by gastropods and by other fossils. The stromatoporoid genera *Actinostromaria* and *Burgundostromaria* are also known from this stage and environment.

Most Upper Cretaceous corals are known from strata of Santonian-Campanian age. Corals from this interval have been found in Slovenia (TURNŠEK, 1978; TURNŠEK & BUSER, 1976), in northern Croatia (TURNŠEK & POLŠAK, 1978; PAŠIĆ, 1951a), and in eastern Serbia (MILOVANOVIĆ, 1939; PAŠIĆ, 1953). In some of these localities many other fossils occur together with corals: *Inoceramus, Acteonella, Hippurites, Gorjanovicia, Bournonia, Radiolites, Kuehnia,* but also *Globotruncana, Praeglobotruncana, Watznaueria, Coccolithus* and others (TURNŠEK, 1978).

Only three coral species have been mentioned so far in the Maastrichtian on the Fruška gora mountains where they are accompanied by *Inoceramus balticus, Loftusia morgani, Gryphaea vesicularis* (PAŠIĆ, 1951b).

A list of the corals with their stratigraphic and paleogeographic position is presented on Fig. 2.

3. Vertical distribution of coral genera in the Yugoslavian Cretaceous

The systematic analysis shows that the Valanginian coral species are equal to the terminal Jurassic ones, while the Hauterivian species are more similar to those of Barremian-Aptian age.

Among the 97 species of Barremian-Aptian age only 4 are the same as in the Upper Jurassic; not a single one ranges into the Upper Cretaceous.

Of the Upper Cretaceous species three are restricted to the Cenomanian and two to the Turonian. Seven species known from the Coniacian continue into the Santonian. The greatest number of species (58) is known from the Santonian-Campanian. Eight species are restricted to the Campanian, 10 have been ascribed to the Campanian-Maastrichtian, and three to the Maastrichtian.

Changes on the species level took place at the beginning of the Hauterivian, at the end of the Aptian, and at the end of the Campanian.

An analysis shows changes at the generic level at different intervals. Of the 62 genera of Barremian-Aptian age, 24 are the same as in the Upper Jurassic, 10 of them continue into the Upper Cretaceous. In the Santonian-Campanian 36 genera occur first, only three of them continue into the Maastrichtian and two into the Tertiary. Major generic changes occurred at the end of the Aptian and again at the end of the Campanian.

4. Distribution of Cretaceous corals on the paleo-geotectonic units of Yugoslavia

The success of corals in the individual regions of the Tethys depends on the interaction of paleogeography, tectonics, water temperature and favorable water currents. Investigations on the present-day distribution of Cretaceous corals have to take into account that numerous reefs at platform margins have been eroded and deposited on secondary sites. It is also possible that soft corals prevailed in some intervals of the Cretaceous period which were not fossilized.

Cretaceous coral localities in Yugoslavia are restricted at different time intervals to certain geotectonic units (combined and re-arranged after HERAK, 1986; KARAMATA, 1988 and BUSER, 1989): Lower Cretaceous corals appear on the Dinaric carbonate platform and in the Carpathian-Balcanid unit. Cenomanian, Turonian and Coniacian corals are mainly resticted to the Drina-Ivanjica unit. Santonian-Campanian and Maastrichtian corals are known only from the northern part of the Inner Dinarids (Fig. 3).

The Dinaric carbonate platform sedimentation was continuous during the Mesozoic. Corals have been found in various Lower Cretaceous levels in patch reefs. On the margin of the platform, reefs must have been abundant. These have been eroded later, which is shown by the Lower Cretaceous corals found in breccias north of the platform. Nearly half of the Lower Cretaceous coral genera are the same as in the Upper Jurassic. It is therefore concluded that the sedimentary environment did not change essentially during this time interval.

The Carpathian-Balcanid unit represents a somewhat different sedimentary environment. The Lower Cretaceous corals are of the same age as those on the Dinaric platform. Species and fossils assemblages are similar. In eastern Serbia, however, numerous rich accumulations of corals have been found. The sediments are as well carbonates as terrigenous. JANKIČEVIĆ (1978) considers this location to have been an unstable continental threshold, where local facies changed rapidly. Corals accumulated on platforms are frequently filled with terrigenous sediments. The facies of these sediments is comparable to the Urgonian facies of southern France and elsewhere.

Within the Inner Dinarids, the Drina-Ivanjica unit represents a shoal between the west Diabase-chert formation and the Vardar zone. It is interpreted by HERAK (1986) as "not persisting land bodies in the course of Alpine orogeny ... but parts of the basement, subsequently uplifted through the ophiolitic nappe system ...". In the Cenomanian, Turonian and Coniacian small patch coral and stromatoporoid reefs were floorishing temporarily on these shoals. The shape of the coral colonies is similar to those of the Dinaric platform, but the reefs are smaller. It can't be excluded that the reefs had been originally larger and were partly eroded after denudation.

In the Inner Dinarids Santonian-Campanian and Maastrichtian corals occur in rocks which are considered to have been deposited at somewhat deeper marine conditions than coral reefs usually occur. This environment was also spreading over the Slovenian trough (BUSER, 1988), the Supradinaricum and partly over the Paradinaricum (HERAK, 1986) where different types of corals have been found. Finally, it occurred in the Fruška gora mountains and in eastern Serbia (MILOVANOVIĆ, 1939; PAŠIĆ, 1951a, b).

The paleoenvironment of northern Slovenian and Croatian occurrences has been interpreted by TURNŠEK (1989). The isolated, biolithitic and brecciated types of coral occurrences are similar to those of the Gosau group in Austria. According to HÖFLING (1985) they were located in the shallow water environments of the back reef and also of the fore reef, where an influence of the open sea was possible. Some of these corals have probably been resedimented and accumulated in secondary sites (Fig. 4).

5. Coral shape and environment

Corals are classified into solitary, phaceloid and massive forms. Among the Lower and "Middle" Cretaceous corals of the Dinaric platform, of the Carpathian-Balcanids and of the Drina-Ivanjica unit, massive forms predominate. Here, solitary corals scarcely exceed 15% of the total assemblage. Among the Upper Cretaceous corals of the Inner Dinarids solitary corals prevail and represent more than 50% of the total assemblage (Fig. 5). The percentage of solitary corals in the Inner Dinarids would even be greater if the massive colonies found in breccias had been separated.

6. Conclusions

During the Cretaceous the geotectonic units described by HERAK (1986), KARAMATA (1988) and for Slovenia by BUSER (1989) have yielded various paleogeographic and sedimentary environments suitable for the growth of corals. The 195 coral species found so far belong to 98 genera and represent one fourth of all known Cretaceous Tethyan scleractinian genera. The highest rates of occurrences are in the Barremian-Aptian and Santonian-Campanian. In all other Cretaceous stages corals are rare, in the Berriasian and Albian no coral has yet been found.

All Lower Cretaceous corals are restricted to the Dinaric carbonate platform and to the Carpathian-Balcanid tectonic unit. Cenomanian, Turonian and Coniacian corals have been found in the Drina-Ivanjica unit. Santonian-Campanian and Maastrichtian corals are only known from the northern part of the Inner Dinarids, and as a whole they represent a Gosau assemblage. Some of them were resedimented.

The success and extinction of the corals can be explained by general environmental changes as well as by local paleogeographic tectonic changes.

Acknowledgements

I thank Milojka HUZJAN for the graphic design, Angelca TURNŠEK for the English translation. I am also grateful to Dr. Heinz KOLLMANN for reviewing the text.

References

ALLOITEAU, J. (1952): Madréporaires post-paléozoiques. In: PIVETAU, J. (ed.), Traité de Paléontologie, **1:** 539–782. Paris.

ALLOITEAU, J. (1958): Monographie des Madréporaires fossiles de Madagascar. – Ann. geol. Madagascar, **25:** 1–218. Paris.

BATALLER, J. R. (1937): La fauna corallina del Cretacico de Catalunya i regions limitrofes. – Arxius de l'escola sup. agricult., N.S., **3:** 1–299. Barcelona.

BAUVAIS, L., & BEAUVAIS, M. (1974): Studies on the world distribution of the Upper Cretaceous corals. – Second int. reef symposium. Proc. 1. Great Barriere Reef committee, 475–494. Brisbane.

BEAUVAIS, M. (1982): Révision systematique des Madréporaires des couches de Gosau (Crétacé supérieur, Autriche). – Trav. Lab. Paleont. Univ. P. & M. Curie. 1–710. Paris.

BUSER, S. (1989): Development of the Dinaric and the Julian carbonate platforms and of the intermediate Slovenian Basin. – Mem. Soc. Geol. It., **40:** 313–320. Roma.

FELIX, J. (1914): Fossilium Catalogus. I. Animalia: 5. Anthozoa palaeocretacea. 6. Anthozoa cenomanica. 7. Anthozoa neocretacea. 1–273. Berlin.

HERAK, M. (1986): A new concept of geotectonics of the Dinarides. – Prirodoslovna istrazivanja 53, Acta Geologica, **16:** 1–42. Zagreb.

HÖFLING, R. (1985): Faziesverteilung und Fossilvergesellschaftungen im karbonatischen Flachwasser-Milieu der alpinen Oberkreide (Gosau-Formation). Münchner Geowiss. Abh., **3:** 1–241. München.

JANKIČEVIĆ, J. (1978): Barrémien et Aptien des parties moyennes des Carpatho-Balcanides dans la Serbie Orientale au point de vue du développement d'Urgonien. – Geol. an. Balkanskog poluostrva, **42:** 103–194. Beograd.

KARAMATA, S. (1988): "The Diabase-Chert Formation" – some genetic aspects. – Bull. Acad. Serbe Sci. Arts, Cl. Sci. nat. mat., **28:** 1–11. Beograd.

KOCHANSKY-DEVIDÉ, V. (1951): Aptian corals of Eastern Serbia. – Geol. an. Balkanskog poluostrva, **19:** 107–112. Beograd.

Kuzmicheva, E. I. (1986): Koralli verhnego mela SSSR (sistematicheski sostav, stratigraficheskoe rasprostranenie i etapi razvitia). – Fanerozoiskie rifi i koralli SSSR. Trudi 5. Vsesojuznogo sympoziuma po korallam i rifam, Dushanbe 1983, 100–103. Moskwa.

Maksimović, B. V., & Marković, B. (1953): Nouvelle contribution à la connaissance de l'Albien et du Cénomanien des environs de Kadina luka et de Rajac. – Zbornik radova SAN 33, Geol. inst., **5:** 183–210. Beograd.

Marković, O. (1951): Polypiers Mésozoiques de Serbie. – I. Polypiers de l'Aptien aux environs du village Sukovo (Serbie Orientale). – Zbornik radova, **16:** 181–193. Beograd.

Milanović, B. (1939): Sur les Polypiers du Senonien supérieur de Vrbovačka reka (Serbie Orientale). – Geol. an. Balkanskog poluostrva, **16:** 113–115. Beograd.

Morycowa, E. (1964): Hexacorallia des couches de Grodziszcze (Neocomien, Carpathes). – Acta Palaeont. Polon., **9:** 1–114. Warszawa.

Morycowa, E. (1971): Hexacorallia et Octocorallia du Crétacé inférieur de Rarau (Carpathes Orientales Roumaines). – Acta Palaeont. Polon., **16:** 1–149. Warszawa.

Pašić, M. (1951 a): Contribution à la division stratigraphique des couches de Gosau de la Craie supérieur dans la zone des horsts du massif Croato-Slavonien et de Fruška gora. – Geol. an. Balkanskog poluostrva, **19:** 97–101. Beograd.

Pašić, M. (1951). Quelques espèces de polypiers des couches sénoniennes à facies de Gosau de Fruška gora. – Zbornik radova, **16:** 173–179. Beograd.

Pašić, M. (1953): Les polypiers du Crétacé supérieur en Serbie. I. Les polypiers indépendants – genre Cunnolites. – Zbornik radova SAN 22, Geol. inst., **5:** 95–110. Beograd.

Pašić, M. (1957): Biostratigraphische Verhältnisse und Tektonik der Oberkreide in der weiteren Umgebung von Kosjerić (Westserbien). – Posebna izdanja Geol. inst. Jovan Zujovic, **7:** 1–208. Beograd.

Radoičić, R. (1960): Microfacies du Crétacé et du Paléogene des Dinarides Externes de Yougoslavie. Paléontologie des Dinarides Yougoslaves. Ser. A: Micropaléontologie, **4/1:** 3–172. Titograd.

Siharulidze, G. (1979): The corals of the Urgonian facies of Georgia. – Geobios, Mém. spéc., **3:** 301–304. Lyon.

Sučić, Z. (1953): Contribution à la connaissance de la constitution géologique des montagnes Ozren et Devica (Serbie Orientale). – Geol. an. Balkanskog poluostrva, **21:** 77–123. Beograd.

Tchechmedjieva, V. (1986): Paléoécologie des Madréporaires du Crétacé supérieur dans le Srednegori de l'ouest (Bulgarie occidentale). – Geol. Balcanica, **16/5:** 55–81. Sofia.

Turnšek, D. (1968): Some hydrozoans and corals from Jurassic and Cretaceous strata of southwestern Yugoslavia. – Razprave SAZU, IV., **11:** 351–367. Ljubljana.

Turnšek, D. (1970): Cretaceous Hydrozoa from the Zlatibor Mountain in western Serbia. – Razprave SAZU, IV., **13:** 193–208. Ljubljana.

Turnšek, D. (1978): Solitary Senonian corals from Stranice and Mt. Medvednica (NW Yugoslavia). – Razprave SAZU, IV., **21:** 61–128. Ljubljana.

Turnšek, D. (1989): Diversification of corals and coral reef associations in Mesozoic paleogeographic units of northwestern Yugoslavia. – Mem. Ass. Australas. Palaeont., **8:** 283–289. Brisbane.

Turnšek, D., & Buser, S. (1974): The Lower Cretaceous corals, hydrozoans and chaetetids of Banjška planota and Trnovski gozd. – Razprave SAZU, IV., **17:** 81–124. Ljubljana.

Turnšek, D., & Buser, S. (1976): Cnidarian fauna from the Senonian breccia of Banjška planota (NW Yugoslavia). – Razprave SAZU, IV., **19:** 37–88. Ljubljana.

TURNŠEK, D., & MIHAJLOVIĆ, M. (1981): Lower Cretaceous cnidarians from eastern Serbia. – Razprave SAZU. IV., **23:** 1–54. Ljubljana.
TURNŠEK, D., & POLŠAK, A. (1978): Senonian colonial corals from Orešje on Mt. Medvednica (NW Yugoslavia). – Razprave SAZU, IV., **21:** 129–180. Ljubljana.
VAUGHAN, T. W., & WELLS, J. W. (1943): Revision of the suborders, families, and genera of the Scleractinia. – Geol. Soc. Am., Spec. Pap., **44:** 1–363. Baltimore.
WELLS, J. W. (1932): Corals of the Trinity Group of the Comanchean of Central Texas. – Journ. Pal., **6/3:** 225–256. Tulsa.
WELLS, J. W. (1933): Corals of the Cretaceous of the Atlantic and Golf Coastal Plains and Western Interior of the United States. – Bull. Amer. Paleont., **18/67:** 1–292. Tulsa.
WELLS, J. W. (1956): Scleractinia. In: Moore, R. C. (ed.): Treatise on Invertebrate Paleontology, Part f, Coelenterata. F328–F444. Lawrence.
WELLS, J. W. (1986): A list of scleractinian generic and subgeneric taxa, 1758–1985. Fossil Cnidaria, **15:** 1–69. Münster.

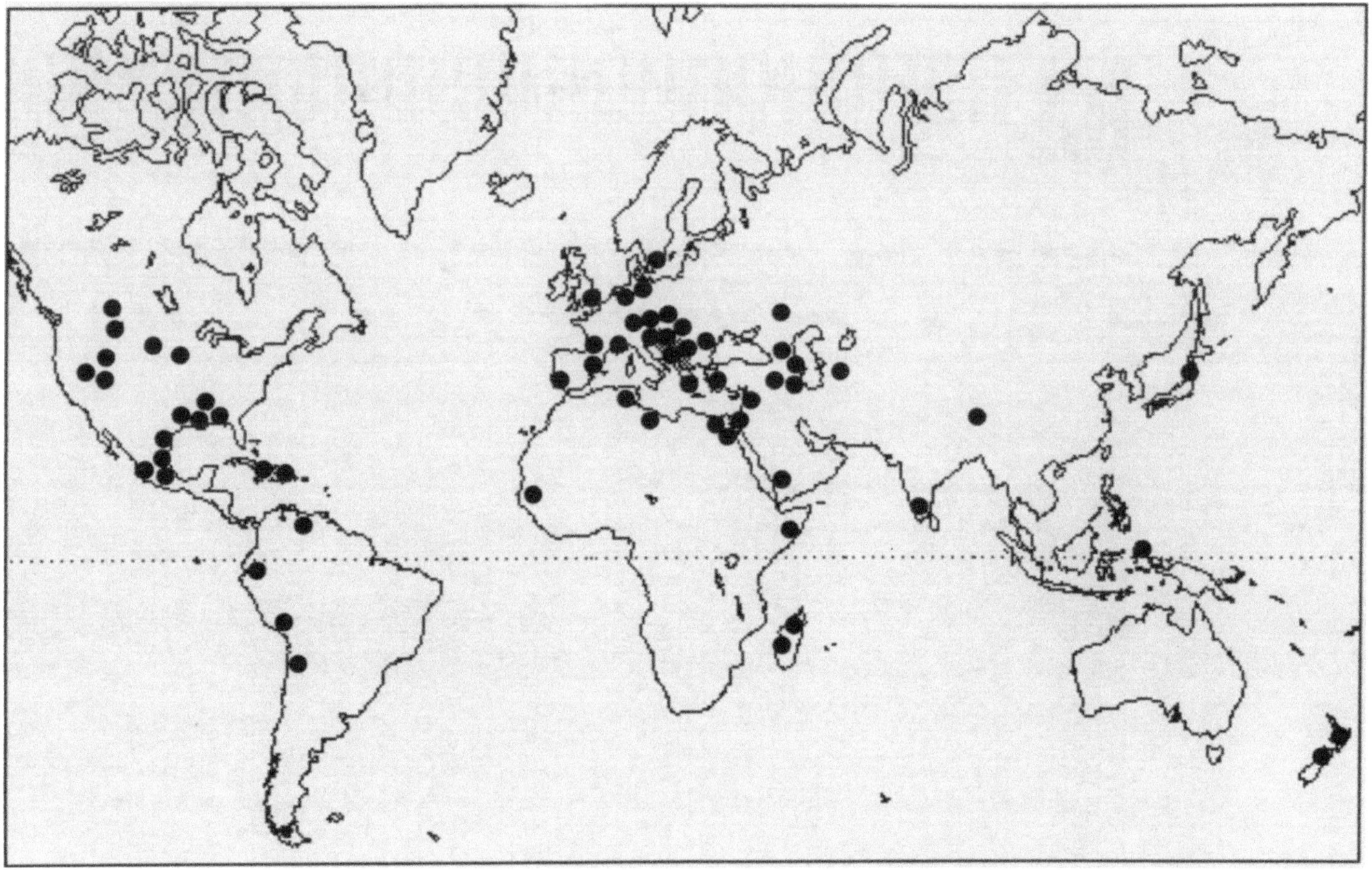

Fig. 1: Cretaceous coral localities in the Tethyan region (partly from BEAUVAIS & BEAUVAIS, 1974).

Lower Cretaceous Species	Localities	Paleogeographical units: Dinarid platform (Local. 1 7 8)	Paleogeographical units: Carpathianr. Balcanid unit (2 4 5 9)	Stratigraphical distribution (Jurassic, Valang., Haute., Barr., Apt., Alb.)
Ironella giseldonensis Starostina & Krasnov 1970	1	●		Valang.
Microphyllia cf. *M.bachmayeri* Geyer 1955	1	●		Jurassic–Valang.
Microphyllia undans Etallon 1858	1	●		Jurassic–Valang.
Eohydnophora picteti (Koby 1897)	2 4 5 7 8 9	●	●	Haute.–Apt.
Latiastraea kaufmanni (Koby 1898)	2 9		●	Haute.–Barr.
Mesomorpha sablensis (Karakash 1907)	2		●	Haute.
Axosmilia cf.A.kobyi (DeAngelis 1905)	2 5 9		●	Haute.–Apt.
Acanthogyra aptiana Turnšek 1981	9		●	Barr.–Apt.
Actinaraea cf. *A.tenuis* Morycowa 1971	9		●	Barr.–Apt.
Actinastraea pseudominima major Morycowa 1971	9		●	Barr.–Apt.
Amphiastraea aethiopica Dietrich 1926	8	●		Barr.–Apt.
Axosmilia bofilli DeAngelis 1905	9		●	Barr.–Apt.
Axosmilia villersensis (Koby 1898)	9		●	Barr.–Apt.
Baryphyllia haimei Fromentel 1857	9		●	Barr.–Apt.
Calamophylliopsis fotisalensis (Bendukidze 1961)	8	●		Barr.–Apt.
Calamophylliopsis sp.	8	●		Barr.–Apt.
Clausastraea bolzei Alloiteau 1960	7 8 9	●	●	Barr.–Apt.
Columnocoenia ksiazkiewiczi Morycowa 1971	9		●	Barr.–Apt.
Cyathophora miyakoensis (Eguchi 1936)	8	●		Barr.–Apt.
Cyathophora pygmaea Volz 1903	7 8 9	●	●	Barr.–Apt.
Cyathophora steinmanni Fritzche 1924	8 9	●	●	Barr.–Apt.
Dermosmilia cretacica Turnšek 1974	7	●		Barr.–Apt.
Diplocoenia saltensis major Morycowa 1971	9		●	Barr.–Apt.
Diplogyra lamellosa eguchii Morycowa 1971	8	●		Barr.–Apt.
Donacosmilia sp.	7	●		Barr.–Apt.
Enallhelia sp.	7	●		Barr.–Apt.
Eohydnophora incerta (Morycowa 1971)	8	●		Barr.–Apt.
Eugyra cotteaui Fromentel 1857	7 8 9	●	●	Barr.–Apt.
Eugyra digitata Koby 1896	4 5 9		●	Barr.–Apt.
Eugyra lanckoronensis (Morycowa 1964)	8 9	●	●	Barr.–Apt.
Eugyra neocomiensis Fromentel 1857	4 5 9		●	Barr.–Apt.
Felixigyra patruliusi tenuiseptata Morycowa 1971	7	●		Barr.–Apt.
Floria planinensis Turnšek 1981	9		●	Barr.–Apt.
Fungiastraea tendagurensis (Dietrich 1926)	8	●		Barr.–Apt.
Fungiastraea sp.	7	●		Barr.–Apt.
Fungiastraeopsis subpolygonalis Morycowa 1971	9		●	Barr.–Apt.
Glenaraea cretacica Počta 1887	7	●		Barr.–Apt.
Gyrodendron serbica Turnšek 1981	9		●	Barr.–Apt.
Heliocoenia actinastrae Turnšek 1981	9		●	Barr.–Apt.
Heliocoenia carpathica Morycowa 1964	9		●	Barr.–Apt.
Heliocoenia rozkowskae Morycowa 1964	9		●	Barr.–Apt.
Holocystis bukowinensis Volz 1903	9		●	Barr.–Apt.
Latusastraea decipiens (Prever 1909)	7 9	●	●	Barr.–Apt.
Latusastraea exiguis (Fromentel 1862)	7	●		Barr.–Apt.
Meandraraea duboisi (Karakash 1907)	9		●	Barr.–Apt.
Meandraraea meandroides Koby 1898			●	Barr.–Apt.
Meandrophyllia cf. *M.lotharinga* (Michelin 1843)	9		●	Barr.–Apt.
Meandroria pirotensis (Toula 1884)	9 5		●	Barr.–Apt.
Mesomorpha excavata (d'Orbigny 1850)	9		●	Barr.–Apt.
Microsolena distefanoi (Prever 1909)	7	●		Barr.–Apt.
Microsolena guttata Koby 1898	7 9	●	●	Barr.–Apt.
Microsolenastraea balcanica Turnšek 1981	9		●	Barr.–Apt.
Ovalastraea polygonalis Alloiteau 1958	9		●	Barr.–Apt.

Fig. 2: Stratigraphic distribution and paleogeographic origin of Cretaceous coral species recorded from Yugoslavia. Part 1.

Lower Cretaceous Species	Localities	Paleogeographical units: Dinarid platform (Local. 1 7 8)	Paleogeographical units: Carpathianr. Balcanid unit (2 4 5 9)	Stratigraphical distribution: Jurassic	Valang.	Haute.	Barr.	Apt.	Alb.
Palaeopsammia zljebinensis Turnšek 1981	9		●				■	■	
Peplosmilia fromenteli DeAngelis 1905	5 7 9	●	●				■	■	
Peplosmilia stutzi (Koby 1896)	5 9		●				■	■	
Phyllocoenia cotteaui Fromentel 1857	5 7	●	●				■	■	
Placophyllia curvata Turnšek 1974	7 9	●	●				■	■	
Plesiofavia dubia (Fromentel 1857)	9		●				■	■	
Pleurophyllia skuviensis Turnšek 1981	9		●				■	■	
Polyphylloseris convexa Fromentel 1857	7 8 9	●	●				■	■	
Polytremacis edwardsana (Stoliczka 1873)	9		●				■	■	
Pseudocoenia annae (Volz 1903)	4 9		●				■	■	
Pseudocoenia hexaphyllia (d'Orbigny 1850)	9		●				■	■	
Pseudocoenia cf. *P.suboctonis* d'Orbigny 1850	9		●				■	■	
Pseudocoeniopsis jurassica (Turnšek 1972)	9		●				■	■	
Pseudopolytremacis spinoseptata Morycowa 1971	7	●					■	■	
Siderastraea senecta Morycowa 1971	9	●					■	■	
Siderofungia irregularis Felix 1891	7		●				■	■	
Smilotrochus tuberosus (M.Edwards & Haime 1850)	9		●				■	■	
Stylina parvistella Volz 1903	9		●				■	■	
Stylina regularis Fromentel 1867	7 9	●	●				■	■	
Stylosmilia alpina Koby 1897	9		●				■	■	
Thamnaraea mammelonata Turnšek 1981	9		●				■	■	
Thamnaraea sp.	9		●				■	■	
Thamnasteria cotteaui Fromentel 1857	9		●				■	■	
Thamnoseris carpathica Morycowa 1971	9		●				■	■	
Trochoidomeandra ovalis Turnšek 1981	9		●				■	■	
Truncoconus inclinatus Turnšek 1981	9		●				■	■	
Baryphyllia barotei Fromentel 1864	5		●					■	
"Convexastraea" cf. *C.almerai* DeAngelis 1905	4		●					■	
"Convexastraea" desori Koby 1890	4		●					■	
"Cryptocoenia" picteti Koby 1898	4		●					■	
"Cryptocoenia" ramosa Toula 1884	5		●					■	
Dimorphastraea bellula d'Orbigny 1849	4		●					■	
"Elipsocoenia hemisphaerica (Fromentel 1857)	4 5		●					■	
Epismilia robusta Koby 1898	5		●					■	
Epismilia sp.	4		●					■	
Eugyra pussila Koby 1898	4 5		●					■	
Hydnophora aff.*H.crassa* Fromentel 1862	5		●					■	
"Phyllocoenia" exculpta (Reuss 1854)	5		●					■	
"Phyllocoenia" picteti Koby 1898	5		●					■	
"Phyllocoenia" cf. *P.zlatarski* Toula 1884	5		●					■	
"Phyllocoenia" sp.	4 5		●					■	
Placosmilia cf. *P.arcuata* M.Edwards & Haime 1849	5		●					■	
Polytremacis urgonensis Koby 1898	4		●					■	
Stylina cf. *S.esmuni* Felix 1909	5		●					■	
Stylina micropora Koby 1898	4		●					■	
Stylohelia sp.	4		●					■	
Thamnasteria urgonensis Koby 1898	4		●					■	
"Thecoseris" pumila Koby 1898	4		●					■	
Thecosmilia tobleri Koby 1898	4		●					■	
"Trochosmilia" neviani DeAngelis 1905	5		●					■	

Fig. 2: Stratigraphic distribution and paleogeographic origin of Cretaceous coral species recorded from Yugoslavia. Part 2.

Upper Cretaceous Species	Paleogeographical units: Drina Ivanjica belt 10 11 12	Paleogeographical units: Inner Dinar. North. Jugosl. 8 14 15 16 17 18 19	Stratigraphical distribution: Cenom.	Turon.	Conia.	Sant.	Camp.	Maast.
Amphiastraea cf. *A.paronai* Prever 1909. 10	●		x					
Thamnoseris hoernesi (Reuss 1854) 10 15	●	●	x			x	x	
"Trochosmilia" sp. 10	●		x					
Thecosmilia minor Prever 1909 11	●			x				
"Trochosmilia" polymorpha Prever 1909 11	●			x				
Actinastrastraea orbignyana (M-Edwards & Haime 1849)	●				x	x		
"Heliastraea" lilli Reuss 1854 11	●				x	x		
Nefophyllia multicincta (Reuss 1854) 11	●				x	x		
Placocoenia dumortieri Fromentel 1879 11 14 17 19	●	●			x	x	x	x
Procladocora tenuis (Reuss 1854) 11 15	●	●			x	x	x	
Corbariastraea junctiseptata (Oppenheim 1930) 11	●				x	x		
Microsolena aff. *M.distefanoi* (Prever 1909) 12	●				x	x	x	
Acrosmilia conica (d'Orbigny 1850) 16		●				x	x	
Actinacis martiniana d'Orbigny 1849 12 15	●	●				x	x	x
Actinacis remesi Felix 1903 8 17		●				x	x	
Actinastraea ramosa (Michelin 1847) 14 15 17		●				x	x	
Actinastraea octolamellosa (Michelin 1846) 15		●				x	x	
Astraraea media (Sowerby 1832) 15 17		●				x	x	
Aulosmilia aspera (Sowerby 1831) 16		●				x	x	
Aulosmilia cuneiformis (M.Edwards & Haime 1849) 16		●				x	x	
Aulosmilia salisbergensis (M.Edwards & Haime 1849) 14		●				x	x	
Columactinastraea pygmaea (Felix 1903) 15		●				x	x	
Columastraea formosa (Goldfuss 1826) 15		●				x	x	
Columastraea striata (Goldfuss 1826) 15		●				x	x	
Columellogyra lomensis Turnšek 1976 8		●				x	x	
Conicosmilotrochus dentatus Turnšek 1978 16		●				x	x	
Conicosmilotrochus stranicensis Turnšek 1978 16		●				x	x	
Conicosmilotrochus strictus Turnšek 1978 16		●				x	x	
Cunnolites (Cunnolites) faecata (Stoliczka 1873) 16 18		●				x	x	
C.(Cunnolites) longifossata Tchechmedjieva 1973 16		●				x	x	
C.(Cunnolites) profunda (Oppenheim 1930) 16 18		●				x	x	
C.(Cunnolites) reussi (Fromentel 1862) 16 18		●				x	x	
C.(Cunnolites) sellata (Quenstedt 1880) 16		●				x	x	
C.(Cunnolites) undulata (Goldfuss 1826) 14 16		●				x	x	
C.(Paracunnolites) scutellum (Reuss 1854) 16 18		●				x	x	
C.(Plesiocunnolites) cycloides (Felix 1903) 16		●				x	x	
C.(Plesiocunnolites) dispar (Quenstedt 1880) 16		●				x	x	
C.(Plesiocunnolites) cf. *P.depressa* (Reuss 1854) 16		●				x	x	
C.(Plesiocunnolites) gosavicus (Oppenheim 1930) 16		●				x	x	
C.(Plesiocunnolites) nummulus (Reuss 1854) 16 18		●				x	x	
C.(Plesiocunnolites) orbignyi (Fromentel 1864) 16		●				x	x	
C.(Plesiocunnolitopsis) robusta (Quenstedt 1880) 16		●				x	x	
Dasmiopsis lamellicostatus (Reuss 1854) 16		●				x	x	
Dermosmiliopsis orbignyi Alloiteau 1957 15		●				x	x	
Dermosmiliopsis tenuicosta (Reuss 1854) 15		●				x	x	
Diploctenium ferrumequinum (Reuss 1854) 16		●				x	x	
Diploctenium cf. *D.pavoninum* Reuss 1854 16		●				x	x	
Elasmophyllia deformis (Reuss 1854) 15		●				x	x	
Ellipsosmilia sp. 15		●				x	x	
Heterocoenia grandis Reuss 1854 8		●				x	x	
Hydnophora styriaca (Michelin 1847) 8 17		●				x	x	
Meandrastraea antiqua (Reuss 1854) 15		●				x	x	
Meandroria konincki (M.Edwards & Haime 1849) 8 17		●				x	x	
Neocoeniopsis lepida (Reuss 1854) 15 17		●				x	x	

Fig. 2: Stratigraphic distribution and paleogeographic origin of Cretaceous coral species recorded from Yugoslavia. Part 3.

Upper Cretaceous Species	Paleogeographical units: Drina Ivanjica belt 10 11 12	Paleogeographical units: Inner Dinar. North. Jugosl. 8 14 15 16 17 18 19	Stratigraphical distribution: Cenom.	Turon.	Conia.	Sant.	Camp.	Maast.
Phyllocoeniopsis pediculata (Deshayes 1831) 15		●				▬	▬	
Phyllocoeniopsis sp. 15		●				▬	▬	
Phyllosmilia sp. 16		●				▬	▬	
Phragmosmilia sp. 16		●				▬	▬	
Pleurocora (?Rhabdophyllia) crassa (Reuss 1854) 15		●				▬	▬	
Pleurocora haueri (M.Edwards & Haime 1848) 15		●				▬	▬	
Procladocora simonyi (Reuss 1854) 8 15		●				▬	▬	
Pseudofavia grandiflora (Reuss 1854) 8		●				▬	▬	
Rennensismilia complanata (Goldfuss 1826) 16		●				▬	▬	
Rennensismilia chondrophora (Felix 1903) 14 19		●				▬	▬	▬
Rennensismilia subinduta (Reuss 1854) 16		●				▬	▬	
Stephanosmilia polydectes Kolosvary 1954 8		●				▬	▬	
Synastraea procera (Reuss 1854) 15		●				▬	▬	
Thamnoseris morchella (Reuss 1854) 8		●				▬	▬	
Astrogyra voraccissima (Oppenheim 1930) 17		●					▬	
Heterocoenia exigua (Michelin 1847) 17		●					▬	
Hydnophora parviconus (Oppenheim 1930) 17		●					▬	
Meandroria konincki typica (Reuss 1854) 17		●					▬	
Placocoenia microcalix Oppenheim 193o 17		●					▬	
Polytremacis partschi Reuss 1854 17		●					▬	
Psilogyra telleri Felix 1903 17		●					▬	
Reussicoenia edwardsi (Reuss 1854) 17		●					▬	
Cunnolites barrerei 18		●					▬	▬
Cunnolites conoidea(Stoliczka 1873) 18		●					▬	▬
Cunnolites filamentosa (Forbes 1846) 18		●					▬	▬
Cunnolites gigantea (d'Orbigny 1850) 18		●					▬	▬
Cunnolites ligeriensis (M-Edwards & Haime 1851) 18		●					▬	▬
Cunnolites macrostomata (Reuss 1854) 18		●					▬	▬
Cunnolites numismalis (Lamarck 1816) 18		●					▬	▬
Cunnolites polymorphus (Goldfuss 1826) 18		●					▬	▬
Cunnolites quenstedti (Oppenheim 1930) 18		●					▬	▬
Cunnolites sororius (Quenstedt 1880) 18		●					▬	▬
Cunnolites ellipticus (Guettard 1770) 19		●						▬
Phragmosmilia psecadiophora (Felix 1903) 19		●						▬
Rennensismilia didyma (Goldfuss 1826) 19		●						▬

Fig. 2: Stratigraphic distribution and paleogeographic origin of Cretaceous coral species recorded from Yugoslavia. Part 4.
Numbers together with species names refer to localities on Fig. 3.

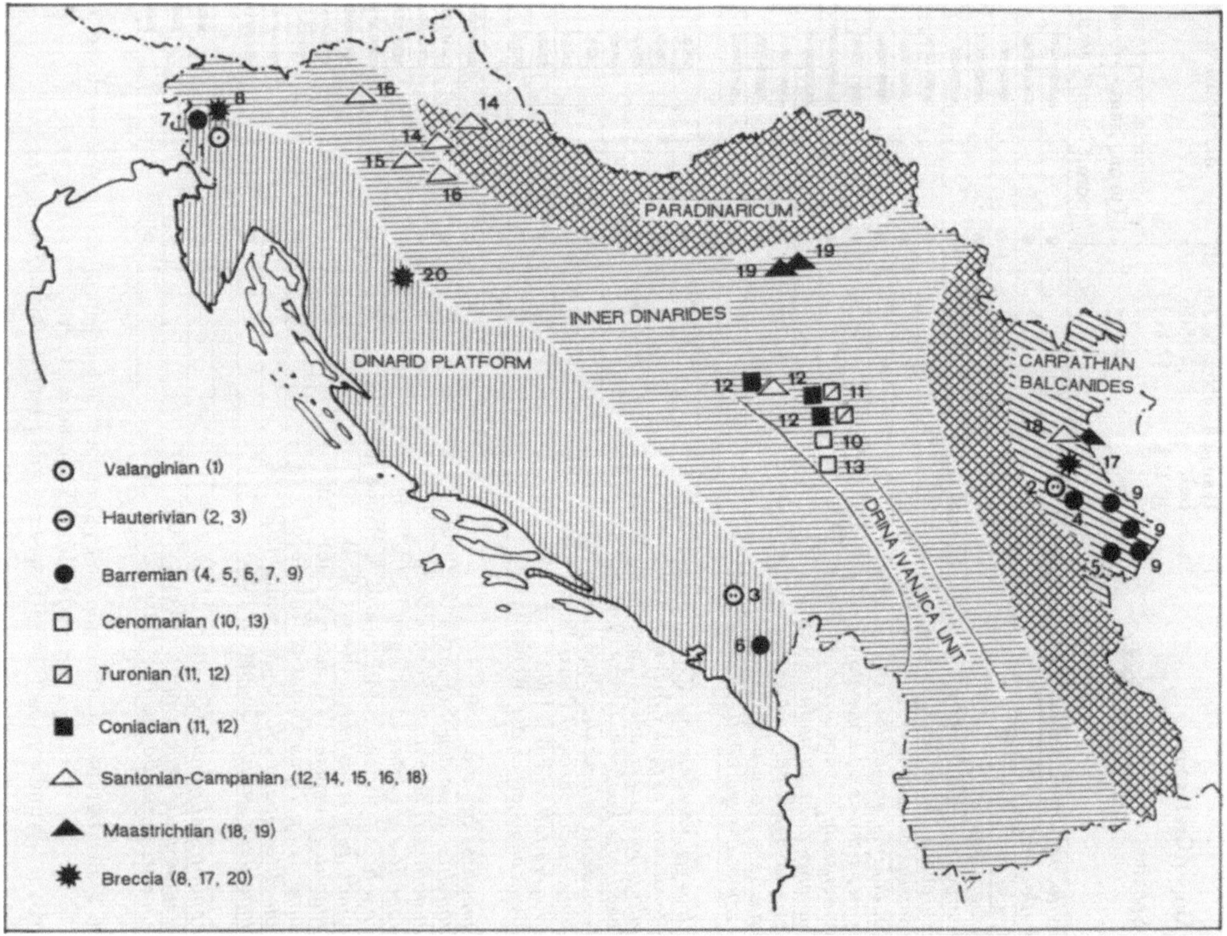

Fig. 3: The age and the distribution of the coral localities on the paleogeotectonic units of Yugoslavia. The numbers 1–19 refer to localities of Fig. 2. Paleogeotectonic units after HERAK (1986), KARAMATA (1988), BUSER (1989).

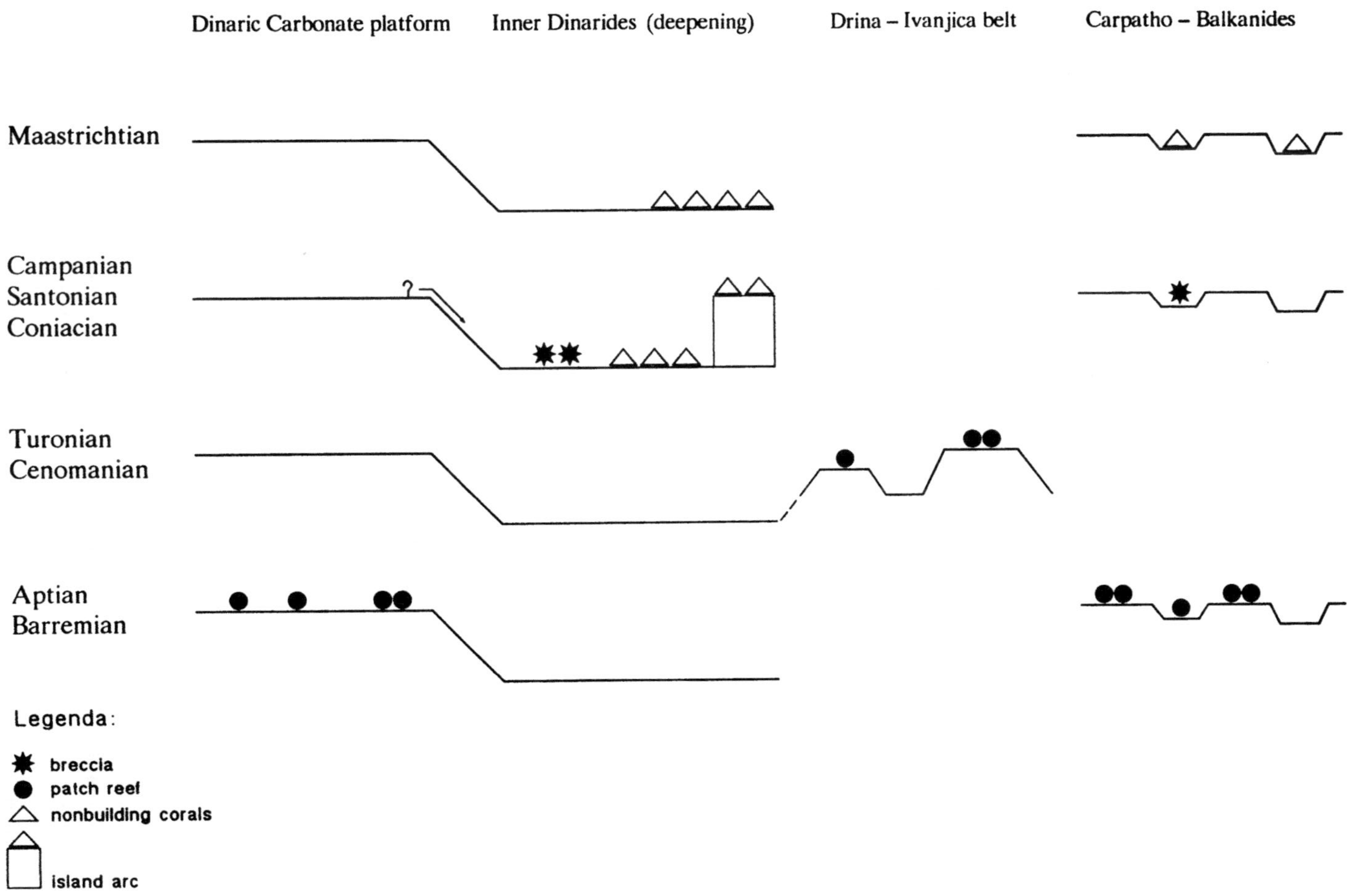

Fig. 4: Generalized palinspastic cross section of Yugoslavia showing the position of coral localities. Note the stable Dinaric Platform containing autochthonous localities, the Inner Dinarids with various forms of coral occurrences and the unstable Carpato-Balkanid Platform with autochthonous occurrences and coral accumulations.

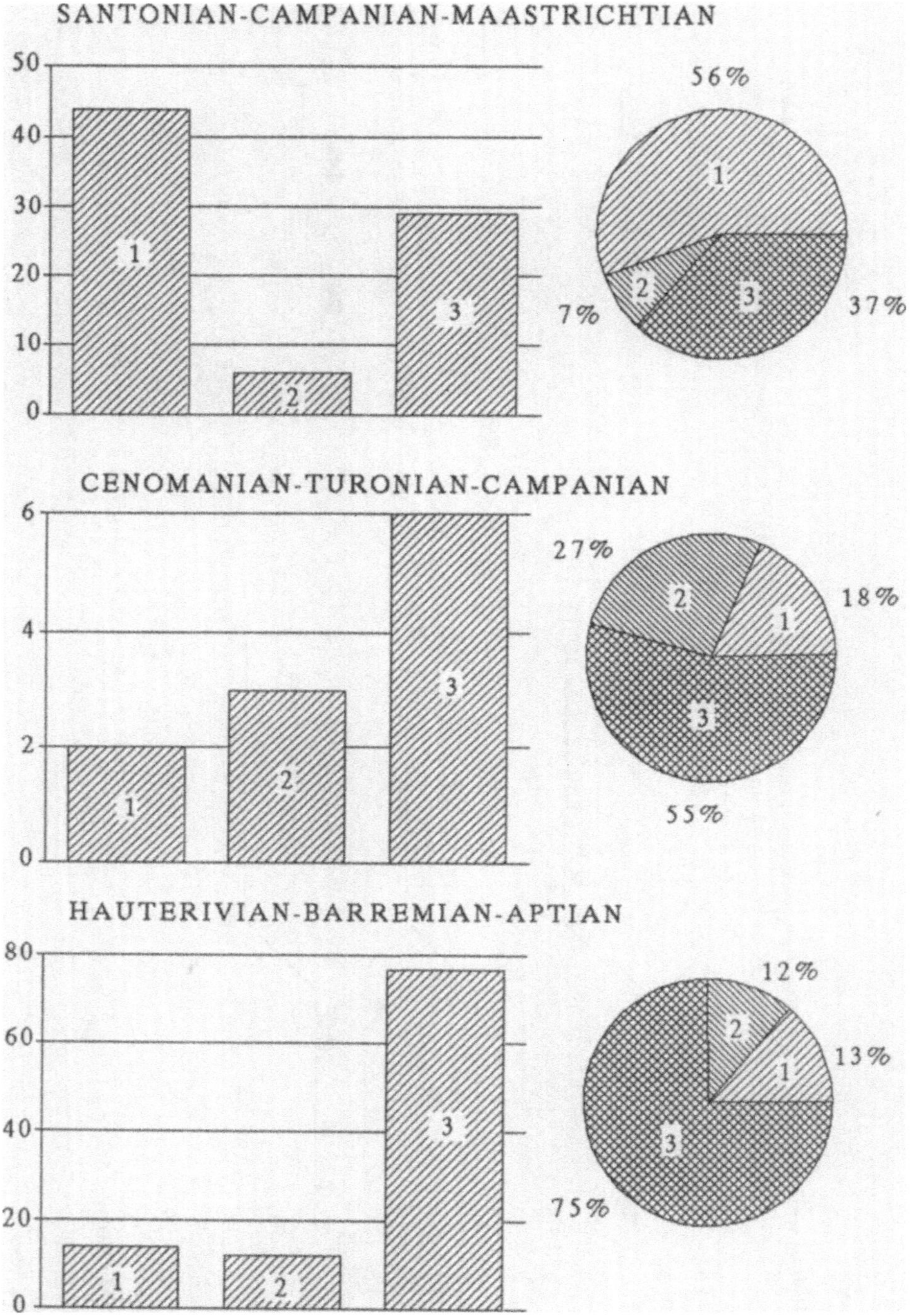

Fig. 5: Relationship between coral shape and environment. 1: solitary corals; 2: phaceloid corals; 3: massive corals. The diagrams on the left show the numbers of species, the diagrams on the right their percentage in the total assemblages. In the Lower and "middle" Cretaceous solitary corals represent ecarcely 15% of the coral assemblages. In the Santonian-Campanian they represent more than 50% of the assemblages and indicate deeper water environments.

New Aspects on Tethyan Cretaceous Fossil Assemblages.
Band 9 Schriftenreihe der Erdwissenschaftlichen Kommissionen der Österreichischen Akademie der Wissenschaften, 171–184.

3.3. Ostracodes du Crétacé moyen et supérieur Téthysien Etat des connaissances – paléogéographie

Par Renée DAMOTTE*)

Résumé

Les principales faunes d'Ostracodes, de l'Aptien au Cénomanien, en régions nord et sud téthysiennes sont passées en revue. Cet état des connaissances conduit à caractériser deux provinces biogéographiques: 1. une provine nordtéthysienne, essentiellement: Ibérie et Sud de la France, les données manquant sur les régions orientales nordtéthysiennes; 2. une province sudtéthysienne qui s'étend de l'Afrique du Nord au Liban. Toutefois à l'Aptien, l'Algarve, province sud du Portugal, présente des affinités avec la province sudtéthysienne.

Abstract

By comparing the major Tethyan ostracode faunas of Aptian to Maastrichtian age two major biogeographic provinces can be distinguished:

1. A north Tethys province which is characterized mainly by faunas recorded in Portugal, Spain and south France. Due to the lack of data the eastern extension of this province is not known. 2. A south Tethys province which is characterized by faunas recorded from North Africa to the Lebanon. The Aptian faunas of the Algarve (south Portugal) show affinities to the south Tethyan province.

Introduction

Dans cet article seront présentés l'état de nos connaissances sur les Ostracodes des régions téthysiennes au Crétacé moyen et supérieur, Aptien à Maas-

*) CNRS et Laboratoire de Micropaléontologie de l'Université Pierre et Marie Curie, 4 Place Jussieu, Paris V°, France.

trichtian, ainsi que des donnés sur leur paléogéographie aux mêmes époques. Toutes les faunes connues ne seront pas envisagées ici, mais seulement celles qui sont connues en plusieurs domaines géographiques et peuvent alors servir de marqueurs téthysiens.

Aptien

Régions Nordtéthysiennes

Les principales régions envisagées sont situées sur les marges sud-ouest de l'Europe: Ibérie, Provence (France) (tableau 1); plus à l'Est les données sont fragmentaires, certaines, assez rares, concernant la Hongrie et la Bulgarie seront traitées séparemment.

Les données existantes proviennent des travaux de OERTLI (1966) sur l'Aptien de Provence; de Baynova (1970), sur la Bulgarie, où des faunes aptiennes sont citées, car si les niveaux y étaient attribués à l'Albien, les zones d'ammonites indiquées sont en réalité aptiennes; de DAMOTTE et al. (1988) sur le Sud du Portugal: Algarve.

En Provence, quelques affinités existent avec les faunes nordeuropéennes, parmi les quelques espéces communes aux deux régions on peut citer:

Schuleridea jonesiana (BOSQUET)
Neocythere mertensi OERTLI
Dolocytheridea intermedia OERTLI
Protocythere derooi OERTLI.

Les faunes de Hongrie sont des faunes à affinités nord européennes (tableau 2).

Par contre, fait interessant, la faune d'Algarve montre des affinités sudtéthysiennes, avec présence d'espèces connues en Afrique du Nord: Algérie et Tunisie:

Paranotacythere luzensis DAMOTTE, RAMALHO & REY
Cythereis (Rehacythereis) cf. *btaterensis btaterensis* BISCHOFF
Cythereis (Rehacythereis) cf. *btaterensis interstincta* BISCHOFF
Cythereis (Rehacythereis) cf. *btaterensis torifera* BISCHOFF.

Régions Sudtéthysiennes (tableau 3)

Des données existent sur les régions orientales: Liban, Israël, Iran et sur les régions occidentales: Algérie, Tunisie. Ces données proviennent des travaux de BISCHOFF (1963–1964) sur le Liban, ROSENFED et RAAB (1984) en Israël, GROSDIDIER (1973) sur l'Iran, BEN YOUSSEF (thèse, 1980) sur la Tunisie, DAMOTTE (1984) sur l'Algérie.

Les principales espèces communes à l'ensemble de ce domaine sudtéthysien sont des espèces décrites par BISCHOFF au Liban:

Cythereis (Rehacythereis) btaterensis btaterensis BISCHOFF

Ostracodes / Pays - Provinces	AQUITAINE	PROVENCE	ESPAGNE	PORTUGAL
Schuleridea jonesiana		Ap A	A C	
Dolocytheridea intermedia		Ap A	A	
Neocythere mertensi		Ap A		
Cythereis buchlerae		Ap A	A	A C
Protocythere aptiensis		A	A	
Bairdia pseudoseptentrionalis	C	C	A	
Cythereis luermannae			A	
Mauritsina soriensis			C	C
Dolocytheridea crassa	C	C	C	A C
Dumontina cenomana	C	C		C
Spinoleberis petrocorica	C	C		C
Trochinus nucalis	C	C		A C
Sarlatina merlensis	C			C
Cythereis ouillierensis		C		C
Curfsina mucronata	C	C		
Paracaudites (D.) grekoffi	C	C		C
Parexophtalmocythere oertlii	C	C		C
Platycythereis minuita	C	C		
Amphicytherura berbiguierensis	C			C
Risaltina aquitanica	C	C	C	
Dolocythere cristata	C	C		C
Costaveenia porrecta	C			C
Pterygocythere gr. pulvinata	C	C	C	
Asciocythere polita	C	C	C	
Dordoniella turonensis		C		

Tableau 1: Répartition des Ostracodes en régions nordtéthysiennes (France – Ibérie): Aptien à Cénomanien.
Ap: Aptien – A: Albien – C: Cénomanien.

Ostracodes / Pays - Provinces	BASSIN ANGLO-PARISIEN	PROVENCE	HONGRIE	BULGARIE
Schuleridea jonesiana	A C		A	Ap A
Dolocythere rara	A			Ap A
Isocythereis fissicostis	A			Ap A
Hechticythere speetonensis	A			Ap A
Neocythere vanveeni	A C			Ap A
Protocythere cf. gaultina	A			Ap A
Cytherelloidea stricta	A			A
Paracypris jonesi	A			A
Pontocyprella harrisiana	A			A
Homocythere harrisiana	A C		A	A
Pontocyprella rara	A		A	
Pontocyprella maynci		Ap A	A C	
Cythereis buchlerae		Ap A	A C	
Cythereis bartensteini		Ap A	A	
Dolocytheridea bosquetiana	A C		A	
Dolocytheridea intermedia		Ap A	A	
Schuleridea derooi		A	A	
Protoveenia florentinensis	A		A	
Matronella matronae	A		A C	
Neocythere lingenensis	A		A C	
Neocythere semilaeva	A		A C	
Bairdia pseudoseptentrionalis	A C		A C	
Cornicythereis gatyensis	A		A C	

Tableau 2: Répartition des Ostracodes en régions nordtéthysiennes (France – Hongrie – Bulgarie): Aptien à Cénomanien.
Ap: Aptien – A: Albien – C: Cénomanien.

Ostracodes / Pays - Provinces	ALGARVE	ALGERIE	TUNISIE	EGYPTE	ISRAEL	LIBAN	IRAN	KOWEIT
Cythereis zoumoffeni	Ap					Ap A		
C. btaterensis btaterensis	Ap	B				A		
C. btaterensis immunicostata	Ap	Ap				A		
C. btaterensis interstincta	Ap	B				A		
C. btaterensis torifera	Ap	B				A		
Cytherelloidea btaterensis	Ap					A		
Paranotacythere luzensis	Ap		Ap					
Cythereis gr. fahrioni		A C				A		
Cythereis malzi			A			A		
Veeniacythereis gr. jezzinensis		A C	C	A C	A C	A C	A C	A C
Centrocythere tunetana		A C	A C		C			
Paracypris mdaouerensis		C T			C T			
Paracypris dubertreti		C T			C	C		
Dolocytheridea atlasica		C T	C	T	C			
Peleriops ziregensis		C	C		C			C
Dordoniella baiderensis					A C	A	A C	
Amphicytherura distincta		C	C		C			
Cythereis aaramtensis		C	C			A		
Cythereis namousensis		C	C	C				
Metacytheropteron berbericus		C	C	C	C		C	
Cytherella gigantosulcata		C	C		C		C	C
Brachycythere gr. sapucariensis		C	C					
Bythoceratina tamarae		C	C		C			
Monoceratina trituberculata		C	C		C			C
Cythereis algeriana		C	C		C	C	A C	C

Tableau 3: Répartition des Ostracodes en régions sudtéthysiennes: Aptien à Cénomanien. B: Barrémien – Ap: Aptien – A: Albien – C: Cénomanien.

Cythereis (Rehacythereis) btaterensis immunicostata BISCHOFF
Cythereis (Rehacythereis) btaterensis interstincta BISCHOFF
Cythereis (Rehacythereis) btaterensis torifera BISCHOFF
Cytherelloidea btaterensis BISCHOFF.

Il faut y ajouter *Paranotacythere luzensis* DAMOTTE, RAMALHO & REY décrite en Algarve et présente en Tunisie.

Albien

Régions Nordtéthysiennes

En France, Provence, la plupart des espèces aptiennes décrites par OERTLI se poursuivent dans l'Albien (tableau 1).

En Espagne, les faunes de l'Albien ont été étudiées par RAMIREZ DEL POSO (1972), mais les figurations y sont difficilement utilisables, et par ANDREU

(Thèse, 1978). En Espagne, les limites Albien-Cénomanien et Cénomanien-Turonien posent encore quelques problèmes; ce qui rend plus aléatoires les comparaisons avec d'autres régions.

Quelques espèces décrites en Provence sont présentes en Espagne:
Cythereis buchlerae OERTLI
Dolocytheridea intermedia OERTLI
Protocythere aptiensis OERTLI
Protocythere triebeli OERTLI.

Des espèces connues en Europe du Nord sont également présentes:
Dolocytheridea bosquetiana (JONES & HINDE)
Cythereis luermannae TRIEBEL
Neocythere vanveeni MERTENS
Schuleridea jonesiana (BOSQUET),
mais ce sont des espèces à très grande extension géographique (Allemagne, Angleterre, Bassin de Paris, Provence . . .).

En Bulgarie (tableau 2), certaines espèces aptiennes sont encore présentes, auxquelles s'ajoutent des espèces albiennes d'Europe du Nord:
Cytherelloidea stricta (JONES & HINDE)
Pontocyprella harrisiana (Jones)
Mandocythere harrisiana (JONES).

En Hongrie (tableau 2), d'après CZASZAR (1976), la partie inférieure de l'Albien, ou argile de Tes, contient, outre des espèces nouvelles, un cortège d'espèces décrites dans le bassin anglo-parisien:
Dolocytheridea bosquetiana (JONES & HINDE)
Mandocythere harrisiana (JONES)
Pontocyprella rara KAYE
Protoveenia florentinensis DAMOTTE
Schuleridea jonesiana (BOSQUET)
et des espèces décrites en Provence:
Cythereis bartensteini OERTLI
Cythereis buchlerae OERTLI
Dolocytheridea intermedia OERTLI
Schuleridea derooi OERTLI.

L'Albien supérieur, marnes de Penzeskut, contient également des formes décrites en Europe du Nord:
Bairdia pseudoseptentrionalis MERTENS
Cornicythereis gatyensis (DAMOTTE & GROSDIDIER)
Cythereis luermannae TRIEBEL
Matronella matronae (DAMOTTE & GROSDIDIER)
Neocythere lingenensis KAYE
Neocythere semilaeava KAYE.

Régions Sudtéthysiennes (tableau 3)

Cet étage est encore assez mal connu en régions sudtéthysiennes. Les faunes du Liban ont été étudiées par BISCHOFF (1962–1964) et par DAMOTTE et SAINT MARC (1970), celles d'Iran par GROSDIDIER (1973), celles de Tunisie par BEN YOUSSEF (1980) et par BISMUTH et al. (1982).

Quelques espèces communes à ces régions existent:

Cythereis fahrioni BISCHOFF

Cythereis malzi BISCHOFF

Veeniacythereis gr. *jezzinensis* (BISCHOFF)

Dordoniella baiderensis DAMOTTE & SAINT-MARC.

Cénomanien

Régions Nordtéthysiennes

France: Dans la province: Aquitaine-Provence (tableau 1), la faune est riche et variée, diversifiée, de type plate forme, avec apparition de nouveaux genres à cachet crétacé supérieur: *Curfsina, Dumontina, Hazelina, Limburgina, Mauritsina, Oertliella, Spinoleberis.* Un certain nombre de ces genres sont également présents en régions sudtéthysiennes mais avec des formes spécifiques différentes.

– Espagne (tableau 1): Des travaux existent dans deux régions du Nord Nord-Ouest du pays: Aragon (ANDREU 1983), Bassin de Soria (GREKOFF et DEROO, 1956; BREMAN 1976).

Quelques affinités avec la Provence ont été reconnues présence de:

Asciocythere polita DAMOTTE

Dolocytheridea crassa DAMOTTE

Pterygocythere pulvinata DAMOTTE

Risaltina aquitanica COLIN et GREKOFF.

Dans le bassin basco-cantabrique (COLIN et al., 1982), l'endémisme est très élevé, avec présence de quelques espèces à affinités boréales.

– Portugal (tableau 1): En Estramadura et dans la région nord du Rio Mondego, sont présentes des espèces communes avec l'Aquitaine et la Provence:

Cythereis ouillerensis BABINOT

Dolocytheridea cristata COLIN

Paracaudites grekoffi (BABINOT)

Parexophtalmocythere oertlii (BABINOT)

Spinoleberis petrocorica (DAMOTTE)

Trochinius nucalis BABINOT

– Hongrie (tableau 2): La plupart des espèces vraconiennes, à affinités boréales, se poursuivent dans le Cénomanien.

Régions Sudtéthysiennes (tableau 3)

De plus nombreux travaux concernent cet étage en région sudtéthysienne, à ceux déjà cités pour l'Albien, on peut ajouter ceux de VAN DEN BOLD (1964) et de COLIN et EL DAKKAK (1975) pour l'Egypte, de BASSOULLET et DAMOTTE (1969) pour l'Algérie, de AL ABDUL RAZZAQ (1977) pour le Koweit, de GARGOURI (1983) pour la Tunisie. En ce qui concerne les publications de REYMENT, en particulier celle de 1982, sur le Maroc, un certain nombre d'attributions spécifiques sont à utiliser avec beaucoup de prudence. Les attributions de plusieurs formes à des espèces décrites en France et en Espagne sont très sujettes à caution et devraient être revues.

Du Moyen Orient à l'Afrique du Nord, le long de la bordure sudtéthysienne, tout un ensemble d'espèces est présent, on peut citer:

Amphicytherura distincta ROSENFELD
Bythoceratina tamarae ROSENFELD
Cytherella gigantosulcata ROSENFELD
Cythereis aaramtensis ROSENFELD
Cythereis algeriana BASSOULLET & DAMOTTE
Cythereis namousensis BASSOULLET & DAMOTTE
Dolocytheridea atlasica BASSOULLET & DAMOTTE
Metacytheropteron berbericus (BASSOULLET & DAMOTTE)
Monoceratina trituberculata ROSENFELD
Peleriops ziregensis (BASSOULLET & DAMOTTE)
Veeniacythereis gr. *jezzinensis* (BISCHOFF).

Turonien

Régions Nordtéthysiennes (tableau 4)

Sud-Est de la France: La Province Aquitaine-Provence subsiste, mais la faune y est très appauvrie.

Espagne: Dans l'Espagne du Centre-Nord (GREKOFF et DEROO 1956; BREMAN 1976), les relations avec la Provence sont nettes, malgré un certain endémisme, on peut citer, en commun dans les deux régions, les espèces suivantes:

Asciocythere polita DAMOTTE
Dolocytheridea crassa DAMOTTE
Dordoniella turonensis DAMOTTE
Pterygocythere pulvinata DAMOTTE
Risaltina aquitanica COLIN & GREKOFF.

Dans les régions asturiennes (SWAIN 1978) et bascocantabriques (COLIN et al., 1982), les affinités sont plus nettes avec le domaine nordeuropéen.

Portugal: D'après J. LAUVERJAT (1981), il y a présence de rares espèces connues en Aquitaine et en Espagne:

Mauritsina soriensis (GREKOFF & DEROO)

Ostracodes / Pays - Provinces	AQUITAINE	PROVENCE	ESPAGNE	PORTUGAL
Dolocytheridea crassa	C	C	T	
Pterygocythere gr. pulvinata	T	T	T	
Risaltina aquitanica	C	C	T	
Asciocythere polita	T	T	T	
Dordoniella turonensis	T	T	T	
Platycythereis minuita	C		T	
Pterygocythere allinensis		T	T	
Mauritsina soriensis		T	T	T
Eucytherura beaussetensis	T	T	S af	
Trochinus nucalis	T	T	S cf	
Pterygocythereis prisca		S	S af	
Limburgina pectinata		T S	S af	
Limburgina damottae		S	S af	
Paracaudites (D.) puncturatoides		S	S af	
Spinoleberis ectypus		T	S af	

Tableau 4 Répartition des Ostracodes en régions nordtéthysiennes: Cénomanien à Sénonien. C: Cénomanien – T: Turonien – S: Sénonien.

Dumontina cenomana (DAMOTTE)

Spinoleberis petrocorica (DAMOTTE) (ces deux dernières espèces ayant été décrites dans le Cénomanien).

Régions Sudtéthysiennes (tableau 5)

Sur l'ensemble de la bordure sudtéthysienne, on peut citer pour le Maghreb les travaux de BASSOULLET et DAMOTTE (1969), de BELLION et al. (1973), de VIVIERE (1985), de BISMUTH et al. (1981), en Israël de ROSENFELD et RAAB (1974), au Koweit de AL ABDUL RAZZAQ (1977), en Egypte de VAN DEN BOLD (1964).

Des espèces sont communes à ces pays:

Amphicytherura yakhinensis ROSENFELD

Cythereis rawashensis VAN DEN BOLD

Haughtinoleberis mdaouerensis (BASSOULLET & DAMOTTE)

Ovocytheridea reniformis VAN DEN BOLD

Paracypris mdaouerensis Bassoullet & Damotte
Spinoleberis yotvatensis Rosenfeld.

Il faut aussi signaler l'existence d'espèces communes au Maghreb et à l'Afrique occidentale:

Haughtinoleberis mdaouerensis (Bassoullet & Damotte)
Perissocytheridea ascalapha (Van den Bold)
Reticulina gr. *vitiliginosa* (Apostolescu)

Reticulina gr. *vitiliginosa* est aussi présente au Maroc, Viviere a mis cette espèce en synonymie avec l'espèce *tarfayensis* décrite par Reyment.

Sénonien (Coniacien à Maastrichtien)

Régions Nordtéthysiennes (tableau 4)

France: La région Provence Languedoc s'isole de l'Aquitaine dès le Santonien, les espèces communes deviennent alors quasi inexistantes. Au Sénonien supérieur, en Provence – Catalogne, les facies deviennent non marins.

Espagne: Dans la région basco cantrabrique, thèse de Rodriguez Lazzaro (1985), quelques rares espéces communes avec l'Aquitaine existent; on peut citer entre autres: *Mauritsina* gr. *hieroglyphica* (Bosquet), décrite originellement en Hollande, *Pterygocythere* gr. *pulvinata* Damotte, *Limburgina* gr. *senonensis* (Damotte), décrites originellement dans le Bassin de Paris. Sur les autres régions, il n'y a pour ainsi dire pas de données.

Régions Sudtéthysiennes (tableau 5)

Afrique du Nord: Les données connues proviennent pour l'Algérie des travaux de Bellion et al. (1973), Viviere (1985), Damotte et Fleury (1989), pour la Tunisie de Donze et al. (1982).

En Israël, on peut citer les travaux de Honingstein (1985, 1986) de Rosenfeld, Raab et Honigstein (1986).

En Jordanie et en Egypte, sont parues diverses publications de Bassiouni, dont en particulier celle de 1980 sur les niveaux maastrichtiens.

Des espèces sont connues en Afrique du Nord et en Egypte:

Brachycythere angulata Grekoff
Cythereis cretaria Van den Bold
Pontocyprella recurva Esker
Protobuntonia numidica Grekoff
Reticulina proteros (Bassiouni);

en Afrique du Nord et en Jordanie:

Doricythereis jordanica nodoreticulata (Bassiouni)
Mauritsina coronata (Esker)
Protobuntonia nakkadi Bassiouni;

en Afrique du Nord et en Israël:

Cythereis cretaria Van den Bold
Protobuntonia nakkadi Grekoff.

Ostracodes / Pays - Provinces	ALGERIE	TUNISIE	EGYPTE	ISRAEL	JORDANIE	AFRIQUE DE L'OUEST
Reticulicosta gr. vitiliginosa	T S	T S				T Co S.
Haughtinoleberis mdaouerensis	T	T				C
Paracypris mdaouerensis	C à Co			C T		
Amphicytherura yakhiniensis	T	T		T		
Cythereis gr. rawashensis	T à K	T Co	T à Sa	T		
Spinoleberis? yotvataensis	T Co	T		T		
Ovocytheridea reniformis	T	T	T			
Cythereis magnei	T	T				
Cytherella cf. gambiensis	T	T				S
Protobuntonia numidica	T K M	S	Sa K	Sa		
Protobuntonia nakkadi	M P	M P			P	
Brachycythere ekpo	Co Sa		K			Co Sa
Reticulina proteros	P	P			P	S P
Doricyt.jord.nodoreticulata	M P	M P			P	
Perissocytheridea ascalapha	Co		T			T
Acanthocythereis? meslei	CoKMP	K M				
Cytherella sarakundaensis	Co Sa					S
Cythereis cretaria	S		K	S		
Brachycythere angulata	Co K	Co K	T K			S M
Pontocyprella recurva	Sa M P	M P	P			
Schizocythere triebeli	Sa					M
Mauritsina? coronata	M P	M P	P		P	

Tableau 5: Répartition des Ostracodes en régions sudtéthysiennes: Turonien à Maastrichtien. C: Cénomanien – T: Turonien – Co: Coniacien – Sa: Santonien – K: Campanien – M: Maastrichtien – S: Sénonien – P: Paléocène.

Dans ces différents pays, à ces espèces communes sont associées des espèces à régionalisme marqué.

Quelques espèces communes à l'Afrique du nord et à l'Afrique de l'Ouest peuvent également être signalées:

Brachycythere angulata GREKOFF
Brachycythere ekpo REYMENT
Brachycythere oguni REYMENT
Reticulina gr. *vitiliginosa* APOSTOLESCU
Schizocythere triebeli REYMENT.

Conclusions

Au Crétacé moyen et supérieur, les faunes d'Ostracodes connues dans les régions nordtéthysiennes sont donc nettement différentes de celles connues dans les régions sudtéthysiennes (Fig. 1).

Les faunes des régions nordtéthysiennes, en particulier Ibérie, France du Sud Est, Aquitaine, sont composées d'espèces qui ne se retrouvent pas en Afrique du Nord. En ce qui concerne les régions orientales de la zone nordtéthysienne, les données manquent, en particulier pour l'Italie, la Grèce, la Turquie, exceptés les travaux récents de BABINOT 1988, sur le Cénomanien de Yougoslavie et de Grèce, présentant une faune endémique aux deux régions étudiées.

En ce qui concerne la zone sudtéthysienne, les faunes d'Ostracodes peuvent caractériser une province sudtéthysienne qui s'étend de l'Afrique du Nord au Moyen Orient. Il faut signaler une exception en ce qui concerne la faune aptienne d'Algarve. Cette faune aptienne d'Algarve comprend des formes identiques à certaines connues en Algérie (Oranie) et en Tunisie et attribuées à des espèces décrites par BISCHOFF au Liban.

Par contre, les espèces, recueillies au Maroc et attribuées par REYMENT à des espèces d'Aquitaine, Provence et Espagne, au vue des figurations données par l'auteur, n'appartiennent vraisemblablement pas à ces espèces européennes.

De même la présence, en Tunisie, de *Amphicytherura* aff. *berbiguierensis,* forme rapprochée de l'espèce décrite par COLIN dans le Cénomanien de Dordogne (France), n'est pas un indice de similitude entre les faunes des deux régions, car il est probable qu'il s'agit de deux espèces différentes.

D'après BABINOT et al. (1978), *Metacytheropteron berbericus,* serait présente dans le Cénomanien du Portugal, mais en 1985, BABINOT lui-même parle de possibilité de convergences dans ce cas. Dans le Maastrichtien du Constantinois, BELLION et al. signale *Mauritsina* cf. *hieroglyphica* s'agit il vraiment de l'espèce décrite en Hollande et retrouvée notamment en Aquitaine, cela est assez douteux.

Une barrière paléogéographique constante a séparé les deux grands ensembles sudeuropéen et nord africain au Crétacé moyen et supérieur. Cette séparation était probablement due à l'existence de masses d'eaux profondes entre les deux zones, ces masses d'eaux faisant obstacles à la migration des Ostracodes dont le mode de vie est généralement benthique. Des causes climatiques semblent moins sûres car, par exemple, actuellement des espèces sont connues en Mer Adriatique et en Mer Pélagienne.

Cette étude comparative entre les faunes nord et sudtéthysiennes pourra être complétée et affinée quand des données complémentaires existeront sur les faunes des régions orientales de la bordure nordtéthysienne, en particulier Grèce, Turquie.

Fig. 1: Répartition géographique des Ostracodes au Cénomanien, d'après la figure 2, DAMOTTE 1985 modifiée.

Espéces d'Ostracodes: A: *Asciocythere polita*, DAMOTTE – B: *Dolocytheridea crassa*, DAMOTTE – C: *Pterygocythere pulvinata*, DAMOTTE – D: *Risaltina aquitanica*, COLIN et GREKOFF – E: *Cythereis ouillerensis*, BABINOT – F: *Dolocytheridea cristata*, COLIN – G: *Paracaudites grekoffi* (BABINOT) – H: *Parexophtalmocythere oertlii* (BABINOT) – I: *Spinoleberis petrocorica* (DAMOTTE) – J: *Trochinius nucalis*, BABINOT

1: *Amphicytherura distincta* (ROSENFELD) – 2: *Bythoceratina tamarae*, ROSENFELD – 3: *Cytherella gigantosulcata*, ROSENFELD – 4: *Cythereis algeriana*, BASSOULLET & DAMOTTE – 5: *Cythereis aaramtensis*, ROSENFELD – 6: *Cythereis namousensis*, BASSOULLET & DAMOTTE – 7: *Dolocytheridea atlasica*, BASSOULLET & DAMOTTE – 8: *Metacytheropteron berbericus* (BASSOULLET & DAMOTTE) – 9: *Monoceratina trituberculata*, ROSENFELD – 10: *Peleriops ziregensis* (BASSOULLET & DAMOTTE) – 11: *Veeniacythereis gr. jezzinensis* (BISCHOFF).

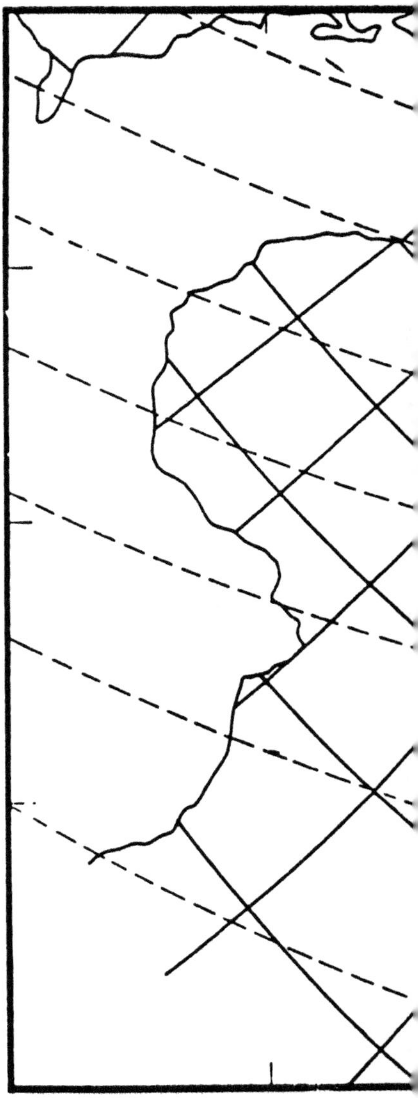

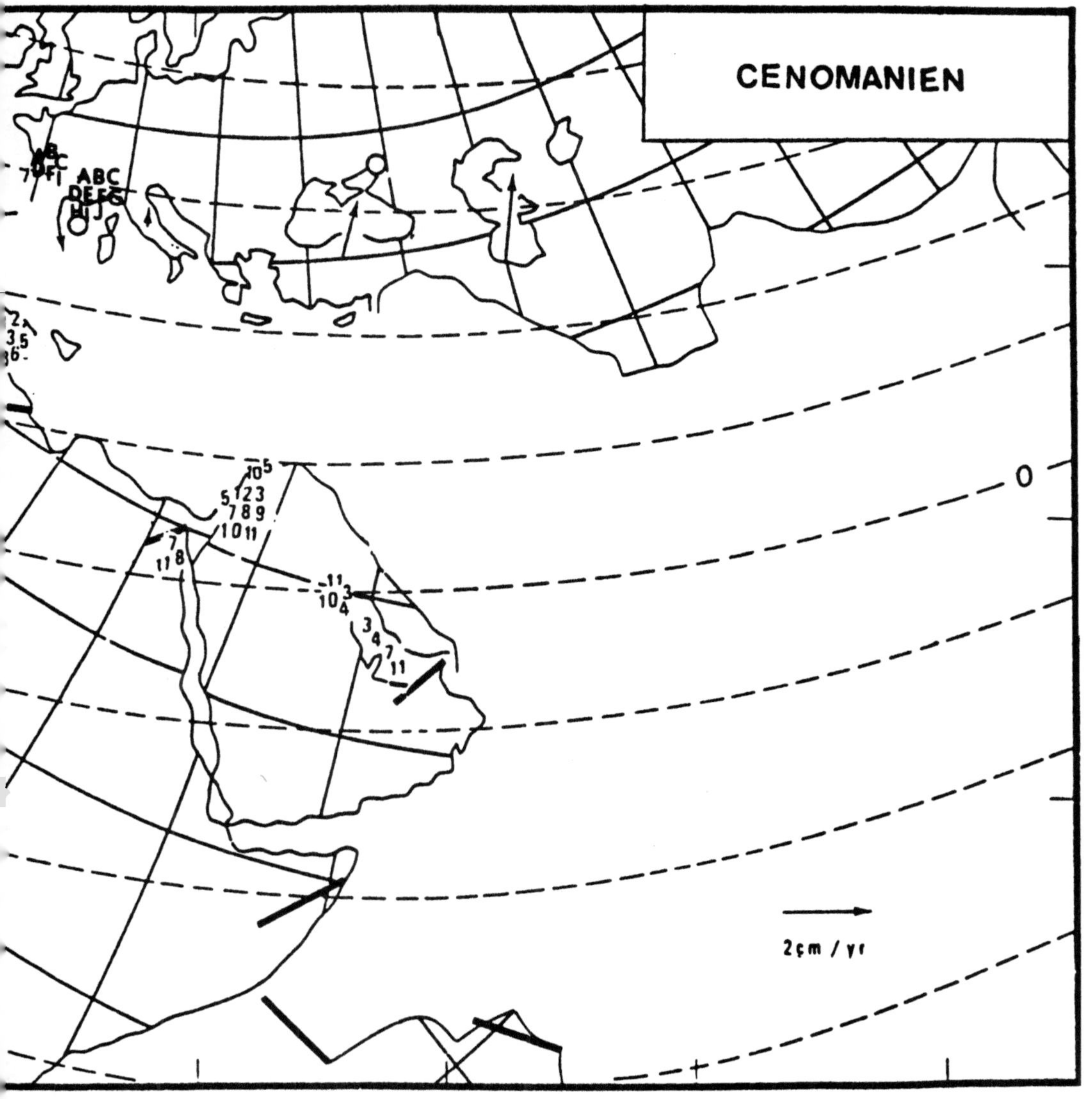
CENOMANIEN
0
2cm / yr

Bibliographie

AL ABDUL RAZZAQ, S. (1977): Study of some Cretaceous Ostracoda of Kuwait. – Thèse, University of Michigan, U.S.A., 2 vol., 421 pp.

AL ABDUL RAZZAQ, S., and GROSDIDIER, E. (1981): Ostracode index species from the Cenomanian of the South shelf of Tethys sea. – Bull. Centres Rech. Explor.-Prod. Elf Aquitaine. **5/2:** 173–191.

ANDREU. B. (1978)· Ostracodes albiens et cénomaniens dans la zone sudpyrénéenne, les chaînes ibériques et aragonaises, de Portugal. – Travaux du Laboratoire de Géologie de l'Université Paul Sabatier. Toulouse. Thèse inédite, 290 pp.

BABINOT. J. F. (1980). Les Ostracodes du Crétacé supérieur de Provence: systématique. biostratigraphie. paléoécologie. – Travaux Laboratoire Géologie historique et Paléontologie. Marseille. **10:** 634 pp.

BABINOT. J. F. (1985) Paléobiogéographie des Ostracodes du Crétacé supérieur des marges ouest-européennes et nord-africaines de la Téthys. – Bull. Soc. Géol. France. (8) I: 739–745.

BABINOT. J. F. (1988) Premières données sur les Ostracodes du Cénomanien de Yougoslavie (Istrie du Sud). – Géobios. **21/1:** 5–15.

BABINOT. J. F.. BERTHOU. P. Y.. COLIN. J. P. LAUVERJAT. J. (1978): Les Ostracodes du Cénomanien du bassin occidental portugais: biostratigraphie et affinités paléogéographiques. – Cahiers de Micropaléontologie. **3:** 11–23.

BASSIOUNI. M. A. (1970): Ostracoda (Mauritsinae und Trachylebridinae) und ihre Bedeutung für die Biostratigraphie des Maastricht und Alttertiär von Jordanien. – Beih. Geolog. Jahrb.. **106:** 5–52.

BASSOULLET. J. P.. et DAMOTTE, R. (1969): Quelques Ostracodes nouveaux du Cénomanien-Turonien de l'Atlas Saharien occidental (Algérie). Revue de Micropaléontologie. **12/3:** 130–144.

BEN YOUSSEF. M. (1980): Etude stratigraphique et micropaléontologique du Crétacé des Djebels Koumine et Kharroub (Tunisie). – Thèse de 3ème cycle, Université de Nice. 104 pp.

BISCHOFF. G. (1963a): Ostracoden-Studien im Libanon. 1: Die Gattung *Cythereis* in der Unterkreide. – Senckenbergiana Lethaea, **44/1:** 1–77.

BISCHOFF. G. (1963b): Ostracoden-Studien im Libanon. 2. Die Gattung *Cypridea* im unteren Aptien. – Senckenbergiana Lethaea, **44/1:** 1–77.

BISCHOFF. G. (1964): Ostracoden-Studien im Libanon. 3: Die Gattung *Cytherelloidea* im oberen Jura und der Unterkreide. – Senckenbergiana Lethaea, **45/1–4:** 1–28.

BISMUTH. H., DONZE, P., LEFEVRE, J., SAINT-MARC, P. (1981): Nouvelles espèces d'Ostracodes dans le Crétacé moyen et supérieur du Djebel Semmama (Tunisie du Centre Nord). – Cahiers de Micropaléontologie, **3:** 51–69.

BISMUTH, H., BOLTENHAGEN, C., DONZE, P., LEFEVRE, J., SAINT-MARC, P. (1982): Le Crétacé moyen et supérieur du Djebel Semmama (Tunisie du Centre Nord); microstratigraphie et évolution sédimentologique. – Bull. Centre Rech. Explor.-Prod. Elf Aquitaine, **5/2:** 193–267.

COLIN, J. P., et EL DAKKAK, M. W. (1975): Quelques Ostracodes du Cénomanien du Djebel Nezzazat. Sinaï. Egypte. – Revista Espanola de Micropaleontologia, num. esp., pp. 49–60.

DAMOTTE. R. (1985): Ostracodes barrémiens-cénomaniens en Algérie occidentale (coupe du Djebel Cheguiga. Monts de Daia. Oranie). – Géologie Méditerranéenne, **11 (1984):** 159–172.

Damotte, R. (1985): Les Ostracodes du Crétacé moyen sud-mésogéen et leur répartition paléogéographique. – Bull. Soc. Géol. France, (8), **I/5:** 733–737.
Damotte, R. (1986): Ostracodes du Crétacé supérieur d'Algérie occidentale (Djebel Tessala, Oranie) – Comparaisons avec le faune d'Afrique occidentale. – Revue de la Faculté des Sciences de Marrakech (Maroc), n° spec. **2:** 155–171.
Damotte, R., Babinot, J. F., Colin, J. P. (1981): Les Ostracodes du Crétacé moyen européen. – Cretaceous Research, **2:** 287–306.
Damotte, R., Fleury, J. J. (1989): Ostracodes maastrichtiens et paléocènes du Djebel Dyr (Constantinois, Algérie). – Géologie Méditerranéenne, **14 (1987):** 87–107.
Damotte, R., Ramalho, M., Rey, J. (1988): Les Ostracodes aptiens de l'Algarve occidental. – Revue de Micropaléontologie, **31/1: 36–48.**
Damotte, R., et Saint-Marc, P. (1972): Contribution à la connaissance des Ostracodes crétacés du Liban. Revista Espanola de Micropaleontologia, **IV/3:** 273–296.
Gargouri Razgallah, S. (1983): Le Cénomanien de Tunisie centrale. Etude paléoécologique, stratigraphique, micropaléontologique et paléogéographique. – Documents et travaux, Institut Géologique Albert Lapparent, Paris, **6: 215.**
Grekoff, N. (1969): Sur la valeur stratigraphique et les relations paléogéographiques de quelques Ostracodes du Crétacé, du Paléocène et de l'Eocéne de l'Algérie orientale. – Proc. Third African Micropal., Coll., NIDOC, Le Caire, p. 227–248.
Grekoff, N., et Deroo, G. (1956): Algunos Ostracodos del Cretacico medio del Norte de Espana. Estudios Geologicos, **31–32:** 215–235.
Glintzböckel, C., et Magne, J. (1959): Répartition des microfaunes à plancton et à Ostracodes dans le Crétacé supérieur de la Tunisie et de l'Est algérien. – Revue de Micropaléontologie, **2/2:** 57–67.
Grosdidier, E. (1973): Associations d'Ostracodes du Crétacé d'Iran. – Revue de l'Institut Français du Pétrole, **XXVIII/2:** 131–169.
Honigstein, A. (1984): Senonian Ostracodes from Israël. – Geological Survey of Israël, bull., **78:** 1–48.
Honigstein, A., Raab, M., Rosenfeld, A. (1985): Manual of Cretaceous Ostracodes. – Israël Geological Survey, Spec. publ., **5:** 1–25.
Ramirez del Poso, J. (1972a): Algunas precisiones sobre la biostratigrafia, paleogeografia y micropaleontologia del Cretacio asturiano (zona de Oviedo, Infiesto, Villaviciosa, Gijon). – Boletin de Geologia y Mineria, **33/2:** 122–162.
Ramirez del Poso, J. (1972b): Algunos datos sobre la estratigrafia y micropaleontologia del Aptense et y Albiense al oeste de Santander. – Revista Espanola de Micropaleontologia, num. esp., p. 59–98.
Reyment, R. A. (1982): Note on Upper Cretaceous Ostracodes from Southwestern Moroco. – Cretaceous Research, **3:** 405–414.
Rosenfeld, A., et Raab, M. (1974): Cenomanian-Turonian Ostracodes from the Judea Group. – Geological Survey of Israël, bull., **62:** 1–64.
Rosenfeld, A., et Raab, M. (1983): Ontogenesis and stratigraphy of the Ostracod *Veeniacythereis jezzinznsis* (Bischoff, 1963). – Journal of Micropaleontology, **2:** 59–65.
Rosenfeld, A., et Raab, M. (1984): Lower Cretaceous Ostracodes from Israël and Sinai. – Israël Journal of Earth Sciences, **33/3:** 86–134.
Van den Bold, W. A. (1964): Ostracoden aus der Oberkreide von Abu Rawash, Ägypten. – Paleontographica, Abt. A, **123:** 11–136.
Viviere, J. L.: 1985: Les Ostracodes de Crétacé supérieur (Vraconien à Campanien basal) de la région de Tébessa (Algérie du Nord Est): Stratigraphie, Paléoécologie, Systématique. Thèse 3ème cycle, Université Pierre et Marie Curie, Paris, 261 p.

4. Biostratigraphy

New Aspects on Tethyan Cretaceous Fossil Assemblages.
Band 9 Schriftenreihe der Erdwissenschaftlichen Kommissionen der Österreichischen Akademie der Wissenschaften, 187–200.

4.1. Albian and Cenomanian foraminifera from the Pieniny Klippen Belt (Carpathians, Poland)

By M. Adam GASINSKI*)

Abstract

A local stratigraphic zonation based on planktonic foraminifera has been established for the almost uninterrupted sequences of Albian to Cenomanian sediments of the Polish part of the Pieniny Klippen Belt. The assemblages resemble those of the Southern Alps and the Umbria-Marche Apennins. Assemblages of agglutinated foraminifera from the Flysch Carpathians situated to the North of the Pieniny Klippen Belt resemble those of the Moroccan Rif. The foraminifera assemblages are Tethyan in their composition. Several anoxic events have been distinguished by dating intercalations of black shales in the sequence.

1. Introduction

Microfauna assemblages from the regional Cretaceous formation stratotypes and sections of the Polish part of the Pieniny Klippen Belt indicate an almost uninterrupted sedimentation during Albian and Cenomanian times. Detailed locality data, lithological descriptions and formal lithostratigraphy of the section have been presented by GASINSKI (1988). Previous investigations have shown that samples from these sections contain abundant and well-preserved foraminifera. Planktonic forms are dominant (GASINSKI, 1983, 1988; BIRKENMAJER & JEDNOROWSKA, 1984, 1987), and allow the establishment of a stratigraphical local zonation (Fig. 2). Several authors have described the Cretaceous foraminiferal assemblages and the stratigraphy of these deposits (ALEXANDROWICZ, 1966, 1979; BIRKENMAJER, 1977, 1987; BIRKENMAJER & JEDNOROWSKA, 1984, 1987; JEDNOROWSKA, 1979, 1980, and others). The nannofossils

*) Jagiellonian University, Institute of Geological Sciences, Oleandry 2a. 30–063 Krakow, Poland.

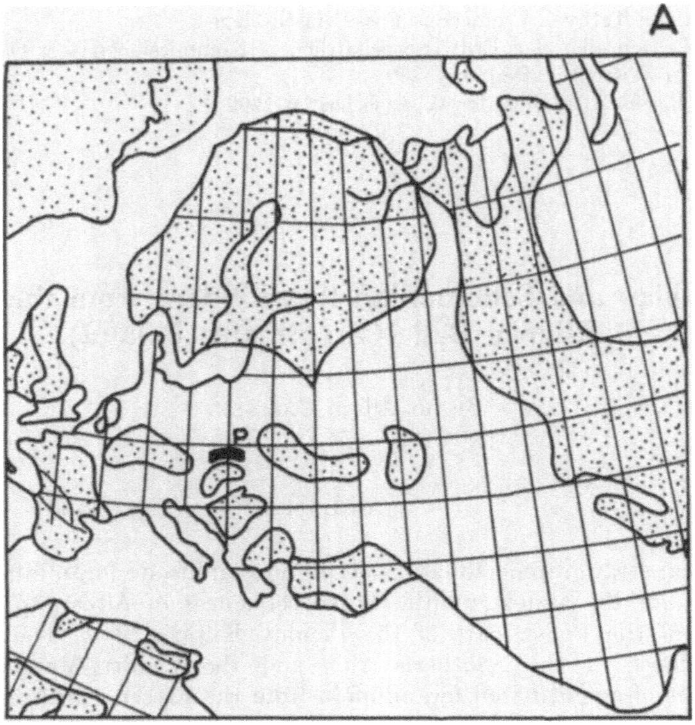

Fig. 1A: Location of the Pieniny Klippen Belt Basin (P.) on the paleogeographic map at 100 m.y. Each time increment is illustrated with a Mercator and a north and south polar Stereographic projection. 30° paleolatitude lines are indicated (after BARRON et al. 1981. Plate 5).

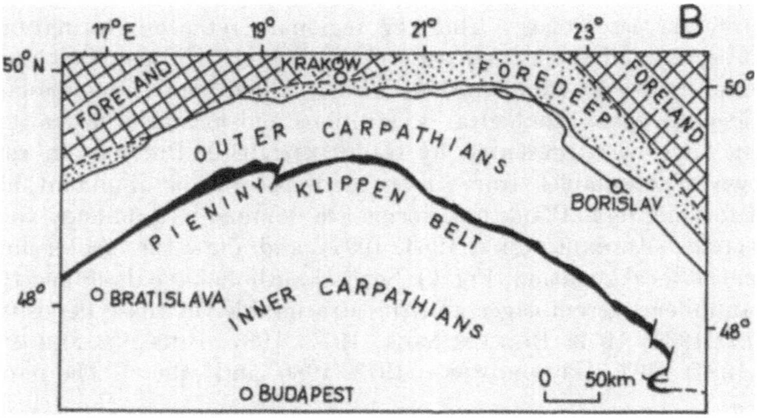

Fig. 1B: Position of the Pieniny Klippen Belt in the Carpathians (after BIRKENMAJER & JEDNOROWSKA, 1987).

have been described by DUDZIAK (1979, 1981 et al.). The analysis of the dinocysts has recently been initiated by LEEREVELD & GASINSKI within IGCP project 262. Megafossils are scarce. The magnetostratigraphy did not yield any results (magnetic long normal zone, see fig. 2).

2. Geological setting

The Pieniny Klippen Belt is situated at the boundary between the Inner and Outer (Flysch) Carpathians (Fig. 1B). A detailed subdivision of this belt, its paleogeographic and palinspastic reconstructions, as well as its tectonic evolution were presented by BIRKENMAJER (1977, 1986). During the Cretaceous the Pieniny Klippen Belt Basin belonged to the northern part of the Tethys (Fig. 1A). Depending on the deposition area in the basin the sediments are developed as marly limestones, marls, shales and turbidites, with intercalating black shales.

The unit, informally called Trawne Beds has been revised recently and was formally named Trawne Member by BIRKENMAJER (1987).

The location and numeration of the investigated samples are given in a previous article (GASINSKI, 1988). Sections without adequate natural outcrops were exposed in trenches.

3. Foraminiferal assemblages

Planktonic foraminifera constitute about 80% of almost all assemblages and are characteristic for the Tethyan realm (REYMENT & BENGTSON, 1986, p. 125). The species belong mainly to *Globigerinelloides, Planomalina, Hedbergella* and *Rotalipora* (GASINSKI, 1983, fig. 5–6; 1988, p. 225, tab. 1). Some of the characteristic foraminifera species are shown on plates 1–3. Detailed descriptions of the taxa were presented by GASINSKI (1983, 1984, 1988). In several samples collected from turbidites a high percentage of agglutinated foraminifera was found.

The following local biostratigraphic zonation based on the first and last occurrence of planktonic index species of foraminifera is proposed (FAD and LAD; Fig. 2):

Hedbergella local assemblage zone (LAZ, informal zone, not on Fig. 2)
FAD of *R. subticinensis?*

Rotalipora subticinensis – Rotalipora ticinensis, local concurrent range zone (LCRZ). Upper boundary: FAD of *P. praebuxtorfi* and LAD of *R. subticinensis.*

Rotalipora ticinensis-Planomalina praebuxtorfi (LCRZ). Upper boundary: FAD of *P. buxtorfi* and LAD of *P. praebuxtorfi.*

Rotalipora ticinensis-Planomalina buxtorfi, local partial concurrent range zone (LPCRZ). Upper boundary: FAD of *R. appenninica.*

Planomalina buxtorfi-Rotalipora appenninica (LCRZ). Upper boundary: LAD of *R. ticinensis* and LAD of *P. buxtorfi.*

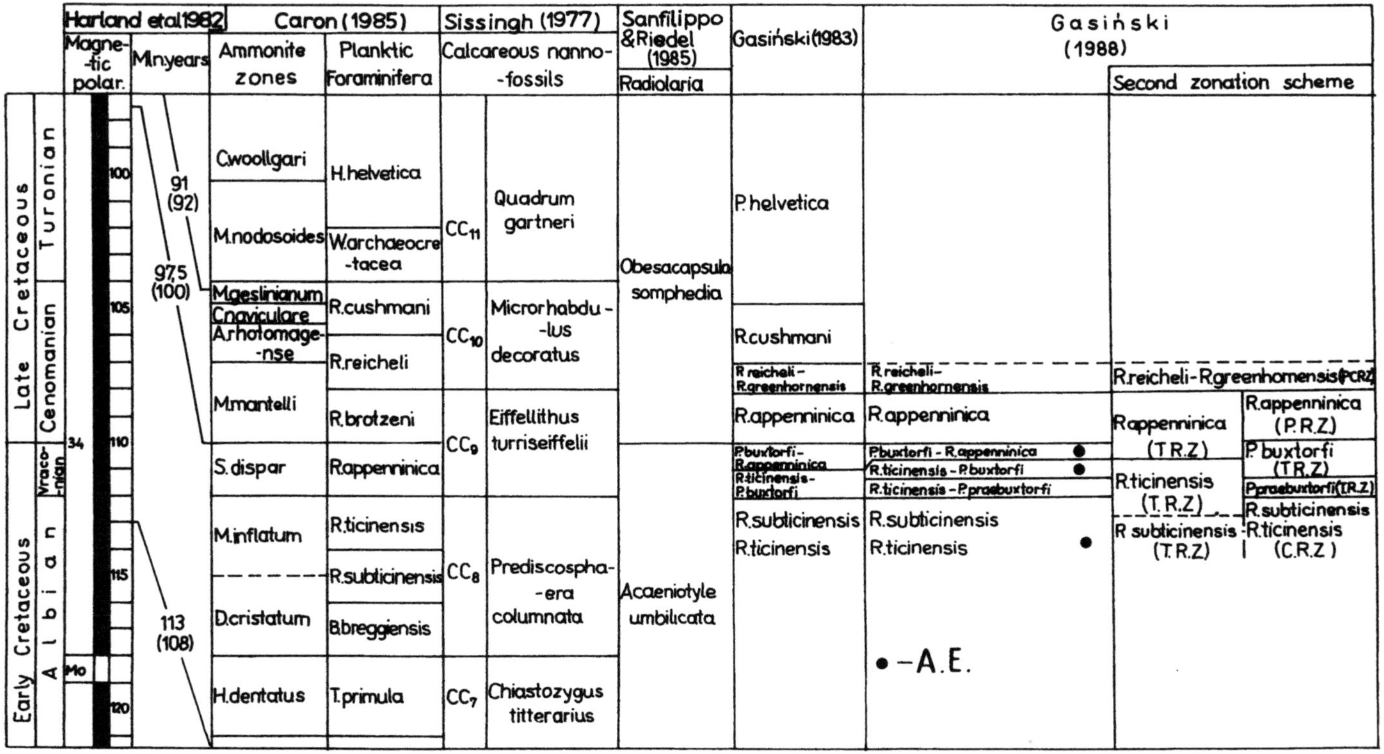

Fig. 2: Local Albian and Cenomanian biostratigraphic zonation in the Pieniny Klippen Belt compared with the ortho- and parastratigraphic zonations of other areas (A.E. anoxic events episodes; after GASIŃSKI, 1988, simplified).

Rotalipora appenninica, local partial range zone (LPRZ). Upper boundary: FAD of *R. reicheli* and FAD of *R. greenhornensis*.

Rotalipora reicheli-Rotalipora greenhornensis (LPCRZ). Upper boundary: FAD of *R. cushmani*.

A second zonation scheme, based on the Taxon Range Zones (TRZ) for facilitating practical application is also proposed (Fig. 2). Descriptions of these local zones with the analyses of assemblages, including correlations and comparisons with other recently proposed zonations were presented. A comparison with the Breggia and Moria sections was discussed (GASINSKI, 1988). BIRKENMAJER & JEDNOROWSKA (1987) have recognized zones established by ROBASZYNSKI &

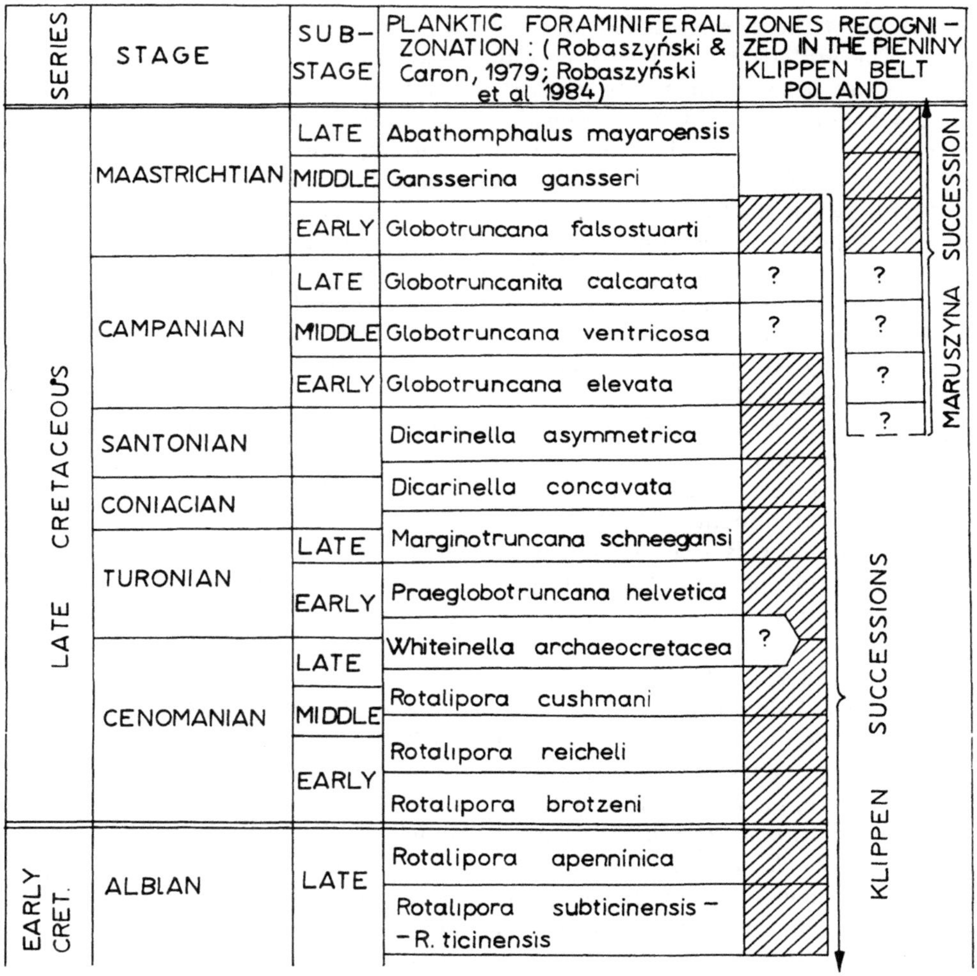

Fig. 3: Foraminiferal zonations by ROBASZYŃSKI & CARON (1979) and ROBASZYŃSKI et al. (1984), recorded in the Pieniny Klippen Belt by BIRKENMAJER & JEDNOROWSKA (1987).

Caron (1979) and ROBASZYNSKI et al. (1984) in the Pieniny Klippen Belt (fig. 3). A comparison of the author's local zones to those established by ROBASZYNSKI & CARON (1979) was presented earlier (GASINSKI, 1983, Fig. 15).

4. Paleoecological remarks

The investigated foraminifera assemblages are similar in composition and species morphology to those of the Southern Alps (Breggia section, Lombardy Basin) and the Umbria-Marche Apennines (Moria section; see GASINSKI, 1988).

Assemblages of agglutinated foraminifera from the Outer (Flysch) Carpathians, situated to the north of the Pieniny Klippen Belt (Fig. 1 B) resemble those of the Moroccan Rif (MORGIEL & OLSZEWSKA, 1982).

The affiliation of the Flysch Carpathian foraminifera assemblages to the Tethys has been confirmed by OLSZEWSKA (1984). Pieniny foraminifera assemblages are definitely different from those described from areas north of the Carpathians especially in the composition of the benthos (HELLER, 1975; PERYT, 1980, 1988 and others). These differences are similar to those recorded by PRICE (1977) between the Tethyan and the Boreal province. The Tethyan character of the investigated microfauna is therefore confirmed. A bathymetric differentiation of the Pieniny Klippen Belt Basin into a deeper part (Pieniny-Branisko furrow) and more shallow part (Czorsztyn s. l. ridge) have been confirmed by foraminiferal assemblages.

For the Albian and the Cenomanian three paleobathymetric associations have been established:

a) "Czorsztyn association", shelf-upper slope, large proportion of nodosarids and miliolids,

b) "Pieniny A association", middle part of slope, oligotaxic planktonic assemblages dominant,

c) "Pieniny B association", bathymetric position similar to "Pieniny A", but with larger portion of agglutinated taxa; characteristic association of pelagic sediments with turbiditic intercalations.

All associations correspond to the various parts of the *"Marssonella"* association of HAIG (1979).

Black shale layers within the sequence are interpreted with SCHLANGER & JENKYNS (1976), ARTHUR & SCHLANGER (1979) and others as reflexions of oceanic anoxic events. Investigations of foraminiferal assemblages in sediments below, above, as well as within the black shales by the author reveal several episodes of anoxia (within OAE1 sensu ARTHUR & SCHLANGER, 1979) in the *R. subticinensis – R. ticinensis, R. ticinensis – P. buxtorfi,* and *P. buxtorfi – R. appeninica* local zones (fig. 2). These episodes have also been observed by the author in the Breggia and Moria sections. The well-known Cenomanian-Turonian anoxic event ("Bonarelli Level") can be also observed in the Pieniny Klippen Belt (BIRKENMAJER & JEDNOROWSKA, 1984, 1987; GASINSKI, 1988). Based on the composition of foraminiferal assemblages the model A (Black Sea model) and model B (mid-water oxygen minimum model) of ARTHUR & SCHLANGER (1979) can be distinguished. Some samples directly overlying black shales in the investigated sections are enriched in Radiolaria. This suggests the existence of upwelling following anoxic episodes (REYMENT & BENGTSON, 1986; GASINSKI, 1988, p. 241).

References

ALEXANDROWICZ, S. W. (1966): Le stratigraphie du Crétacé superieur et moyen dans la partie polonaise de la zone des Klippes Piènines. – Zeszyty Naukowe AGH, Geologia, Rozprawy, **78:** 1–142 (In Polish, French abstract).

ALEXANDROWICZ, S. W. (1979): Albian Foraminifera of the Czorsztyn Series, Chmielowa Formation of the Pieniny Klippen Belt. – Ann. Soc. Gel. Pol., **49:** 165–183 (In Polish, English abstract).

ARTHUR, M. A., & SCHLANGER, S. O. (1979): Cretaceous "Oceanic Anoxic Events" as causal factors in development of reef-reservoired giant oil fields. – AAPG Bull., **63:** 870–885.

BARRON, E. J., HARRISON, CH. G. A., SLOAN, J. L., & HAY, W. W. (1981): Paleogeography. 180 million years ago to the present. – Ecl. geol. Helv., **74:** 443–470.

BIRKENMAJER, K. (1977): Jurassic and Cretaceous lithostratigraphic units of the Pieniny Klippen Belt, Carpathians. – Stud. geol. pol., **45:** 1–158.

BIRKENMAJER, K. (1986): Stages of structural evolution of the Pieniny Klippen Belt, Carpathians. – Stud. geol. pol., **88:** 7–32.

BIRKENMAJER, K. (1987): The Trawne Member (Upper Albian – Upper Cenomanian) a flysch development in the Branisko Nappe, Pieniny Klippen Belt, Carpathians. – Stud. geol. pol., **92:** 29–40.

BIRKENMAJER, K., & JEDNOROWSKA, A. (1984): Upper Cretaceous stratigraphy in the Pieniny Nappe at Sromowce Nizne, Pieniny Klippen Belt, Carpathians. – Stud. geol. pol., **83:** 25–50.

BIRKENMAJER, K., & JEDNOROWSKA, A. (1987): Late Cretaceous foraminiferal biostratigraphy of the Pieniny Klippen Belt (Carpathians, Poland). – Stud. geol. pol., **42:** 7–28.

CARON, M. (1985): Cretaceous planktic Foraminifera. In: Plankton stratigraphy (eds Bolli, H. M., Saunders, J. B., & Perch-Nielsen, K.), pp. 17–86.

DUDZIAK, J. (1979): Nannoplankton from the Upper Cretaceous of the Pieniny Klippen Belt of Poland, Carpathians. Stud. geol. pol., **61:** 77–103.

DUDZIAK, J. (1981): Upper Cretaceous nannoplankton from the Czorsztyn Succession, Pieniny Klippen Belt, Carpathians. A scanning microscope study. Stud. geol. pol., **70:** 55–65.

GASIŃSKI, M. A. (1983): Albian and Cenomanian planktic Foraminiferida from the Trawne Beds (Pieniny Klippen Belt, Polish Carpathians). Cret. Res., **4:** 221–249.

GASIŃSKI, M. A. (1984): Selected benthic Foraminiferida from the so-called Trawne Beds (Pieniny Klippen Belt, Polish Carpathians). Stud. geol. pol., **83:** 51–65.

GASIŃSKI, M. A. (1988): Foraminiferal biostratigraphy of the Albian and Cenomanian sediments in the Polish part of the Pieniny Klippen Belt, Carpathian Mountains. Cret. Res., **9:** 217–247.

HAIG, D. W. (1979): Global distribution patterns for Mid-Cretaceous foraminiferids. Jour. Foram. Res., **9:** 29–40.

HARLAND, W. B., COX, A. V., LLEWELLYN, P. G., PIKTON, C. A. G., SMITH, A. G., & WALTERS, R. (1982): A geologic time scale. 131 pp. (Cambridge Univ. Press).

HELLER, I. (1975): Microbiostratigraphy of the Cretaceous deposits in the southern part

of Łódź Synklinorium (Central Poland). Ann. Soc. Geol. Pol., **45:** 233–254 (In Polish, English Summary).
JEDNOROWSKA, A. (1979): Microfauna and age of Upper Cretaceous Pustelnia Marl Member, Pieniny Klippen Belt of Poland. Stud. geol. pol., **61:** 37–76.
JEDNOROWSKA, A. (1980): Microfauna and age of the Malinowa Shale Formation, Upper Cretaceous, in the Pieniny Klippen Belt, Carpathians. Stud. geol. pol., **67:** 8–21.
MORGIEL, J., & OLSZEWSKA, B. (1982): Uniformity of the Tethyan Faunas from Cretaceous and Paleogene as shown by foraminifers from Morocco and Polish Flysch Carpathians. Cah. Micropal., **3:** 45–53.
OLSZEWSKA, B. (1984): A paleoecological interpretation of the Cretaceous and Paleogene foraminifers of the Polish Outer Carpathians. Biul. IG., **346:** 7–62 (In Polish, English Summary).
PRICE, R. J. (1977): The stratigraphical zonation of the Albian sediments of north-west Europe, as based on foraminifera. Proc. Geol. Assoc., **88:** 65–91.
PERYT, D. (1980): Planktic Foraminifera zonation of the Upper Cretaceous in the Middle Vistula River Valley, Poland. Palaeontol. Pol., **41:** 3–101.
PERYT, D. (1988): Paleoecology of middle and late Cretaceous foraminifers from the Lublin Upland /SE Poland/. Rev. Paleobiol. vol. spec., **2:** (Benthos '86) 311–321.
REYMENT, R., & BENGTSON, P. (COMPILERS). (1986): Events of the Mid-Cretaceous. Phys. Chem. of the Earth., **16: 210 pp. (Pergamon Press, Oxford).**
ROBASZYŃSKI, F., & CARON, M. (Coordinators) (1979): Atlas de Foraminiferes planctoniques du Crétacé moyen (mer Boréale et Tetyhs). I et II. Cah. Pal. I: 1–185, II: 1–181.
ROBASZYŃSKI, F., CARON, M., GONZALES DONOSO, J. M., & WONDERS, A. A. H. (1984): Atlas of the Late Cretaceous Globotruncanids. Rev. Micropal., **26:** 145–305.
SCHLANGER, S. O., & JENKYNS, H. C. (1976): Cretaceous oceanic anoxic events– causes and consequences. Geol. Mijn., **55:** 179–184.
SISSINGH, W. (1977): Biostratigraphy of Cretaceous calcareous nannoplankton. Geol. Mijn., **56:** 37–65.

Plate 1.

Fig. 1. *Tritaxia gaultina* (Morozova). Sample FL-26, *R. ticinensis – P. buxtorfi* LPCRZ (x 100).

Fig. 2. *Dorothia oxycona* (Reuss). Sample Cz-3, *R. ticinensis – P. buxtorfi* LPCRZ (x 120).

Fig. 3. *Biticinella breggiensis* (Gandolfi). Sample Fl-4, *R. subticinensis – R. ticinensis* LCRZ (x 200).

Fig. 4. *Globigerinelloides bentonensis* (Morrow). Sample FL-16, *R. ticinensis – P. buxtorfi* LPCRZ (x 200).

Fig. 5. *Globigerinelloides bentonensis* (Morrow). Sample Kp-5, *R. subticinensis – R. ticinensis* LPCRZ (x 300).

Figs. 6–7. *Hedbergella delrioensis* (Carsey). Sample Fl-7, *R. subticinensis – R. ticinensis* LCRZ (x 300).

Figs. 8–9. *Hedbergella delrioensis* (Carsey). Sample Fl-14, *R. subticinensis – R. ticinensis* LCRZ (x 200).

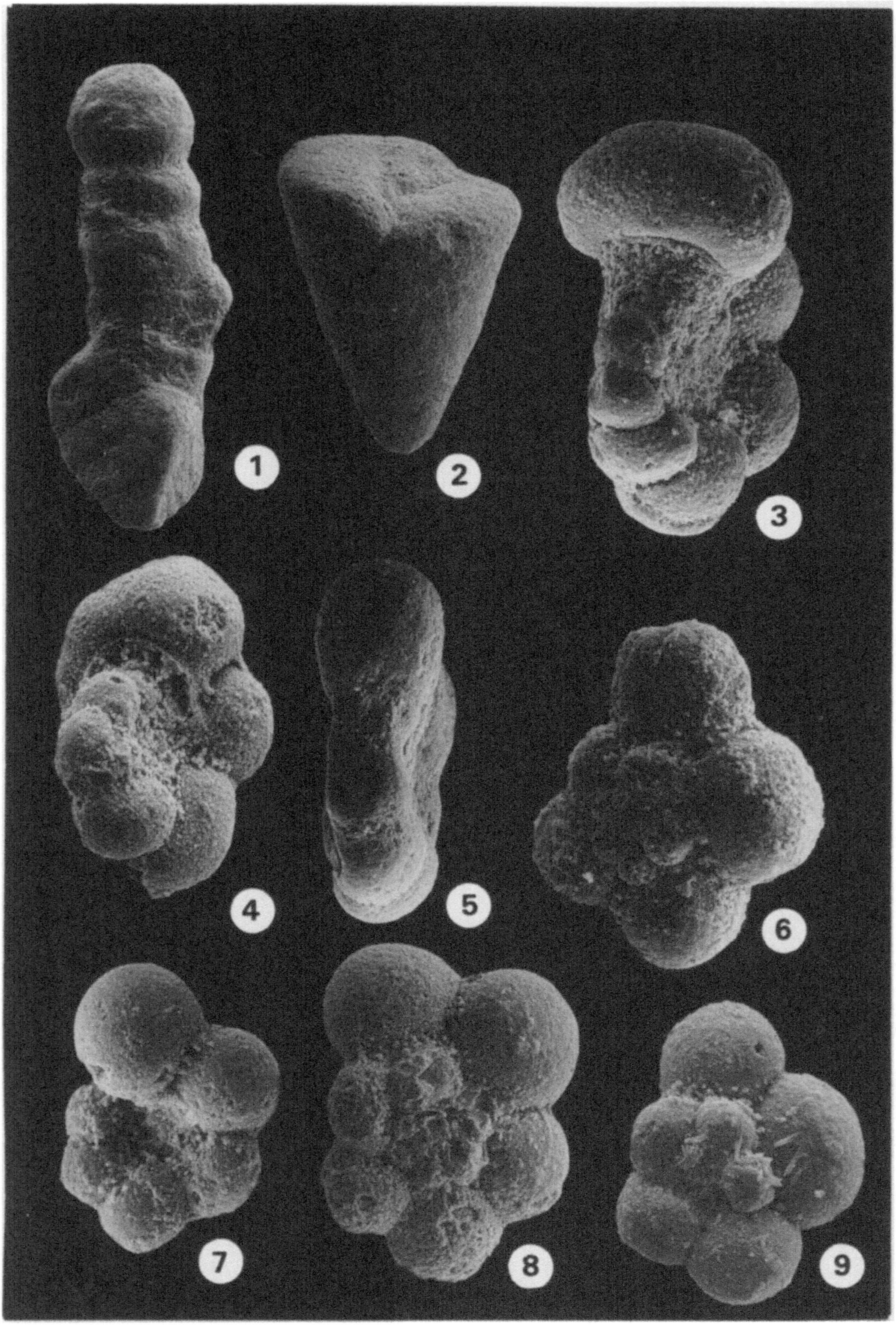
1
2
3
4
5
6
7
8
9

Plate 2

Figs. 1–2. *Hedbergella delrioensis* (Carsey). Sample Kp-4, *R. ticinensis* – *R. praebuxtorfi* LCRZ (1 : x 200; 2 : x 240).
Fig. 3. *Hedbergella delrioensis* (Carsey). Sample Cz-3, *R. ticinensis* – *P. buxtorfi* LPCRZ (x 300).
Fig. 4. *Hedbergella delrioensis* (Carsey). Sample Kp-7, *P. buxtorfi* – *R. appeninica* LCRZ (x 200).
Fig. 5. *Hedbergella simplex* (Morrow). Sample Fl-34, *P. buxtorfi* – *R. appenninica* LCRZ (x 320).
Fig. 6. *Hedbergella planispira* (Tappan). Sample Fl-8, *R. subticinensis* – *R. ticinensis* LCRZ (x 160).
Fig. 7. *Hedbergella simplex* (Morrow). Sample H-7, *R. reicheli* – *R. greenhornensis* LPCRZ (x 260).
Fig. 8. *Planomalina buxtorfi* (Gandolfi). Sample Fl-28, *P. buxtorfi* – *R. appenninica* LCRZ (x 160).

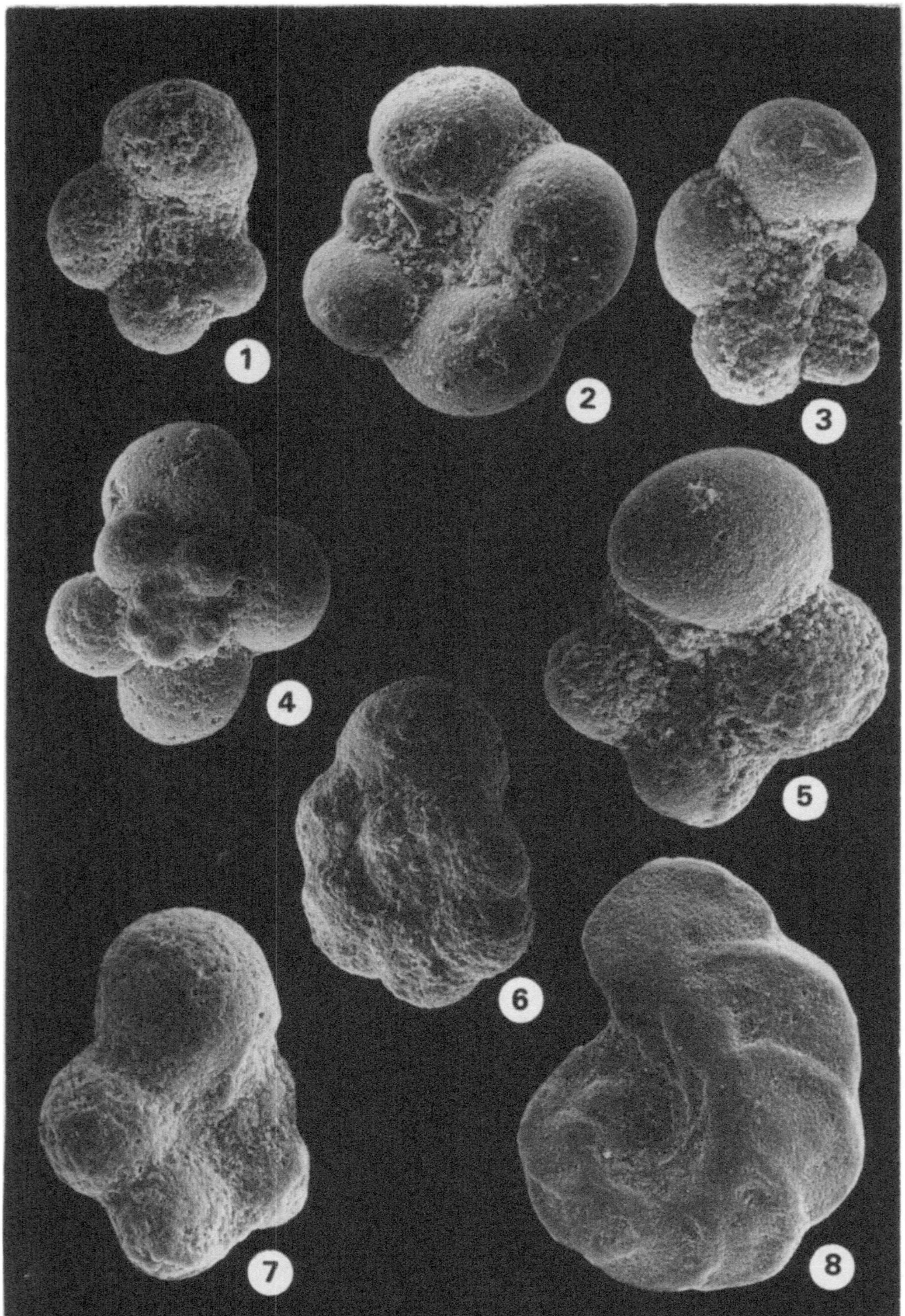
1
2
3
4
5
6
7
8

Plate 3

Fig. 1. *Rotalipora subticinensis* (GANDOLFI). Sample Kp-1, *R. subticinensis* – *R. ticinensis* (LCRZ (x 150).
Fig. 2. *Rotalipora subticinensis* (GANDOLFI). Sample Fl-5. *R. subticinensis* – *R. ticinensis* LCRZ (x 150).
Fig. 3. *Rotalipora ticinensis* (GANDOLFI). Sample Fl-8. *R. subticinensis* – *R. ticinensis* LCRZ (x 130).
Fig. 4. *Rotalipora ticinensis* (GANDOLFI). Sample Kp-1, *R. subticinensis* – *R. ticinensis* LCRZ (x 130).
Fig. 5. *Rotalipora appenninica* (RENZ). Fl-33, *P. buxtorfi* – *R. appenninica* LCRZ (x 160).
Fig. 6. *Radiolaria*, gen. indet. Sample Cz-3 (x 120).
Fig. 7. *Archaeodictyomitra* sp. Sample Fl-19 (x 240).
Fig. 8. *Dictyomitra koslovae* FOREMAN. Sample Kp-6, *R. ticinensis* – *P. buxtorfi* LPCRZ (x 240).
Fig. 9. *Rotalipora appenninica* (RENZ). SAMPLE KP-9, *R. appenninica* LPRZ (x 160).

1
2
3
4
5
6
7
8
9

New Aspects on Tethyan Cretaceous Fossil Assemblages.
Band 9 Schriftenreihe der Erdwissenschaftlichen Kommissionen der Österreichischen Akademie der Wissenschaften, 201–221.

4.2. Stratigraphy and biozonation of a reference Aptian – Albian p.p. Tethyan carbonate platform succession: The Sierra del Carche series (oriental Prebetic zone – Murcia, Spain)

By Jean Pierre MASSE*), Consuelo ARIAS**) and Lorenzo VILAS**).

Abstract

The Sierra del Carche succession represents the Aptian-Albian p.p. of the Prebetic zone. It is characterized by platform carbonates and terrigenous sediments. Two formations can be distinguished: the stratigraphically lower formation was deposited on a carbonate platform during the Bedoulian-Early Gargasian. The stratigraphically higher formation was deposited on a mixed carbonate-siliciclastic platform during the Late Gargasian-Late Albian p.p. Vertical ranges of benthic foraminifera, calcareous green algae and rudists previously recorded in this region show that many taxa can be used either alone, in association, or according to the order of their stratigraphic appearance as indices for biozonation. Thus, 10 biozones have been distinguished in the Bedoulian p.p. – Upper Albian p.p. interval; each one is interpreted with regard to its chronological value. As many of the taxa involved in this framework have a large paleobiogeographic distribution, the proposed biozonation may proof valuable for the stratigraphic correlation of carbonate platform successions wide apart in the Tethyan realm.

1. Introduction

In the circum-mediterranean areas important paleogeographic changes occurred during Aptian-Albian times which affected especially the carbonate platforms. The indispensable basis for detailed studies of these changes are

*) Laboratoire de Stratigraphie et de Paléoécologie, URA, 1208 du C.N.R.S., Université de Provence, 3 Pl. V. Hugo. F-13331 Marseille Cedex 3. France.

**) Dpto. de Estratigrafia. Univ. Complutense y U.E.I. de Correlaciones estratigraficas y Paleogeografia C.S.I.C. 28040 Madrid, Spain.

regional correlations and a precise biostratigraphy of the series. The objective of the present work is to illustrate a detailed local succession which is considered as a reference for the biozonation of three biological groups: foraminifera, calcareous algae and rudists, each being important in "urgonian facies".

The choice of the La Sierra del Carche section was based on:

– good outcrop conditions and access, which are favourable for the observation, despite certain tectonic complications of the whole succession,

– the three above-mentioned groups occurring in a subcontinuos carbonate sequence,

– abundance, diversity and good preservation of the palaeontological content, especially of rudists.

The Aptian-Albian series of the Iberic and Betic regions have been the subject of numerous biostratigraphic studies. Several biozonations, mainly based on foraminifera have been proposed (FOURCADE, 1970; FOURCADE et al., 1972; GARCIA-HERNANDEZ, 1978; AZEMA et al, 1979; GARCIA-HERNANDEZ, 1981).

Detailed investigations of many of the local series (ARIAS et al., 1987, and work in progress) allow us to focus on the Sierra del Carche. This is considered to be a valuable reference section for the biostratigraphy in the meridional zone of the Iberian Peninsula.

The Sierra del Carche is situated in the north of the province of Murcia, East of the village Jumilla (Fig. 1). It is located near the northern boundary of the Inner Prebetic tectonic zone sensu GARCIA-HERNANDEZ et al. (1980) and shows strong tectonic deformation and prominent faults. According to authors (RODRIGUEZ ESTRELLA, 1979; BAENA, 1981) its base is affected by a thrust with a sense of movement toward the north. Cretaceous rocks are the main constituent of the Sierra del Carche; these rocks indicate the development and disappearance of a carbonate platform and a mixed platform during the Aptian and Albian p.p.

2. General Stratigraphy

The exposed base of the Cretaceous in the Sierra del Carche is formed by basinal sands and marls with fragments of *Spitidiscus* aff. *intermedius* (D'ORB.), dated by AZEMA (1977) as Barremian. In addition, we have found fragments of ammonites classified as *Leptoceras* sp., *Barremites* sp. and *Pulchellia* sp. This association indicates a Middle-Upper Barremian age. A carbonate platform and a mixed terrigenous-carbonate succession ranging into the Upper Albian, overlies this unit. It is followed above by a sequence of terrigenous deposits and dolomites. We have found *Neorbitolinopsis conulus* (DOUVILLÉ) and *Caprina choffati* DOUVILLÉ in limestones intercalated at the base of this sequence. The lowermost levels of the terrigenous sequence are dated as Upper Albian-Vraconian by these faunal elements.

We have distinguished several divisions within the carbonates and the terrigenous-carbonate succession. They are principally based on stratigraphic dis-

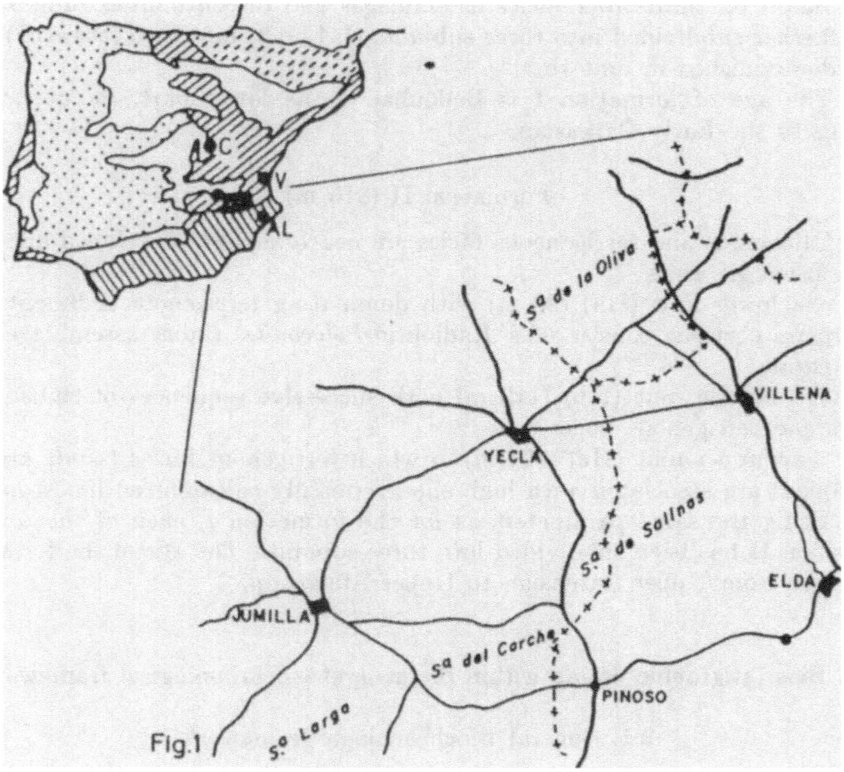

Fig. 1: The geographical and geological position of the Sierra del Carche.

continuities which form boundaries between the sedimentary sequences, and on palaentological and lithological characteristics (Fig. 2).

According to the stratigraphic position two major formations can be distinguished, each is composed of several units and subunits:

Formation I (434 m)

Limestones with both, biocalcarenitic and rudist facies, are dominant. Terrigenous sediments are poorly represented. Two units can be distinguished:

– a lower unit (Ia) (248 m) made of well sorted bioclastic to oolitic limestones, orbitolinid to coral packstones, caprinid packstones-wackestones, requienid *(Toucasia)* beds, wackestones-packstones and marly limestones rich in bivalves (ostreids and pholadomyids).

– an upper unit (Ib) (186 m) consisting of both terrigeneous and calcareous facies: sands, clays and marls with some plant remains intercalating with requienid limestones, Bacinella wackestones-packstones with some branching corals, *Toucasia* and black-pebble horizons.

Based on third-order facies assemblages and discontinuities, unit Ia has been further subdivided into three subunits (Ia1, Ia2, Ia3), two (Ib1–Ib2) have been distinguished in unit Ib.

The age of formation I is Bedoulian in its lower part, its upper part belongs to the Early Gargasian.

Formation II (316 m)

Calcareous and terrigeneous facies are nearly equally developed and form three lithologic units.

– a lower unit (IIa) (99 m) with dominating terrigenous sediments, the calcareous portions consist of a Radiolitid-*Polyconites* rudist assemblage with some corals.

– a middle unit (IIb) (113 m) with successive sequences of rudist limestones and terrigenous facies,

– an upper unit (IIc) (104 m) in which terrigenous facies (sands are well developed) are associated with high energy (mainly calcarenitic) limestones.

Using the same parameters as for the formation I, each of the units of formation II has been subdivided into three subunits. The age of the formation II ranges from Upper Gargasian to Upper Albian pp.

3. Biostratigraphic dating within the general biochronological framework

3.1. General Biochronologic framework

Champetier and Moullade (1968), Busnardo et al. (1968), and Fourcade (1970) called attention to the biochronological value of taxa which can be used for biozonation: *Iraquia simplex* (Henson), *Orbitolinopsis reticulata* Moullade and Peybernes (= *O. carochi* in Fourcade), *Mesorbitolina parva* (Douglass), *Pseudochoffatella cuvillieri* Deloffre, *Simplorbitolina manasi* Ciry and Rat, *Mesorbitolina texana* (Roemer) and *Neorbitolinopsis conulus* (Douvillé). The authors proposed six biozones from the Bedoulian p.p. to the Albian p.p. in the Iberic and Betic regions. In 1976, Peybernes recognized five biozones characterized by the following species (from base to top): *Mesorbitolina parva, Pseudochoffatella cuvillieri, Simplorbitolina* cf. *manasi* (= *S. chauvei* Fourcade), *Simplorbitolina manasi, Simplorbitolina conulus* Schroeder. This author was the first to propose that the Clansayesian is represented by the first appearance of *Mesorbitolina texana.* Azema et al. (1979), introduced a zone with *Simplorbitolina chauvei* according to Fourcade (1970) between to zones with *Pseudochoffatella cuvillieri* and *Simplorbitolina manasi* although the object of their work was not the subdivision of the Clansayesian.

In 1979, Peybernes revised his former zonal scheme by transferring *Simplorbitolina chauvei* into the Lower Clansayesian and *Simplorbitolina manasi* into the Middle and Upper Clansayesian. In 1981, Garcia-Hernandez, distinguished a certain hierarchy between these species and concluded that some were valuable for defining a biozone while others were limited to various horizons within one or more biozones.

The repartition of species by PASCAL (1984) agrees with that of PEYBERNES. However, in the majority of cases, the vertical distribution of the index forms is larger and shows some overlapping. For this reason PASCAL does not use a formal biozonation in his work. PASCAL puts *Simplorbitolina chauvei* and *Simplorbitolina manasi* (in the absence of *Simplorbitolina conulus*) together with *Pseudochoffatella cuvillieri* into the same chronological interval and mentions their coexistence in the Clansayesian. However, this does not agree with N'DA LOUKOU (1984), who puts *Pseudochoffatella cuvillieri* into the Upper Gargasian and dates the beginning of the Clansayesien with the first appearance of *Mesorbitolina texana*. N'DA LOUKOU's scheme was adopted by ARNAUD-VANNEAU (1986).

The chronological interpretation of these previously established biozonations and, which is of less importance, of individual distributions assigned to the taxa by various authors, is different in the Iberic and the Pyreneo-Aquitanic regions. While many different biozonation concepts exist for the Upper Aptian, there is general agreement regarding the Albian; this is exemplified by the vertical range of the succession *Simplorbitolina manasi*, *Simplorbitolina conulus* and *Neorbitolinopsis conulus* respectively represented in the Lower Albian p.p., the Upper and/or Middle Albian, and the Upper Albian (REY et al., 1977; BERTHOU and SCHROEDER, 1978; PEYBERNES, 1979; GARCIA-HERNANDEZ, 1981; ARNAUD-VANNEAU et al., 1985).

3.2. Biostratigraphical description and biochronological interpretation

In this chapter, the main taxa recognized in the lithological units and subunits of the Sierra del Carche are presented. The units/subunits are dated according to the general biostratigraphic framework discussed before.

Formation I

Lower unit (Ia). – The lower subunit (Ia1) is characterized by the association of: *Palorbitolina lenticularis* (BLUMENBACH), *Orbitolinopsis cuvillieri* MOULLADE, *O. buccifer* ARNAUD-VANNEAU & THIEULOY, *O. briacensis* ARNAUD-VANNEAU, *Dictyoconus? vercorii* ARNAUD-VANNEAU, *Rectodictyoconus? giganteus* SCHROEDER, *Choffatella decipiens* SCHLUMBERGER, *Debarina hahounerensis* FOURCADE et al., and *Neotrocholina* aff. *aptiensis* (IOCHEVA). In the southeast of France this assemblage indicates the Lower Aptian (MASSE, 1976; ARNAUD-VANNEAU, 1980). The middle subunit (Ia2) contains in its lower part elements of the preceding association. In its upper part *Iraquia* simplex has its first appearance together with a rudist fauna of *Offneria* sp., *Caprina douvillei* PAQUIER and *Pachytraga paradoxa* (PICTET and CAMPICHE). This rudist fauna is characteristic for the Lower Aptian (MASSE, 1976–1985). In the upper subunit (Ia3) the Dictyoconinae are represented exclusively by *Iraquia simplex*. *Choffa-*

tella decipiens is still present; the rudists are dominanted by the genus *Toucasia* associated locally with *Monopleura, Matheronia,* very scarce primitive radiolitids and *Horipleura* cf. *baylei* PAQUIER. This unit can be dated as Lower Aptian, too.

Upper unit (Ib). – The lower subunit (Ib1) contains *Praeorbitolina wienandsi* SCHROEDER and *Mesorbitolina parva.* We consider this association as uppermost Lower Aptian in age. The rudists associated are *Agriopeura* sp. and *Horiopleura* cf. *baylei.* The upper subunit (Ib2) includes in its middle part *Mesorbitolina parva* and *Dictyoconus? pachymarginalis* SCHROEDER, and at the top *Mesorbitolina minuta* (DOUGLASS). According to studies on the stratigraphical distribution of this species, we have assigned a Lower Gargasian age to this sequence.

Formation II

Lower unit IIa. – The lower subunit (IIa1) is characterized by *Orbitolinopsis reticulata,* associated with scarce *Orbitolinopsis aquitanica* SCHROEDER & POIGNANT which are of Upper Gargasian age. Typical *Eoradiolites* are associated with this microfauna as well as *Polyconites* cf. *verneuili* (BAYLE) and *Horiopleura lamberti* BAYLE and COQUAND. The middle subunit (IIa2) lacks a characteristic microfauna. The rudists are dominated by *Pseudotoucasia* cf. *santanderensis* DOUVILLÉ. The upper subunit (IIa3) is well characterized by the presence of *Pseudochoffatella cuvillieri* associated with *Mesorbitolina texana,* relatively evolved radiolitids and forms belonging to the *Horiopleura-Polyconites* group. This fauna is can be dated as Upper Gargasian (Clansayesian pp.?).

Middle unit IIb. – Whereas the Lower subunit (IIb1) contains a insignificant microfauna, the overlying subunit IIb2 is characterized by *Simplorbitolina manasi* associated with Radiolitidae *(Eoradiolites* and *Praeradiolites)* and with the group *Horiopleura – Polyconites;* this suggests a Clansayesian pp. – Lower Albian age. *Simplorbitolina conulus,* defining the period from Lower Albian pp. to Middle Albian appears first in the upper subunit (IIb3).

Upper unit IIc. – In the lower subunit (IIc1) *Simplorbitolina conulus* continues. *Hensonina lenticularis* (HENSON) appears in the middle part of this subunit. The middle subunit (IIc2) contains a similar microfauna. *Neorbitolinopsis conulus* and *Neoiraquia* sp. cf. *convexa* DANILOVA are recorded in the upper subunit (IIc3) accompanied by the previous species. Rudists are mainly represented by costulated monopleurids (*Agriopleura? darderi* ASTRE) associated with evolved *Eoradiolites* and numerous *Polyconites.* According to recent studies, the subassemblage IIc can be attributed to the Upper Albian p.p.

4. Discussion of the stratigraphical distribution of principal taxa

4.1. Calcareous Algae

Among the dasycladaceans we have found *Helioporella cylindrica* SOKAC and NIKLER. This form is well represented in the Outer Dinarids of Montenegro

(SOKAC and NIKLER, 1973) and occurs exclusively in the Bedoulian. In the Upper Aptian-Albian, *Cylindroporella ivanovici* (SOKAC) appears which has a wide stratigraphic distribution. This form, which is the type species of the genus *Korkyrella* SOKAC and VELIC 1981, as redefined by SOKAC in 1987, is assigned to the genus *Cylindroporella* JOHNSON, (MASSE and LUPERTO-SINNI, studies in progress.) It closely resembles *Cylindroporella? barnesii* JOHNSON but differs from this species by its external mace-like morphology. Its sterile and fertile parts are clearly differentiated; furthermore, the fertile ampoulae are much more numerous. SOKAC and VELIC originally included this species into *Cylindroporella? texana* JOHNSON; however, although the morphology is very similar, it can be distinguished by its general organization, which conforms to the genus *Cylindroporella*. We have observed rare specimens of *Montiella?* cf. *elitzae* (BAKALOVA) as well, hitherto only known from Hauterivian to Bedoulian beds. *Carpathoporella fontis* (PATRULIUS) and forms like *Neomeris* which pose specific determination problems and others of uncertain taxonomic status such as *Terquemella*, may be added. Finally, various types of the latter, some of which can be compared with the calcified cysts of *Triploporella*, have been recorded. The scarcity of dasycladaceans in the Upper Aptian-Albian interval is not surprising. This is the case in the entire perimediterranean area (MASSE, in press). It is rather astonishing that dasycladaceans are so rare in the Bedoulian. At this time, the general ecological conditions appear much more favourable. This may be concluded from the common occurrence of benthic foraminifera such as Dictyoconinae and Miliolidae and calcified Chlorophyceae of the genus *Boueina*. The last-mentioned genus, represented by various types, is also very frequent in the Upper Aptian-Albian.

Together with these taxa, some species of wide stratigraphic range occur. These include *Marinella lugeoni* PFENDER, *Polystrata alba* PFENDER, *Permocalculus*, as well as diverse porostromatic structures: *Cayeuxia*, *Bacinella-Lithocodium* and *Girvanella*.

4.2. Foraminifera

Our analysis of the studied interval adds a number of new aspects to previous work on the Iberic region. Recent progress in the micropalaeontology of eocretaceous Miliolidae (see mainly the works of NEAGU, 1984–1986 and ARNAUD-VANNEAU, 1980) has enabled us to examine the stratigraphic distribution and inventory of this family; however, our results have to be considered as provisional. The majority of the forms – *Cornuloculina tenuae* NEAGU, *Moesiloculina danubiana* NEAGU, *Massilina* cf. *ostroviana Neagu, Rumanoloculina* cf. *robusta* NEAGU and *Istriloculina eliptica* IOCHEVA – have a distribution which covers the Barremian-Bedoulian interval. Nevertheless the following should be remarked:

– In accordance with the observations of N'DA LOUKOU (1984), *Derventina filipescui* NEAGU does not extend beyond the limit of the Bedoulian-Gargasian.

– The Miliolidae assemblage changed significantly in the Upper Albian; because a detailed taxonomic review of these forms is lacking, this observation is mainly subjective and needs further evaluation.

Referring to the Orbitolinidae, five points have to be noted:

1. We did not distinguish the different morphotypes of *Iraquia simplex:* This name comprises all types agreeing with the original description of Henson (1984), as well as those designated by Bassoullet and Moullade (1962) under the name *Iraquia rubiensis.*

2. In the stratigraphic succession, *Simplorbitolina conulus* precedes *Neorbitolinopsis conulus;* this agrees with the observations of our predecessors (see mainly the review of R. Schroeder in Arnaud-Vanneau et al., 1985). Until now it was generally accepted that these two species were excluding each other in their stratigraphic range. *Simplorbitolina* was thought to occur essentially in the Lower to Middle Albian, *Neorbitolinopsis conulus* in the Upper Albian. We have observed the two species together in the uppermost beds of formation II (Upper Albian pp.). This confirms Peybernes (1979) who has pointed out the possibility of a coexistence of both species and has emphasized their close evolutionary relationship. The same observations can be made regarding the coexistence of *Simplorbitolina manasi* and *Simplorbitolina conulus* as suggested by Pascal (1984).

3. The presence of the genus *Praeorbitolina* together with *Mesorbitolina parva* in the same level, already reported by Pascal (1984) in the Cantabrian area, is confirmed here.

4. The horizon with *Praeorbitolina* and *Mesorbitolina,* which precedes the *Iraquia simplex* biozone in the Western Prebetic zone (Garcia-Hernandez, 1981) and in the meridional Iberic zone (Arias et al., 1987), appears to be inverted in the Sierra del Carche. We think this is due to local ecological reasons: *Iraquia* is closely connected to the facies with Requieniidae, while the *Praeorbitolina-Mesorbitolina* association is usually found in facies without rudists.

5. *Mesorbitolina texana* appears simultanously with *Pseudochoffatella cuvillieri.* This raises questions regarding the value of both species for defining the Upper Gargasian in relation to the Clansayesian in carbonate platform facies. This species, vaguely defined by Roemer (1849) and revised by Douglass (1960) and Schroeder in 1979, has been the object of numerous interpretations. The name *Mesorbitolina texana* has been used for *"Mesorbitolina"* with sub- and supra-embrionary zones showing a well developed network of septa (Douglass, 1960). Moullade and Saint-Marc (1975) considered under this denomination forms with hemispheric proloculi, flat bases, and quite large size. For Schroeder (1979), who has revised the specimens collected and figured by Roemer, *O. texana* has a sub-spheric, small to medium size "protoconch" with a generally flattened upper part.

Consequently, the forms mentioned by Douglass, 1960 and Moullade and Saint-Marc, 1975, such as *O. (M.) texana* and *O. (M.) minuta* Douglass, belong after Schroeder respectively to *O. (M.) subconcava* (Leymerie) and *O. (M.) texana* Roemer.

The results of the review of SCHROEDER with respect to *M. texana* appear to be correct. However, SCHROEDER's invalidation of *M. minuta* DOUGLASS is not clear, as it is based on the characteristics of the embryonal system and not on the structure of the marginal zone; the taxonomical importance of the latter structure was emphasized by DOUGLASS. Thus the identity of the taxa *O. texana* and *O. minuta* is not yet well demonstrated. After SCHROEDER's attempt to rehabilitate the concept of *O. (M.) subconcava* LEYMERIE, PEYBERNES (1982) published a review of the type material. PEYBERNES showed that it was impossible to give formal validity to this taxon and proposed to designate *O. leymeriei* as neotype of *O. subconcava* as it was assigned by SCHROEDER (PEYBERNES, 1982). However, after this author *O. leymeriei* does not include the forms with hemispheric embryos and flat bases (pro *"O. subconcava"*, sensu SCHROEDER and *"Orbitolina* sp. *A"* sensu FOURCADE and RAOULT 1973) and could not be related to *O. texana*.

This opinion is in contradiction with the data presented in SCHROEDER's review in 1979. Recognizing these nomenclatural difficulties and considering the work of various authors on the Iberic region and the southwest of France, we have placed the mesorbitolinids with embryos, showing numerous chamberlets, into two groups:

– *O. (M.) minuta* comprises forms with subspheric embryos which are in agreement with the description of DOUGLASS. It could include, p.p., *O. (M.) texana* ROEMER, sensu SCHROEDER's review in 1979.

– *O. (M.) texana* for the forms with hemispheric embryos and a flat base, which obviously do not correspond to the *O. (M.) texana* "type" but to the *M. leymeriei* PEYBERNES – *"Orbitolina" pervia* DOUGLASS group, not yet reviewed in detail.

Finally it must be noted that we haven't found the genus *Archaeoalveolina* in the Upper Aptian-Albian interval, as recorded in the Baleares (AZEMA et al., 1979) nor *Coskinolinella*, well known in the Aquitanic-Pyreneic area and its Cantabrian extension (ARNAUD-VANNEAU et al., 1985).

4.3. Rudists

Rudists are well represented in the whole succession and show a characteristic vertical distribution.

Caprinidae: *Offneria* sp. (large form with developed canals all around the valves and therefore attributed to the group *O. rhodanica* PAQUIER), and *Caprina douvillei*, are restricted to the Bedoulian like *Pachytraga paradoxa*. This stratigraphic position agrees with that known from the southeast of France (MASSE, 1976). No Caprinidae are present in the Upper Aptian and in most of the Albian; the genus *Caprina* re-appears at the end of the Albian. It is represented by *C. choffati* DOUVILLÉ in the basal part of the dolomites, overlying the studied succession (see p. 4).

Radiolitidae, represented mainly by *Eoradiolites*, increase in number notably in the middle of the Gargasian (levels with *Orbitolinopsis reticulata*,

O. aquitanica and *Mesorbitolina minuta*). These forms are small with thin and only partially cellular shells. In the Upper Gargasian larger forms with well developed cellular structures occur. Other forms of *Eoradiolites* with relatively large size and thick, cellular shells are recorded from the Clansayesian and the Albian; they are associated with *Praeradiolites*. As the general systematic framework of this group of primitive Radiolitidae is unsatisfactory we have voluntarily abstained from attributing these diverse forms to species described in the literature. A detailed taxonomic work is needed.

Like the Radiolitidae the *Polyconites-Horiopleura* group, identified in the uppermost Bedoulian and represented here by small forms, shows considerable development in the Upper Gargasian; *Horiopleura* cf. *lamberti* and *Polyconites* of the *verneuili* groups are very common and persist into the Clansayesian-Albian.

Monopleuridae are present in the whole succession, although they generally play a minor role from a quantitative point of view. Nevertheless they are particularily abundant in the Middle and Upper Albian. Costulated tubular forms belonging to the *Agriopleura? darderi* group form important biostromes. Requienidae are dominated by the genus *Toucasia* which is widely distributed abroad. *Pseudotoucasia* develops in the middle part of the Gargasien. *Pseudotoucasia santanderensis* is very abundant in both, the Upper Gargasian and the Clansayesian.

It becomes evident from this brief description, that the Sierra del Carche locality is essential for the knowledge of the Spanish Aptian-Albian rudists; until now, these were poorly known and stratigraphically only partially evaluated. The possibility of analyzing the evolution of primitive Radiolitidae within a precise biostratigraphic framework is exceptional in this area. Finally, our analysis suggests a certain number of modifications regarding the vertical partitions proposed by palaeontologists who have worked in the Aquitanic-Pyreneic domain and the revisions by Peybernes, in 1979. The stratigraphic appearance of the principal Gargasian-Albian taxa is generally younger than previously accepted.

5. Conclusions

The studied succession covers a time interval from the Lower Aptian p.p. to the Upper Albian p.p. It is divided into two main formations, each subdivided respectively into two and three units. Five subunits are recognized in the lower formation and nine in the upper, each showing distinct facies associations. Based essentially on foraminifera the time interval is subdivided into 10 biozones, whereas rudists are used as complementary indices. These biozones are defined by:

- the total range of a single species,
- the total range of a group of taxa (cenozone).

In addition, the limits of some zones are defined by the appearance of a particular taxon which may pass the boundary of the time interval (for example: *Mesorbitolina*).

The following biozones have been recorded:

1. *Orbitolinopsis cuvillieri* and *Rectodictyoconus? giganteus*, together with the *Caprina-Offneria-Pachytraga* group.

2. *Iraquia simplex*.

3. *Praeorbitolina-Mesorbitolina*.

In zones 2 and 3 the first Radiolitidae and primitive *Horiopleura* appear.

4. *Dictyoconus pachymarginalis*, with the initial appearance of *Mesorbitolina minuta*.

5. *Orbitolinopsis reticulata* and *O. aquitanica*.

In zones 4 and 5 large *Polyconites*, *Horiopleura* and *Pseudotoucasia* develop together with Radiolitidae.

6. *Pseudochoffatella cuvillieri*. Initial appearance of *Mesorbitolina texana*.

7. *Simplorbitolina manasi*. *Praeradiolites* appears together with *Hensonina lenticularis*.

8. *Simplorbitolina manasi* and *Simplorbitolina conulus*.

In zones 6, 7 and 8, Radiolitidae of large size develop.

9. *Simplorbitolina conulus* and *Neorbitolinopsis conulus*.

10. *Neorbitolinopsis conulus* and *Neoiraquia*.

In zones 9 and 10, Radiolitidae and *Agriopleura? darderi* develop, forming biostromes.

This biozonation framework lacks the *Coskinolinella* and *Archaeoalveolina* species which may be considered as valuable index fossils, as well as *Simplorbitolina chauvei*.

Some of the recorded taxa are linked to the northwestern part of the "European continent", especially to the "Iberic plate" and adjacent regions (i.e. of the Aquitaine). Many species are known from European and African regions or from both. Consequently, most of the proposed biozones are valuable for biostratigraphical correlations within the Tethyan realm.

Acknowledgements

We gratefully acknowledge the determination of the Barremian ammonites by the Rvd. G. Conte of the University of Marseille.

We also acknowledge the help of Miss B. Bartolomé and Miss R. Ojeda who kindly typed the manuscript, of Mr. J. L. Gonzalez-Pachón who printed the photographs, of Mr. J. L. Gonzales-Galán and Mr. G. Herrero who made the thin sections and Mr. C. Sanchez who drew the illustrations.

This study was supported by C.S.I.C.-C.A.I.Y.T. 452 project and also by the Spanish-French Action n. 90/168.

References

ARIAS, C., MASSE, J. P., and VILAS, L. (1987): Modalités d'installation et de développement des plates-formes carbonatées urgoniennes à la charnière des domaines ibérique et bétique (Espagne meridionale) durant l'Aptien inférieur. – Mém. Géol. Univ. Dijon, **11:** 213–223.

ARNAUD-VANNEAU, A. (1980): Micropaleontologie, paleoécologie et sédimentologie d'une plate-forme carbonatée de la marge passive de la Tethys. – Geol. Alpine. Grenoble, **11:** 874 p.

ARNAUD-VANNEAU, A. (1986): Episodes transgressifs et renouvellements des Foraminifères benthiques sur les plates-formes du Cretacé inferieur du Sud de la France. – Bull. Centres Rech. Explor.-Prod. Elf-Aquitaine, **10/2:** 405–420.

ARNAUD-VANNEAU, A., CHERCHI, A., CHIOCCHINI, M., DE CASTRO, P., FOURCADE, E., HAMAOUI, M., LUPERTO SINNI, E., NEUMANN, H., PRESTAT, B., and SCHROEDER, R. (1985): Les grands Foraminiféres du Crétacé moyen de la région méditerraneenne. – Géobios. Lyon. Mem. spec., **7:** 160 p.

AZEMA, J. (1977): Etude géologique des zones externes des Cordillères bétiques aux confins des provinces d'Alicante et de Murcie (Espagne). – Thèse Sciences Univ. Paris VI, 393 pp.

AZEMA, J., CHABRIER, G., CHAUVE, P., and FOURCADE, E. (1979): Nouvelles donées stratigraphiques sur le Jurassique et le Crétacé du Nord-Ouest d'Ibiza (Baleares, Espagne). – Geologica Rom., **18:** 1–21.

BAENA, J. (1981): Mapa Geológico de España E. 1/50.000. n° 869, Jumilla. IGME. Madrid.

BASSOULET, J. P., and MOULLADE, M. (1962): Les Orbitolinidae du Crétacé inferieur de la Sierra du Montsech, Province de Lèrida (Espagne). – Rev. Micropaleont., **5/2:** 104–114.

BERTHOU, P. Y., and SCHROEDER, R. (1978): Les Orbitolinidae et Alveolinidae de l'Albien superieur-Cenomanien inferieur et le problème de la limite Albien-Cénomanien dans le sud-ouest de la region de Lisbonne (Portugal). – Cahiers Micropaleont., Paris, **3:** 51–104.

BUSNARDO, R., CHAMPETIER, Y., FOURCADE, E., et MOULLADE, M. (1968): Etude stratigraphique des faciès à Orbitolinidés et à Rudistes de la Sierra Mariola (province d'Alicante, Espagne). – Géobios, **1:** 165–185.

CHAMPETIER, Y., et MOULLADE, M. (1968): Sur la présence de Barrémien et d'Aptien à Orbitolinidae dans la partie septentrionale de la feuille d'Alcira (prov. de Valence, Espagne). – C. R. Somm. Soc. Géol. France, **1:** 12–14.

DOUGLASS, R. C. (1960): The foraminiferal genus Orbitolina in North America. – Geol. Surv., Prof. pap., **333:** 52 p.

FOURCADE, E. (1970): Le Jurassique et le Crétacé aux confins des Chaînes Bétiques et Ibériques (SE de l'Espagne). Thése Sciences Univ. Paris VI, 468 pp.

FOURCADE, E., JEREZ, L., RODRIGUEZ, T., and JAFFREZO, M. (1972): El Jurásico terminaly y el Cretácico inferior de la Sierra de la Muela (provincia de Murcia). Consideraciones sobre las biozonas con Foraminiferos del Albense-Aptense del Sureste de España. – Rev. Esp. Micropaleont. No extraordinario, 215–248.

FOURCADE, E., et RAOULT, J. F. (1973): Crétacé du Kef Hahouner et position stratigraphique d' *«Ovalveolina» reicheli* P. de CASTRO (série septentrionale du môle néritique du Constantinois, Algérie) – Rev. Micropaléont., **15/4:** 227–246.

GARCIA-HERNANDEZ, M. (1978): El Jurásico terminal y el Cretácico inferior en las Sierras de Cazorla y del Segura (zona Prebética). – Tesis Doct. Universidad de Granada. 344 p.

GARCIA-HERNANDEZ, M. (1981): Biozonation du Cretacè inferieur à l'aide des Foraminifères benthiques et des algùes Dasycladacées dans le Prebetique Occidental (Cordilléres betiques, Espagne). – Géobios, **14:** 261–267.

GARCIA-HERNANDEZ, M., LOPEZ GARRIDO, A. C., RIVAS, P., SANZ DE GALDEANO, C., and VERA, J. A. (1980): Mesozoic Paleogeographic evolution of the External Zones of the Betic Cordillera. – Geol. Mijnbouw, **59:** 155–168.

HENSON, F. R. S. (1948): Larger imperforate Foraminifera of South-Western Asia, Families Lituolidae, Orbitolinidae and Meandropsinidae. – British Museum (Nat. Hist.), London, 127 p.

MASSE, J. P. (1976): Les calcaires urgoniens de Provence (Valanginien-Aptien inferieur): stratigraphie, paleontologie: les paleoenvironnements et leur evolution. Thèse Sciences, Marseille, 445 p.

MASSE, J. P. (1985): Paleobiographie des Rudistes du domaine perimediterranéen à l'Aptien inferieur. – Bull. Soc. géol. France, (8), **1:** 715–721.

MASSE, J. P. (in lit.) Relations entre modifications biologiques et phenomènes geologiques sur les plates-formes carbonatées du domaine peri-mediterranéen au passage Bedoulien-Gargasien. Géobios.

MOULLADE, M., et SAINT-MARC, P. (1975). Les «Mésorbitolines»: révision taxonomique, importance stratigraphique et paléobiogéographique. – Bull. Soc. géol. France, (7), **XVII/5:** 828–842.

N'DA LOUKOU, V. (1984): L'Urgonien des Pyrenées Occidentales. Thesè Doctorat 3e cycle, Univ. de Pau, 318 p.

NEAGU, TH. (1984) Nouvelles données sur la morphologie du test, sur la systématique et la nomenclature des Miliolidés agatisthegues du Mesozoïque. – Rev. Española Micropaleont., **16:** 75–90.

NEAGU, TH. (1986): Barremian Lower Aptian Miliolid Fauna in Southern Dobrogea (Romania). – Rev. Española Micropaleont., **28:** 313–348.

PASCAL, A. (1984): Les Systemes biosédimentaires urgoniens (Aptien-Albien) sur la marge Nord Iberique. – Thèse Sciences Dijon, 561 p.

PEYBERNES, B. (1976): Le Jurassique et le Crétacé inferieur des Pyrénées franco-espagnoles entre la Garonne et la Mediterranée. – Thèse Univ. Toulouse, 459 pp.

PEYBERNES, B. (1979): L'Urgonien des Pyrenées; essai de synthèse. – Géobios, Mém. spec., **3:** 79–87.

PEYBERNES, B. (1982): Les Orbitolinidés Crétacés d'Afrique: essai de synthèse. – Cah. de Micropaleont., **2:** 13–28.

REY, J., BILOTTE, M., et PEYBERNES, B. (1977): Analyse biostratigraphique et paleontologique de l'Albien marin d'Estremadura (Portugal). – Géobios, **10/3:** 369–393.

RODRIGUEZ ESTRELLA, T. (1979): Geología e Hidrogeología del sector de Alcaraz-Lietor-Yeste (prov. de Albacete). – Mem. IGME., **97.** (I y II): 566 pp.

ROEMER, F. (1849): Texas. Bonn. 464 p.

SCHROEDER, R. (1979): Les Orbitolines de l'Aptien: definition, origine et évolution. Colloque sur l'Urgonien des Pays Méditerranéens. Grenoble. – Géobios, Mem. sp., **3:** 289–299.

SOKAC, B. (1987): On some controversial Dasyclad genera and species and their stratigraphic position in the Lower Cretaceous deposits of the Dinarides. – Geol. Vjesnik. Zagreb, **40:** 9–38.

SOKAC, B., and NIKLER, L. (1973): Calcareous algae from the Lower Cretaceous of the environs of Niksic, Crna. Gora (Montenegro). – Paleont. Jugosl., **13:** 57 p.

SOKAC, B., and VELIC, I. (1981): Korkyrella n. gen. (Dasycladaceae) from the Upper Barremian and Lower Aptian of the island of Korcula. – Paleont. Jugosl., **24:** 1–12.

Plate I

Fig. 1 & 2: *Iraquia simplex* (Henson).
1. Oblique vertical section.
2. Horizontal section.

Fig. 3 & 6: *Simplorbitolina manasi* Ciry and Rat.
3. Axial section.
6. Horizontal section.

Fig. 4, 5 & 10: *Simplorbitolina conulus* Schroeder.
4. Vertical section.
5. Horizontal section.
10. Axial section.

Fig. 7: *Neorbitolinopsis conulus* (Douvillé).
Horizontal section.

Fig. 8: *Neoiraquia convexa* Danilova.
Axial section.

Fig. 9 & 11: *Neoiraquia* cf. *convexa* Danilova.
Oblique vertical sections.

Magnification for all sections x 40.

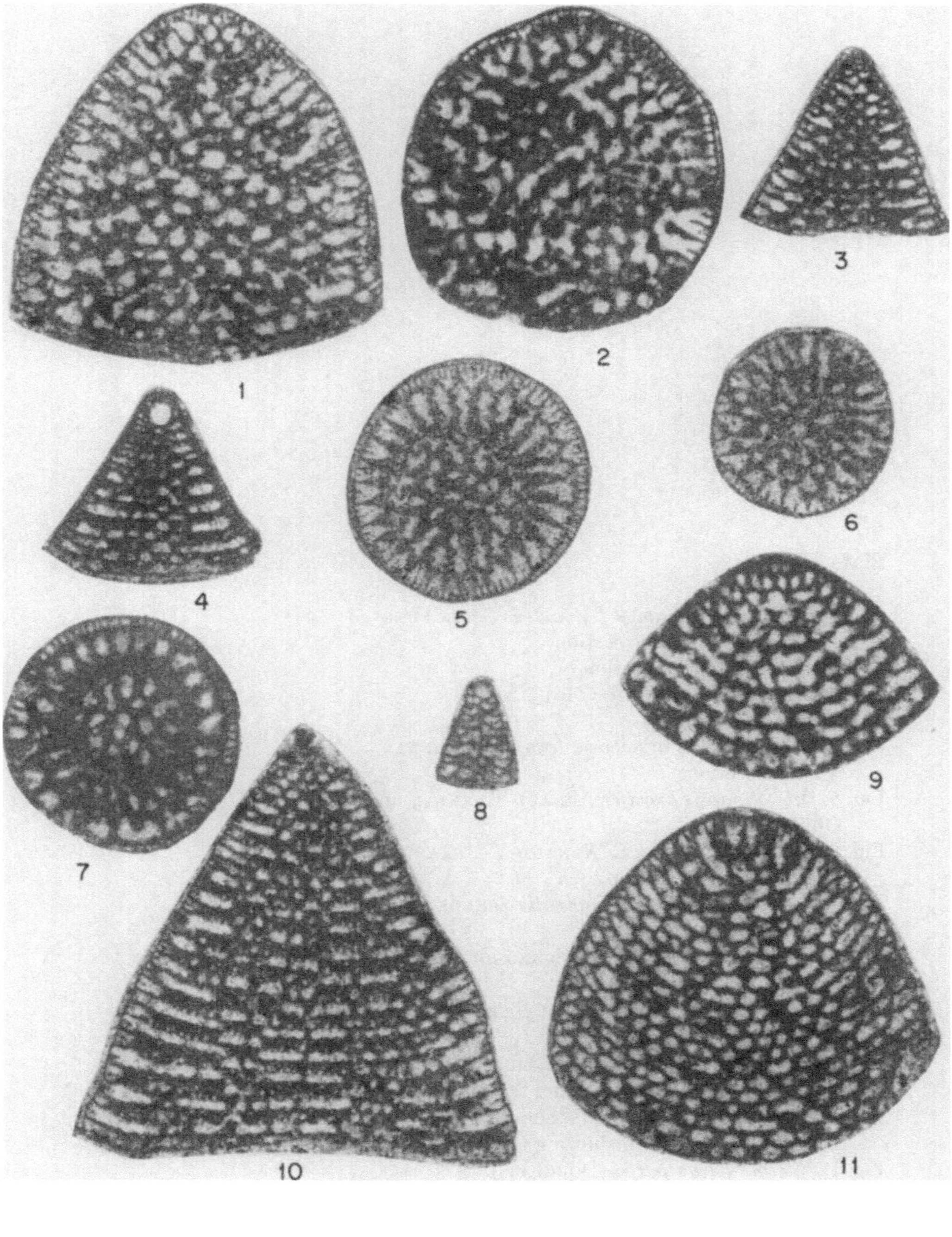
1
2
3
4
5
6
7
8
9
10
11

Plate II

Fig. 1 & 2: *Rectodictyoconus ? giganteus* SCHROEDER
1. Horizontal oblique section.
2. Vertical oblique section.

Fig. 3: *Orbitolinopsis cuvillieri* MOULLADE.
Vertical section.

Fig. 4: Orbitolinopsis briacensis ARNAUD-VANNEAU.
Vertical section.

Fig. 5: *Orbitolinopsis buccifer* ARNAUD-VANNEAU and THIEULOY.
Oblique section.

Fig. 6: *Dictyoconus ? vercorii* ARNAUD-VANNEAU.
Oblique vertical section.

Fig. 7: *Dictyoconus ? pachymarginalis* SCHROEDER.
Oblique vertical section.

Fig. 8: *Orbitolinopsis aquitanica* SCHROEDER and POIGNANT.
Oblique vertical section.

Fig. 9: *Orbitolinopsis reticulata* MOULLADE.
Oblique vertical section.

Fig. 10: *Palorbitolina lenticularis* (BLUMENBACH).
Apical oblique section showing the macrospheric embryonic apparatus.

Fig. 11: *Praeorbitolina wienandsi* SCHROEDER.
Apical oblique section showing the macrospheric embryonic apparatus.

Fig. 12: *Mesorbitolina parva* (DOUGLASS).
Subaxial oblique section showing the macrospheric embryonic apparatus.

Fig. 13: *Mesorbitolina texana* (ROEMER).
Axial section.

Magnification for all sections x 40.

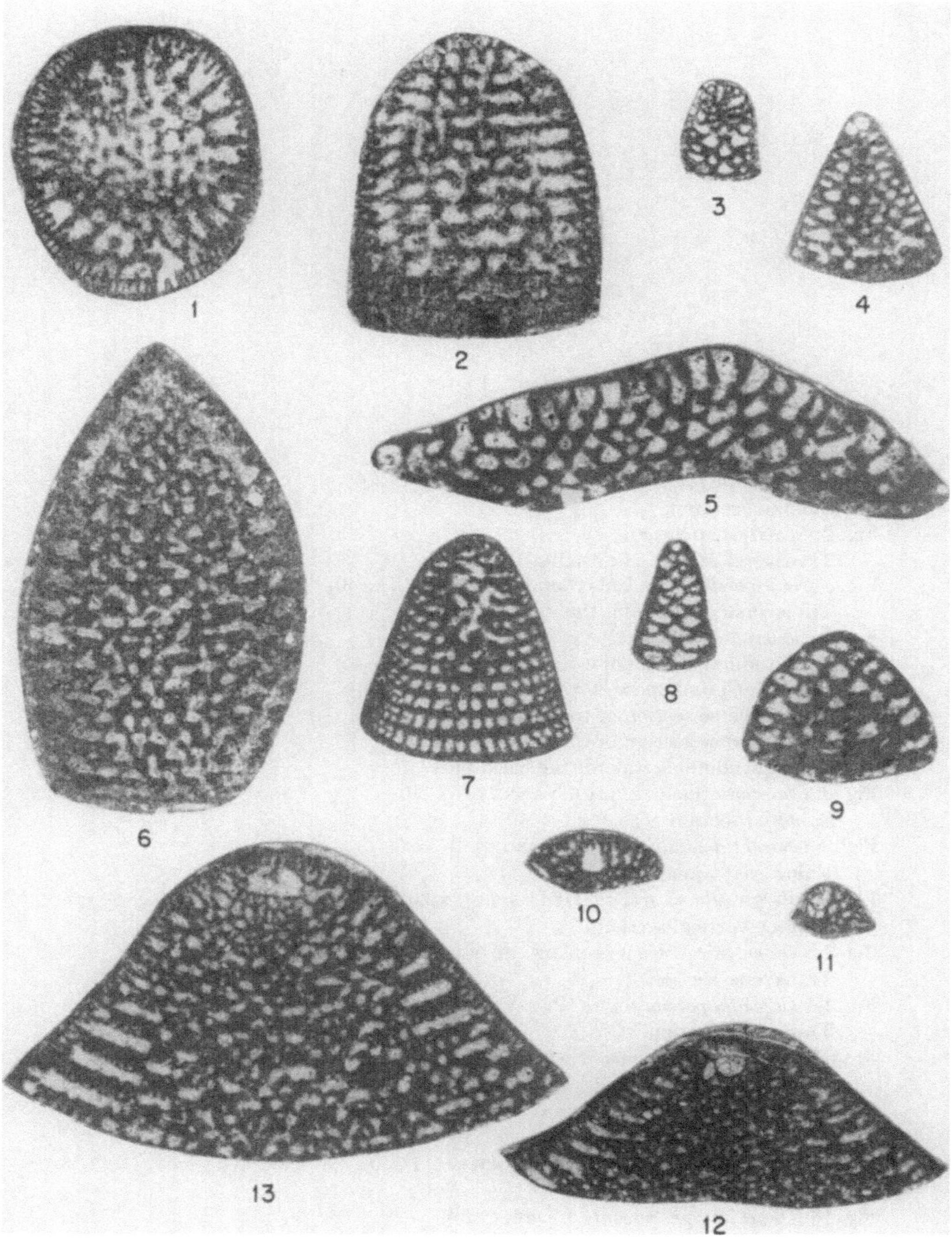
1
2
3
4
5
6
7
8
9
10
11
12
13

Plate III

Fig. 1: *Pseudochoffatella cuvillieri* DELOFFRE (x 20)
Subaxial section.
Fig. 2: *Neomeris* sp. (x 50)
Transverse section of a thallus fragment.
Fig. 3: *Helioporella cylindrica* SOKAC and NIKLER (x 40)
Transverse section of a thallus fragment.
Fig. 4: *Boueina* (x 40)
Longitudinal section of a thallus fragment.
Fig. 5, 6 & 7: *Cylindroporella ivanovici* (SOKAC) (x 50)
5. Transverse section of the fertile part.
6. Transverse section of the sterile basal part.
7. Longitudinal section of the basal part.
Fig. 8: *Charentia nana* ARNAUD-VANNEAU (x 50)
Subaxial section.
Fig. 9: *Charentia cuvillieri* NEUMANN (x 40)
Equatorial transverse section.
Fig. 10: *Vercorsella scarcellai* (DE CASTRO) (x 50)
Oblique vertical section.
Fig. 11: *Glomospira urgoniana* ARNAUD-VANNEAU (x 50)
Transverse section.
Fig. 12: *Carpathoporella fontis* (PATRULIUS) (x 50)
Transverse section.
Fig. 13: *Derventina filipescui* NEAGU (x 50)
Subaxial section.
Fig. 14: *Nautiloculina bronnimanni* ARNAUD-VANNEAU and PEYBERNES (x 50)
Subaxial section.
Fig. 15: *Neotrocholina* aff. *aptiensis* (IOCHEVA) (x 50)
Axial section.
Fig. 16: *Cuneolina* gr. *pavonia* D'ORB. (x 50)
Axial section.
Fig. 17: *Choffatella decipiens* SCHLUMBERGER (x 50)
Subaxial section.
Fig. 18: *Montiella* cf. *elitzae* (BAKALOVA) (x 40)
Longitudinal section.

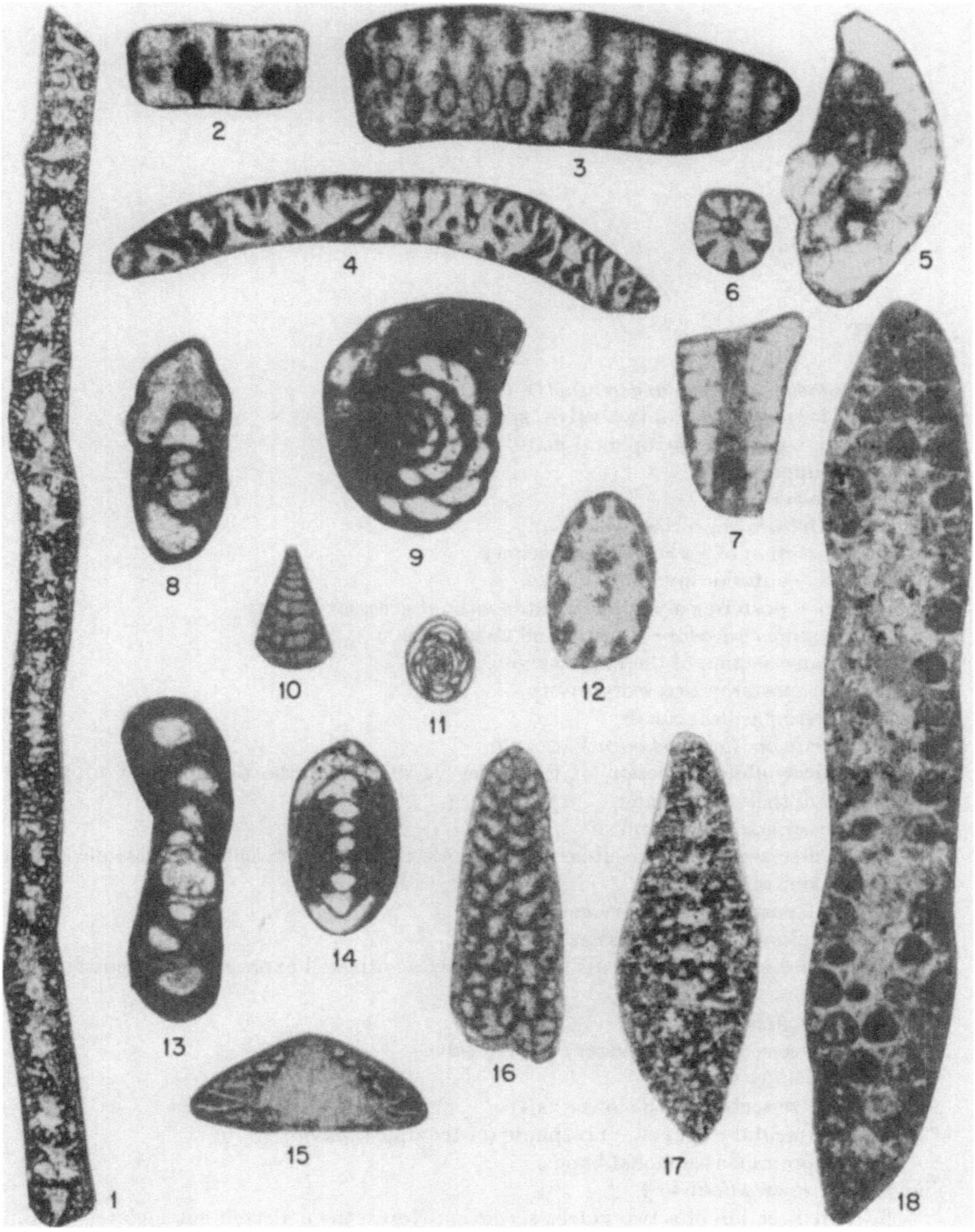
1
2
3
4
5
6
7
8
9
10
11
12
13
14
15
16
17
18

Plate IV

Fig. 1: *Pseudotoucasia santanderensis* (DOUVILLE)
Transverse section of a two valves specimen.
p. m. p. – posterior myophoral plate,
u. v. – upper valve,
l. v. – lower valve.

Fig. 2: *Horiopleura* sp. (primitive form).
Radial section of a two valves specimen.
a. m. p. – anterior myophoral plate,
p. m. p. – posterior myophoral plate (without accessory cavity).

Fig. 3: *Pachytraga paradoxa* (PICTET and CAMPICHE)
Transverse section of the lower valve.
p. a. c. – posterior accessory cavity,
c. – anterior palleal canals.

Fig. 4: *Offneria* sp. (gr. *rhodanica* PAQUIER)
Transverse oblique section of the lower valve (notice the development of palleal canals on the ventral side).

Fig. 5: *Caprina douvillei* PAQUIER
Transverse section of the upper valve (notice the pyriform palleal canals developed on the ventral side).
p. a. c. – posterior accessory cavity.

Fig. 6: *Agriopleura ? darderi* (ASTRE)
Transverse section of a cluster of lower valves with well expressed outer costal ornamentation.

Fig. 7: *Eoradiolites* sp.
Transverse section of a cluster of lower valves.

Fig. 8: *Eoradiolites* sp.
Transverse section of the lower valve.
L. – ligamental ridge, m. – myophore (of the upper valve),
S.-E. – prominent siphonal bands.

Fig. 9 & 10: *Praeradiolites* sp.
9. Radial section of a two valves specimen. Notice the deflected outer plates (p) and the myophore of the upper valve (m).
10. Transverse section of the lower valve showing the great development of the outer plates on the ventral side and the myophoral apparatus (of the upper valve).
L. – ligamental ridge, m. – myophores.
For all the sections: G. – body cavity.

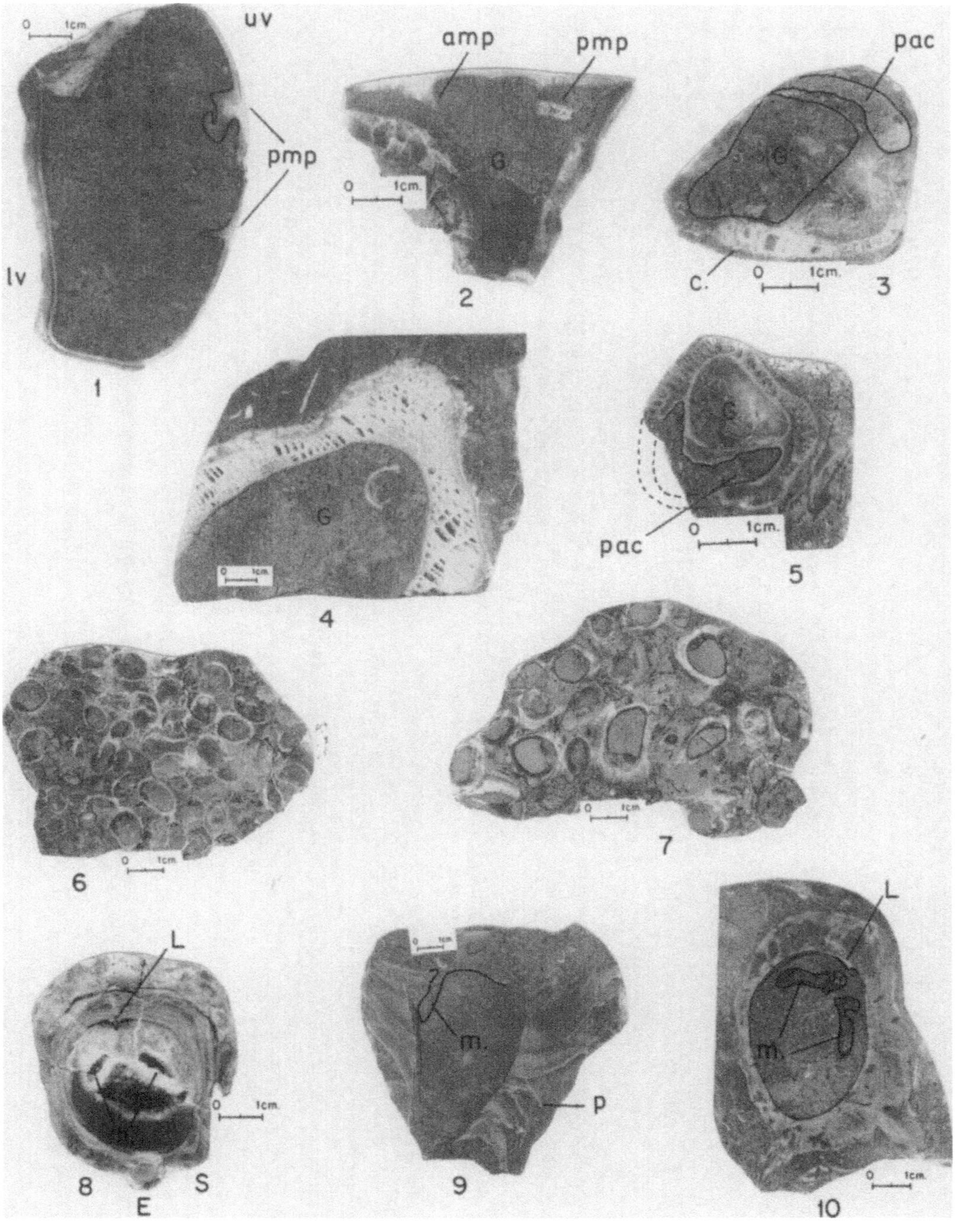
0 1cm
uv
amp
pmp
pac
pmp
G
G
lv
c.
1
2
3
G
G
pac
4
5
6
7
L
L
m.
m.
p
8
9
10
E
S

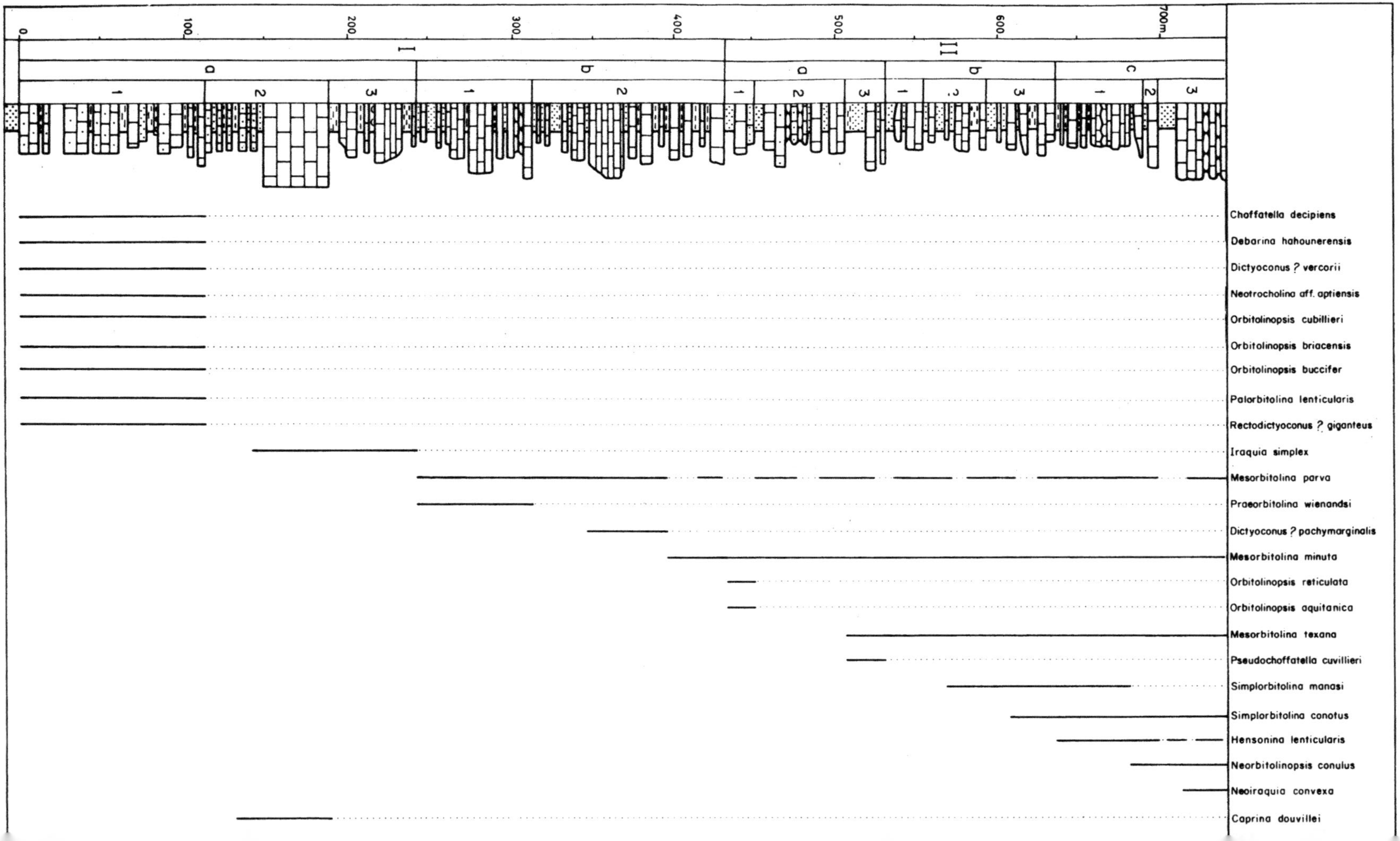

Choffatella decipiens
Debarina hahounerensis
Dictyoconus ? vercorii
Neotrocholina aff. aptiensis
Orbitolinopsis cubillieri
Orbitolinopsis briacensis
Orbitolinopsis buccifer
Palorbitolina lenticularis
Rectodictyoconus ? giganteus
Iraquia simplex
Mesorbitolina parva
Praeorbitolina wienandsi
Dictyoconus ? pachymarginalis
Mesorbitolina minuta
Orbitolinopsis reticulata
Orbitolinopsis aquitanica
Mesorbitolina texana
Pseudochoffatella cuvillieri
Simplorbitolina manasi
Simplorbitolina conotus
Hensonina lenticularis
Neorbitolinopsis conulus
Neoiraquia convexa
Caprina douvillei
0
100
200
300
400
500
600
700m
I
II
a
b
c
1
2
3

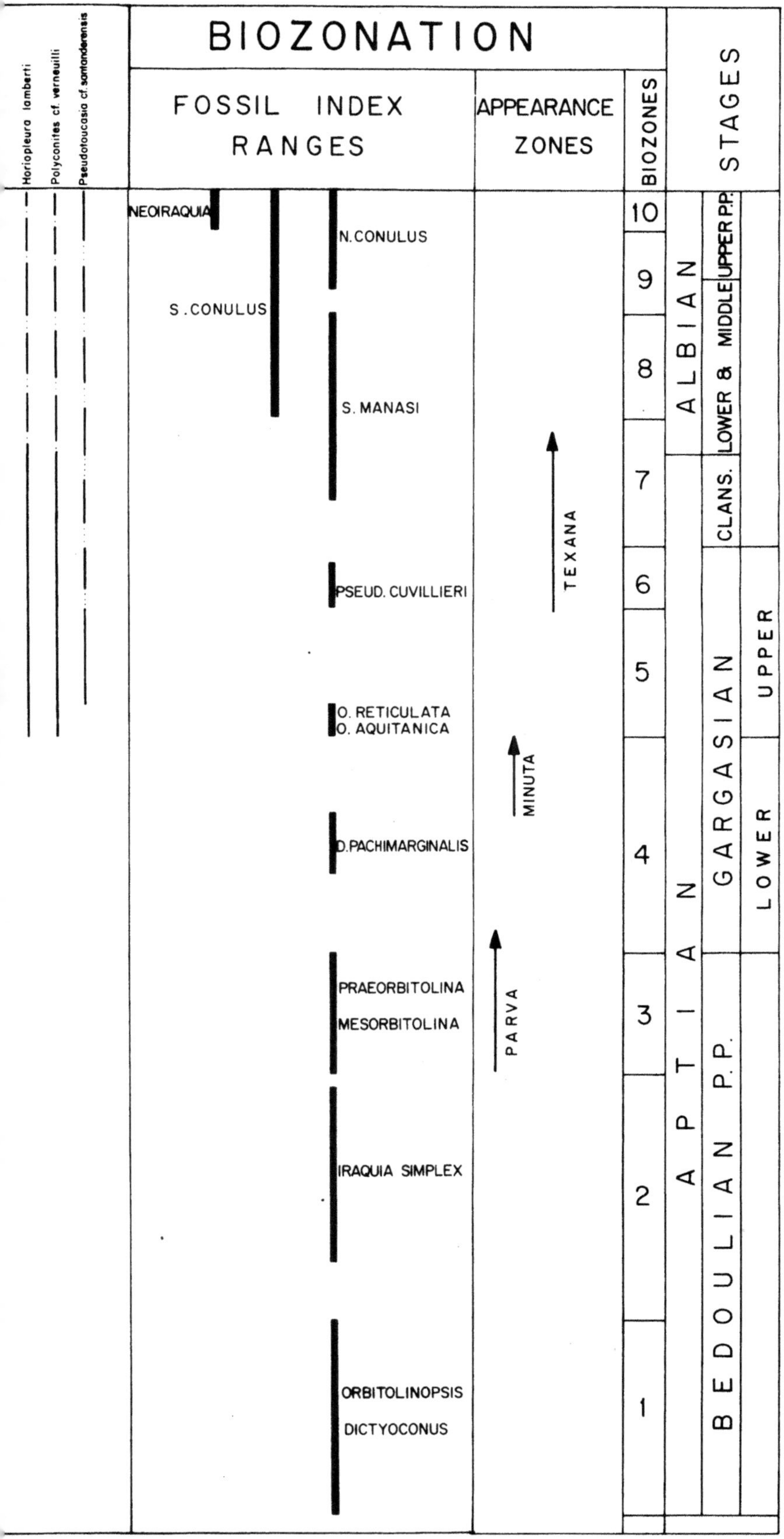

Fig. 2: Lithological and biostratigraphical subdivision of the Sierra del Carche sequence.

New Aspects on Tethyan Cretaceous Fossil Assemblages.
Band 9 Schriftenreihe der Erdwissenschaftlichen Kommissionen der Österreichischen Akademie der Wissenschaften, 223–230.

4.3. Les successions d'ammonites et la subdivision des étages du Crétacé inférieur téthysien

Par G. Mandov et T. Nikolov*)

Abstract

In general, the ammonite successions of Bulgaria agree with those of the rest of the Mesogée. However, some particularities make it evident that these faunas form the link between those of Western Europe and the Crimea-Caucasus region. This has to be considered when biogeographical questions concerning the Lower Cretaceous are discussed.

Each of the numerous biostratigraphical zonations contains a certain speculative element and is therefore not very satisfactory for the geological correlation and for the solution of still open biostratigraphical problems. Only physically existing sections may reveal the vertical distribution of ammonites and are therefore suitable for solving these problems. The Bulgarian sections are continuous and rich in ammonites. They are therefore important for solving the biostratigraphical problems of the Lower Cretaceous.

Introduction

Au siècle passé, les ammonites ont joué, en France et en Suisse, un rôle déterminant pour la division des étages du Crétacé inférieur. Leur importance lors de la subdivision et de la correlation des dépots a assuré la stabilité des unités stratigraphiques. Des unités telles que l'étage facial Urgonien caractérisé avec d'autres groupes d'organismes et qui peu après ne faisaient plus partie de la nomenclature stratigraphique, font exception. Meme de nos jours, malgré la valeur archistratigraphique des ammonites il existe des problèmes sur la subdivision des étages et des sous-étages et sur la division zonale des dépots du Crétacé inférieur dans la Tethys.

*) L'Université de Sofia, Bulgarie.

Cet exposé présente une brève récapitulation de la stratigraphie ammonitique en mettant l'accent sur deux points: Le premier a trait aux problèmes existant encore concernant la subdivision et des limites de la série du Crétacé inférieur. Le deuxième à certaines particularités des successions d'ammonites du Crétacé inférieur en Bulgarie, dont le territoire occupe selon nous une place importante dans le panorama du Crétacé inférieur dans la Mésogée.

Berriasien

La limite Jurassique-Crétacé, ou la limite inférieur du Berriasien, a été longement discutée par de nombreux spécialistes dans leur publications et lors de forums internationaux (LE HÉGART, 1971; NIKOLOV, 1987). Cela est du au fait que dans les couches voisines de la limite Jurassique-Crétacé les ranges-zones des genres du Tithoniques terminal, *Paraulacosphinctes, Parapalasiceras, Aspidoceras, Glochiceras, Pseudargentiniceras,* etc., recouvrent partiellement celles d'un nombre de genres dont le dévelloppement principal se situe dans le Berriasien: *Pseudosubplanites (Pseudosubplanites), Pseudosubplanites (Hegaratella), Berriasella (Berriasella), Berriasella (Picteticeras), Berriasella (Strambergella), Delphinella, Fauriella, Malbosiceras, Pomeliceras (Mazenoticeras), Jabronella, Subalpinites, Substeuroceras, Protocanthodiscus, Corongoceras* etc. (Fig. 1). Ce recouvrement coincide avec la sous-zone *Malbosiceras chaperi* (NIKOLOV, 1967 = zone *Berriasella jacobi* de LE HÉGARAT, 1973). L'appartenance de cette dérnière au sommet de l'étage Tithonique est indubitable, si l'on considère son équivalence avec le sommet de la zone *Paraulacosphinctes transitorius* du Tithonique terminal. Cette situation exige que le problème de la limite Jurassique-Crétacé, et donc de la limite infèrieure du Berriasien, soit résolu d'une part d'une manière conventionelle et de l'autre au niveau d'espèces des faunes ammonitiques. Les monographies de LE HÉGARAT (1973) et de NIKOLOV (1982) démontrent le caractère universel de la zone susjacente *Pseudosubplanites grandis* dans la Téthys et son importance pour la corrélation des dépots. Pour cette raison son choix en tant que base de l'étage Berriasien est indiscutable.

Valanginien

Quant à la succession dans la répartition des genres d'ammonites, la limite Berriasien-Valanginien est plus nettement marquée. D'une part cela est prouvé par la disparition de la plupart des répresentants de la sous-famille Berriasellinae: *Berriasella* (s. l.), *Delphinella, Dalmasiceras, Fauriella, Euthymiceras, Jabronella* (s. l.), *Retowskites,* etc. D'autre part ce sont les prémières manifestations et la dévelloppement abondant des genres *Busnardoites, Kilianella, Thurmanniceras, Neocomites* (s. l.), et plus tard des genres *Luppovella, Saynoceras, Valanginites* etc. qui individualisent bien l'étage Valanginien. La limite infèrieur de l'étage Valanginien en Bulgarie correspond à la limite inférieure de la zone

Kilianella roubaudiana, interprétée selon la première proposition de Kilian (1907–1913) et identifiée pour la première fois par Nikolov (1960). En Bulgarie cette zone coincide avec le volume du Valanginien inférieur. Malgré la présence de presque tous les genres qui existent en France, les trois zones d'ammonites dans le Valanginien inférieur, proposées par Busnardo, Thieuloy, Moullade et al. (1979) ne peuvent pas etre identifiées actuellement en Bulgarie. Les communications de Company et Tavera (1980) et Company (1980) pour la présence de *Thurmanniceras otopeta* en Espagne, ainsi que la découverte de la mème ammonite dans les dépots du Valanginien dans le Prébalkan central en Bulgarie (G. Mandov, données non-publiées) démontrent que des couches contenant la meme ammonite sont largement répandues dans une grande partie de la Téthys. Pourtant, notamment en Bulgarie, des trouvailles isolées ne permettent pas pour le moment l'appréciation de sa répartition verticale, des ammonites coexistantes et des possibilités de séparation de la zone *Thurmanniceras otopeta.* La coincidence partielle des range-zones locales de *T. otopeta* avec celles de *Berriasella calisto, Fauriella boissieri, Fauriella donzei* et *Tirnovella alpillensis* en France (Busnardo, Thieuloy, Moullade et al., 1979) et en Espagne (Company, 1982) donne raison à Hoedemaeker (1984) qui maintient son opinion de 1982 selon laquelle une partie de couches à *Berriasella calisto* et *Tirnovella alpillensis* appartiendraient à la base du Valanginien. La répartition en Bulgarie des ammonites sus-mentionnées démontre que la limite Berriasien-Valanginien est bien fixée au sommet de la zone *Fauriella boissieri,* et donc de la sous-zone *Berriasella calisto* (Nikolov, 1982, 1987).

Hauterivien

La limite inférieur de l'étage Hauterivien est bien caracterisée dans toute le Mésogée de la meme façon qu'en Bulgarie. L'apparition de rares représentants du genre *Acanthodiscus* permet l'identification de la zone basale de l'étage Hauterivien. Pourtant, pour les travaux stratigraphiques, l'apparition et le développement abondant du genre *Crioceratites,* dont les représentants caractérisent la base ainsi que l'étage Hauterivien entier, sont d'une plus grande importance. Les genres *Thurmanniceras, Busnardoites, Luppovella* et *Saynoceras* ont complètement disparu. La limite Valanginien-Hauterivien se caractérise de meme par l'apparition des genres *Lyticoceras, Oosterella, Phyllopachyceras* ainsi que par des espèces isolées des genres *Olcostephanus, Valanginites, Eleniceras* etc. Pour les terres bulgares la coincidence entre la disparition du genre *Saynoceras* et l'apparition abondante du genre *Crioceratites* représente une particularité caractéristique. D'autre part, des genres tels que *Himantoceras* et *Criosarasinella* qui se recontraient dans le sud-est de la France dans la partie terminale du Valanginien (Thieuloy, 1973, 1977) apparaissent en Bulgarie avec des représentants abondants du genre *Crioceratites* (Mandov, 1976). Il en est de meme pour l'espèce caractéristique *Teschenites callidiscus* Thieuloy qui en Bulgarie se trouve à la base du Hauterivien. Pour cette raison leur répartition dans les coupes, sur le territoire bulgare, démontre des différences dans leur position stratigraphique. D'autre part, en Bulgarie on ne peut identifier les zones *Teschenites callidiscus* et *H., trinodosum* établies dans le Sud-Est de la France. Cela exige selon nous une plus grand attention et que l'on discute la proposition de HOEDEMAEKER (1984) de tracer la limite Valanginien-Hauterivien immédiatement au-dessus de la zone *Saynoceras verrucosum*. A part cela, la répartition réelle des ammonites dans les Balkanides occidentaux (Bulgarie occidentale) démontre des différences avec les zones classique du Hauterivien (NIKOLOV, 1969) ainsi qu'avec celles établies par MOULLADE et THIEULOY (1967) dans la fosse Vocontienne. Pour cette raison MANDOV (1976) utilise pour cette région une zonation purement empirique qui n'est pas propre à une large corrélation, mais elle est plus proche de l'image naturelle.

Barrémien

Les genres d'ammonites du Barrémien sont très bien représentés en Bulgarie et sont une base importante pour la délimitation de l'étage et sa subdivision (DIMITROVA, 1967; BRESKOVSKI, 1975; NIKOLOV, 1987). Il s'agit des genres apparaissant dès la base de l'étage: *Carstenia, Pulchellia, Holcodiscus, Torcapella, Valdedorsella, Anahamulina, Protacrioceras, Uhligia* etc. Au-dessus viennent les genres *Jaubertites, Paraspinoceras, Eoleptoceras, Leptoceras, Hamulina, Materonites, Costidiscus, Heteroceras, Imerites, Nicklesia, Silesites, Pseudohaploceras* etc. Il s'y ajoute cependant aussi des genres tels que *Crioceratites, Barremites* (s. l.), *Plesiospitidiscus, Astieridiscus* qui s'avèrent traversants par rapport à la limite inférieure du Barrémien. Voila pourquoi en Bulgarie, ainsi que dans diverses régions de la Téthys la limite inférieure de l'étage est fixée différemment. Les principales contradictions sont réduites essentiellement à trois conceptions, géneralisées par BIRKELUND et al. (1984). En réalité, ce sont les couches à *Pseudothurmannia* qui sont à la base de ces désaccords et la limite fixée par rapport à des couches, c'est-à-dire au-dessus d'elles, au-dessous ou à travers les couches elles-mêmes. D'après nous, le problème survient lors de l'analyse des ammonites dans les couches voisins de la limite Hauterivien-Barrémien. Deux notions sont confondues: celle de la «zone *Pseudothurmannia angulicostata*» et celle des «couches à *Pseudothurmannia*». Les représentants du genre *Pseudothurmannia* s'avèrent aussi traversants par rapport à la limite inférieure du Barrémien. Cela signifie qu'à leur base ils sont accompagnés de représentants nettement hauteriviens. Ce sont des espèces des genres *Crioceratites, Plesiospitidiscus, Astieridiscus* etc. De cette façon apparait un volume de couches, qui peuvent etre identifiées entièrement à la zone *Pseudothurmannia angulicostata* de la partie terminale de étage Hauterivien dans le sens de la proposition faite au Colloque de Lyon (1963). Les chercheurs ne doivent pas etre embarassés par le fait que léspèce-indice de la zone se recontre rarement et est difficilement reconnaissable. Pour ce qui est des couches à *Pseudothurmannia* qui viennent plus en haut et se caractérisent avec des genres et espèces typiquement barrémiens, elles doivent etre rapportées naturellement à l'étage Barrémien. Voilà pourquoi il est difficile d'accepter l'idée de BUSNARDO et VERMEULEN (1986) proposant de fixer la limite Hauterivien-Barrémien d'après la disparition des representants de *Pseudothurmannia*. En d'autres termes, nous considérons que la limite entre les deux étages doit etre fixée d'après la première manifestation des ammonites barrémiennes.

Aptien

La répartition stratigraphique des genres d'ammonites dans l'étage Aptien représente une base favorable pour sa délimitation et sa subdivision detaillée. A la base de l'étage, les premiers qui appareissent sont les genres *Procheloniceras, Kutatissites, Prodeshayesites, Ancyloceras*. Un peu plus tard, dans les coupes de la Bulgarie du Nord-Est où les dépots aptiens sont le mieux développés et les plus riches en fossiles, viennent *Deshayesites, Paradeshayesites, Dufrenoyia, Tropeum, Cheloniceras (Cheloniceras)*. Dans la partie médiane et supérieure de l'étage apparaissent et disparaissent peu après des genres caractéristiques tels que *Gargasiceras, Tonohamites, Cheloniceras (Epicheloniceras), Colombiceras, Acanthoplites, Diadochoceras, Nodosohoplites*, etc.

La délimitation de l'étage Aptien du Barrémien en Bulgarie grace aux genres d'ammonites caractéristiques n'est d'aucune difficulté. L'intérêt revient ici à l'emplacement des couches à *Colchidites* (NIKOLOV, STOYKOVA, IVANOV, 1983). Les couches conténant des ammonites de ce genre et considérées comme typiquement barémiennes dans la partie méridionale de l'URSS et en Bulgarie du Nord démontrent une succession intéressante d'ammonites barrémiennes et aptiennes. Ainsi, dans la partie inférieure des couches à *Colchidites*, on rencontre les genres *Heteroceras, Hamulina* et *Anahamulina*, considérés comme typiquement barrémiens. Cependant dans la partie supérieure de ces couches, avec les *Colchidites* viennent les ammonites des genres *Procheloniceras, Kutatissites* et *Prodeshayesites*. De cette façon, en Bulgarie du Nord (au nord-ouest de la ville de Popovo), les couches à *Colchidites* sont analogues au cas des «couches à *Pseudothurmannia*» voisines de la limite Hauterivien-Barrémien. Ici aussi la manifestation des ammonites typiquement aptiennes est considéré comme la base de distinction de deux étages.

Albien

Les ammonites albiennes montrent aussi une répartition verticale limitée. Presque toutes commencent et finissent leur développement dans le cadre de l'étage, même dans le volume de ses sous-étages. Voilà pourquoi il n'y a aucune difficulté dans la délimitation et la subdivision détaillée de l'étage qui est répandu en Bulgarie du Nord-Ouest et dans le Nord de la partie centrale de la Bulgarie (IVANOV, 1981; IVANOV, STOYKOVA, NIKOLOV, 1982). Dans la partie inférieure de l'étage se manifeste tout d'abord le sous-genre *Leymeriella (Leymeriella)* et un peu plus tard aussi *Leymeriella (Neoleymeriella)*. L'abondance d'espèces et d'exemplaires permet l'identifaction de la zone *Leymeriella tardefurcata,* même sa subdivision en sous-zones et en meme temps la délimitation de l'étage Albien de l'Aptien. Parmi les relativement peu nombreuses ammonites s'avèrent traversantes par rapport à la limite Aptien-Albien, le genre *Hypacanthoplites* est seul assez important. Plus en haut apparaissent progressivement les genres *Kossmatella, Protanisoceras, Douvilleiceras, Hoplites, Anahoplites, Hamites* (s. l.), *Anisoceras, Hysteroceras, Mortoniceras, Mariella, Stoliczkaia, Lechites, Arraphoceras.*

La limite supérieure de l'étage Albien en Bulgarie présente le plus souvent une surface erodée. Par endroits, en Bulgarie du Nord-Ouest où les coupes sont continues, la limite avec le Cénomanien est fixée d'après les premières manifestations du genre Mantelliceras.

Conclusions

Ce bref exposé de la répartition des genres d'ammonites dans les dépots du Crétacé inférieur en Bulgarie démontre que:

1. Les successions d'ammonites en Bulgarie s'accordent en général avec celles de la région mésogéenne. Cependant, les quelques particularités susdites mettent en évidence la place importante des terres bulgares entre l'Europe du Sud-Ouest et la région Crimée-Caucase. Voilà pourquoi ce fait doit etre pris en considération en cherchant une solution aux questions discutables au sujet de la stratigraphie du Crétacé inférieur. Les coupes riches en ammonites de tous les étages du Crétacé inférieur bulgare sont un argument de plus.

2. Les nombreux schémas zonaux dans la litérature géologique qui contiennent toujours un certain élément de spéculation, ne se prètent guère à la corrélation concrète et à la solution des problèmes discutés. C'est pour lorsqu'on cherche à résoudre les problèmes de la subdivision du Crétacé inférieur il faut étudier et comparer des coupes concrètes à ammonitofaune et la répartition verticale réelle de celle-ci.

References

Birkelund, T., Hancock, J. M., Hart, M. B., Rawson, P. F., Remane, J., Robaszynski, F., Schmid, F., Surlyk, F. (1984): Cretaceous Stage Boundaries – Proposals. – Bull. geol. Soc. Denmark, **33**: 3–20.

Breskovski, S. (1975): Les zones et sous-zones ammonitiques dans l'étage Barrémien en Bulgarie du Nord-Est. – Geol. Balc., **5/2**: 47–66.

Busnardo, R., Thieuloy, J.-P., Moullade, M. et al. (1979): Hypostratotype mésogéen de l'étage Valanginien (Sud-Est de la France). C.N.R.S. 143 p.

Busnardo, R., Vermeulen, J. (1986): La limite Hauterivien-Barrémien dans la région stratotypique d'Angles (Sud-Est de la France). – C. R. Acad. Sc. Paris, **302/2/7:** 457–459.

Company, M. (1982): Contribucion a la bioestratigrafia del Valanginiense en las Cordilleras Béticas. – Cuadernos Geologia Ibérica, **8:** 665–676.

Company, M., Tavera, J. M. (1982): Los ammonites del trànsito Berriasense-Valanginiense en la region de Cehegin (Prov. de Murcia, SE de Espana). – Cuadernos Geologia Ibérica, **8:** 651–664.

Hoedemaeker, Ph.-J. (1982): Ammonite biostratigraphy of the uppermost Tithonian, Berriasian, and lower Valanginian along the Rio Argos (Caravaca, SE Spain). – Scripta geol., **65:** 1–81.

Hoedemaeker, Ph.-J. (1984): Proposals for the stratigraphic positions of the Berriasian-Valanginian and the Valanginian-Hauterivian boundaries. – Bull. geol. Soc. Denmark, **33:** 139–146.

Hoedemaeker, Ph.-J. (1987): Correlation possibilities around the Jurassic-Cretaceous boundary. – Scripta Geol., **84:** 1–55.

Ivanov, M. (1981): Notes on the Biostratigraphy of the Albian Stage between Botunja and Iskar River. – Rev. Bulg. Geol. Soc., **42/3:** 352–360 (Bulg.).

Ivanov, M., Stoikova, K., Nikolov, T. (1982): Biostratigraphical studies of the Albian Stage in the Northern part of Pleven district. – Ann. Univ. Sofia, **72/1** Geol.: 79–87 (Bulg.).

Kilian, W. (1907–1913): Lethaea geognostica. II: das Mesozoicum, 3. Kreide. Stuttgart, Schweizerbart'scher Verlag. 398 p.

Le Hégarat, G. (1971): Le Berriasien du Sud-Est de la France. – Doc. Lab. Geol. Fac. Lyon, **43/1–2:** 1–576.

Mandov, G. (1974): Représentants du genre *Himantoceras* Thieuloy, 1964 (Ammonoidea) dans les dépôts du Crétacé inférieur en Bulgarie. – Rev. Bulg. Geol. Soc., **35/2:** 143–148 (Bulg.).

Mandov, G. (1976): L'étage Hauterivien dans les Balkanides Occidentales (Bulgarie de l'Ouest) et sa faune d'ammonites. – Ann. Univ. Sofia, Fac. Géol. et Géogr., **67/1** – Géologie: 11–99 (Bulg.).

Moullade, M., Thieuloy, J.-P. (1967): Les zones d'Ammonites du Valanginien supérieur et de l'Hauterivien vocontien. – C. R. somm. Soc. Géol. France, **6:** 228–229.

Nikolov, T. (1960): La faune d'ammonites dans le Valanginien du Prébalkan oriental. – Trav. Géol. Bulgarie sér. Paléont., **2:** 143–261 (Bulg.).

Nikolov, T. (1969): Stratigraphy of the Lower Cretaceous in a part of North-Eastern Bulgaria. – Bull. Geol. Inst., ser. Strat. and Litol., **18:** 31–71 (Bulg.).

Nikolov, T. (1969): Le Crétacé inférieur en Bulgarie. – BSGF, **7/11:** 56–68.

Nikolov, T. (1970): On the boundary between the Aptian and the Albian Stages. – Bull. Geol. Inst., Ser. Strat. and Lytol., **19:** 61–78 (Russ.).

Nikolov, T. G. (1979): Le cadre paléobiogéographique du Crétacé inférieur bulgare. Aspecte der Kreide Europas. JUGS A, **6:** 425–433.

Nikolov, T. G. (1982): Les ammonites de la famille Berriasellitae Spath, 1922. Tithonique supérieur – Berriasien. – Acad. bulg. sci., 269 p.

Nikolov, T. G. (1987): The mediterranean Lower Cretaceous. – Geologica Balc., Ser. op. sing., **2:** 1–269.

Nikolov, T., Stoykova, K., Ivanov, M. (1983): Ammonite occurrence in the Barremian-Aptian boundary beds in North-Eastern Bulgaria. – C. R. Bulg. Acad. Sci., **36/8:** 1081–1084.

Stoikova, K. (1983): Ammonites from the Aptian Stage in Northeast Bulgaria. – Rev. Bulg. Geol. Soc., **44/1:** 77–90 (Bulg.).
Thieuloy, J.-P. (1973): Le genre *Saynoceras* (Ammonoidea) et la zonéographie valanginienne dans l'Arc subalpin de Castellâne (Alpes-de-Haut-Provence). – C. R. Acad. Sc. Paris, **276:** 925–927.
Thieuloy, J.-P. (1977): La zone à *Callidiscus* du Valanginien supérieur vocontien (S. E. de la France). Lithostratigraphie, ammonitofaune, limite Valanginien-Hauterivien, corélations. – Geologie Alpine, **53:** 83–143.

Fig. 1. Répartitions stratigraphique des genres d'ammonites les plus importants dans le Crétacé inférieur en Bulgarie

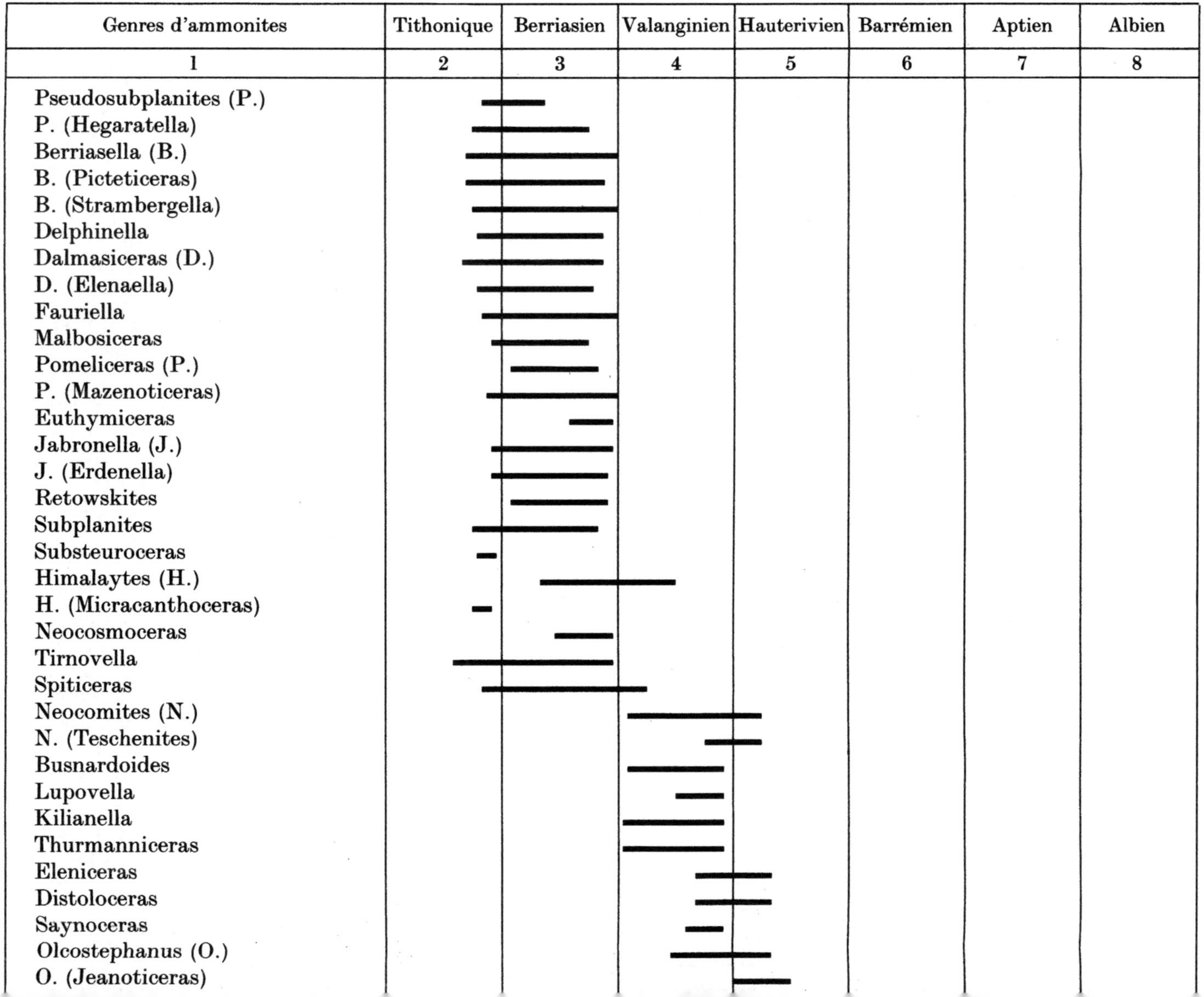

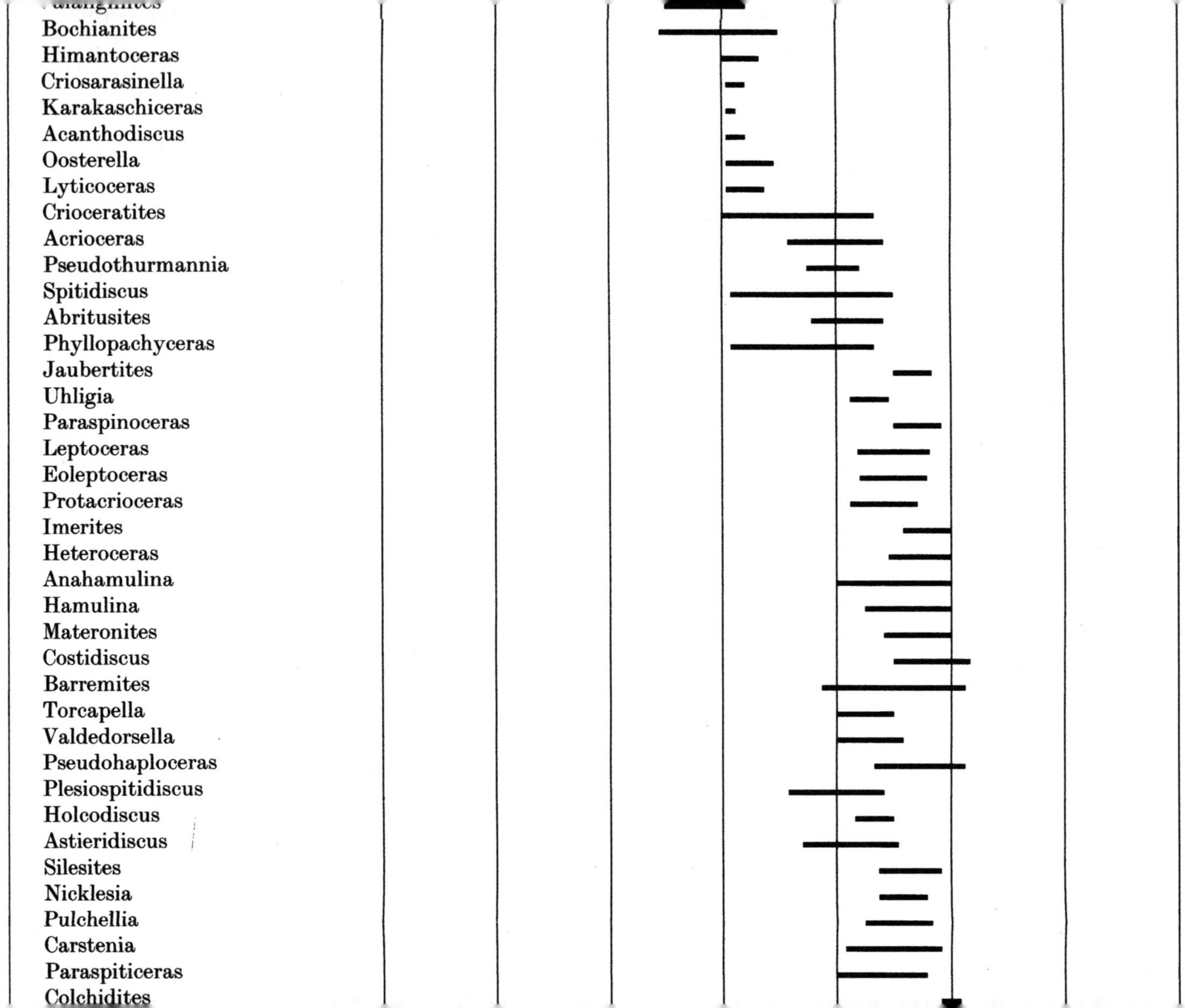
Bochianites
Himantoceras
Criosarasinella
Karakaschiceras
Acanthodiscus
Oosterella
Lyticoceras
Crioceratites
Acrioceras
Pseudothurmannia
Spitidiscus
Abritusites
Phyllopachyceras
Jaubertites
Uhligia
Paraspinoceras
Leptoceras
Eoleptoceras
Protacrioceras
Imerites
Heteroceras
Anahamulina
Hamulina
Materonites
Costidiscus
Barremites
Torcapella
Valdedorsella
Pseudohaploceras
Plesiospitidiscus
Holcodiscus
Astieridiscus
Silesites
Nicklesia
Pulchellia
Carstenia
Paraspiticeras
Colchidites

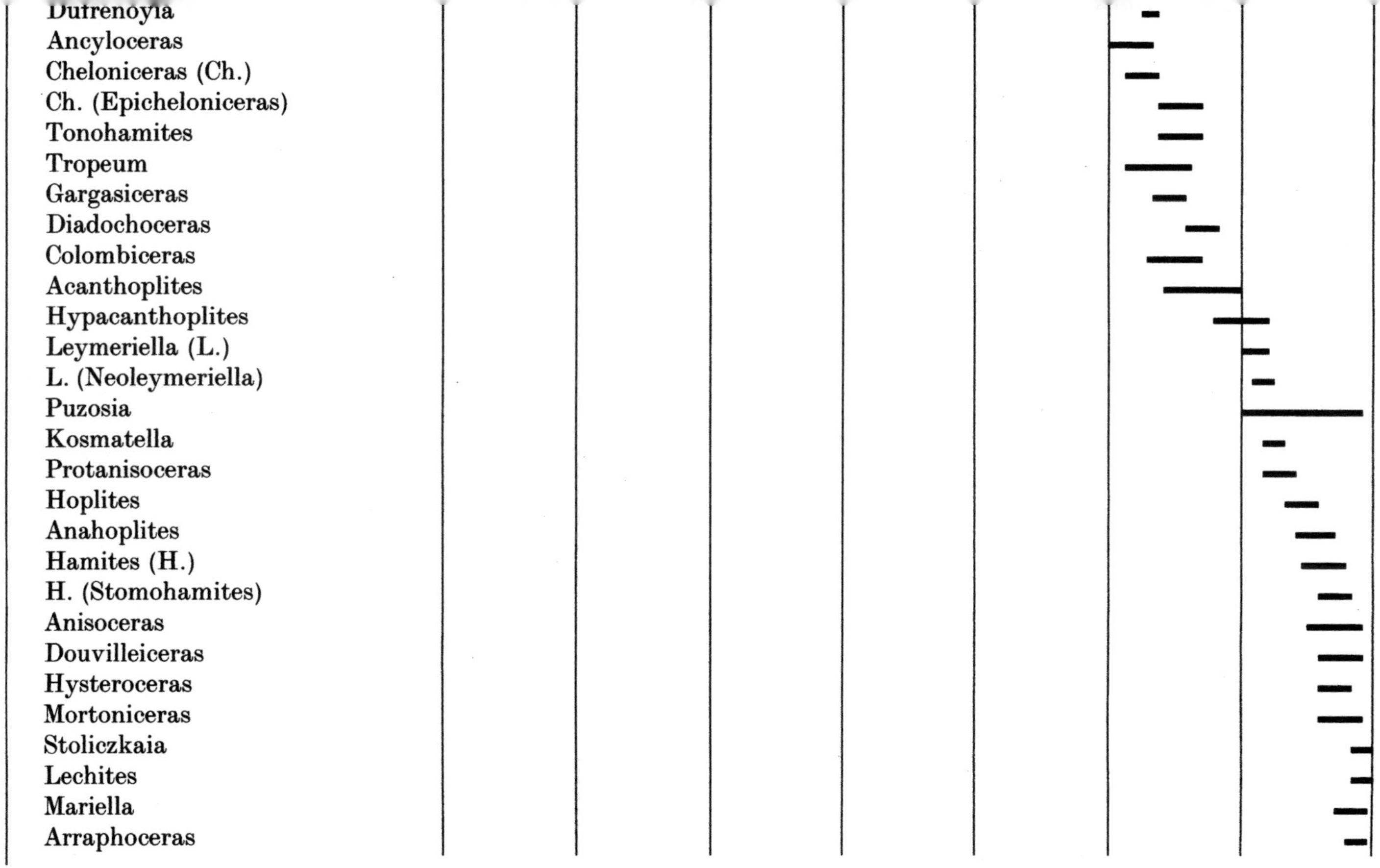

Dufrenoyia
Ancyloceras
Cheloniceras (Ch.)
Ch. (Epicheloniceras)
Tonohamites
Tropeum
Gargasiceras
Diadochoceras
Colombiceras
Acanthoplites
Hypacanthoplites
Leymeriella (L.)
L. (Neoleymeriella)
Puzosia
Kosmatella
Protanisoceras
Hoplites
Anahoplites
Hamites (H.)
H. (Stomohamites)
Anisoceras
Douvilleiceras
Hysteroceras
Mortoniceras
Stoliczkaia
Lechites
Mariella
Arraphoceras

New Aspects on Tethyan Cretaceous Fossil Assemblages.
Band 9 Schriftenreihe der Erdwissenschaftlichen Kommissionen der Österreichischen Akademie der Wissenschaften, 231–239.

4.4. Rudists and Larger Foraminifera below the Cretaceous-Tertiary Boundary in the Dolenja Vas Section

By Mario PLENIČAR*), Katica DROBNE**), Bojan OGORELEC***)

Summary

The Vreme beds are part of the Liburnian Formation. At Dolenja vas which is situated on the Trieste-Komen platform, the Upper Maastrichtian age has been proved with co-occurring rudistids and foraminifera.

Introduction

Carbonate series passing continuously from the Cretaceous into the Tertiary are observable only in a few places in the Dinarids. A presumably continuous section is this of Dolenja vas near Senožeče (Fig. 1). Here, the carbonates of the Trieste-Komen plateau which have been deposited from the Lower Cretaceous to the end of Cretaceous are overlayn by the Tertiary (Paleocene) carbonates of the Vipava synclinorium. Cuisian flysch is overlying the Paleocene carbonate rocks.

The Maastrichtian open marine platform sediments of the Trieste-Komen plateau are developed as grey limestones with rudists (rudistid limestones). In restricted environments (partly lagoons of the carbonate platform) dark grey, grey brown or black micritic limestones or limestone breccias have been deposited. Brackish or fresh water intercalations may occur in the marine sediments. This development is typical for the Trieste-Komen plateau. It extends from the Cretaceous into the Paleocene and was named Liburnian Formation by STACHE

*) Katedra za geologijo in paleontologijo FNT, Askerčeva 12, pp. 311, SLO-61001 Ljubljana.

**) Paleontološki inštitut I. Rakovca ZRC SAZU, Novi trg 3–5, pp. 323, SLO-61001.

***) Geološki zavod Ljubljana. Dimičeva, SLO-61000.

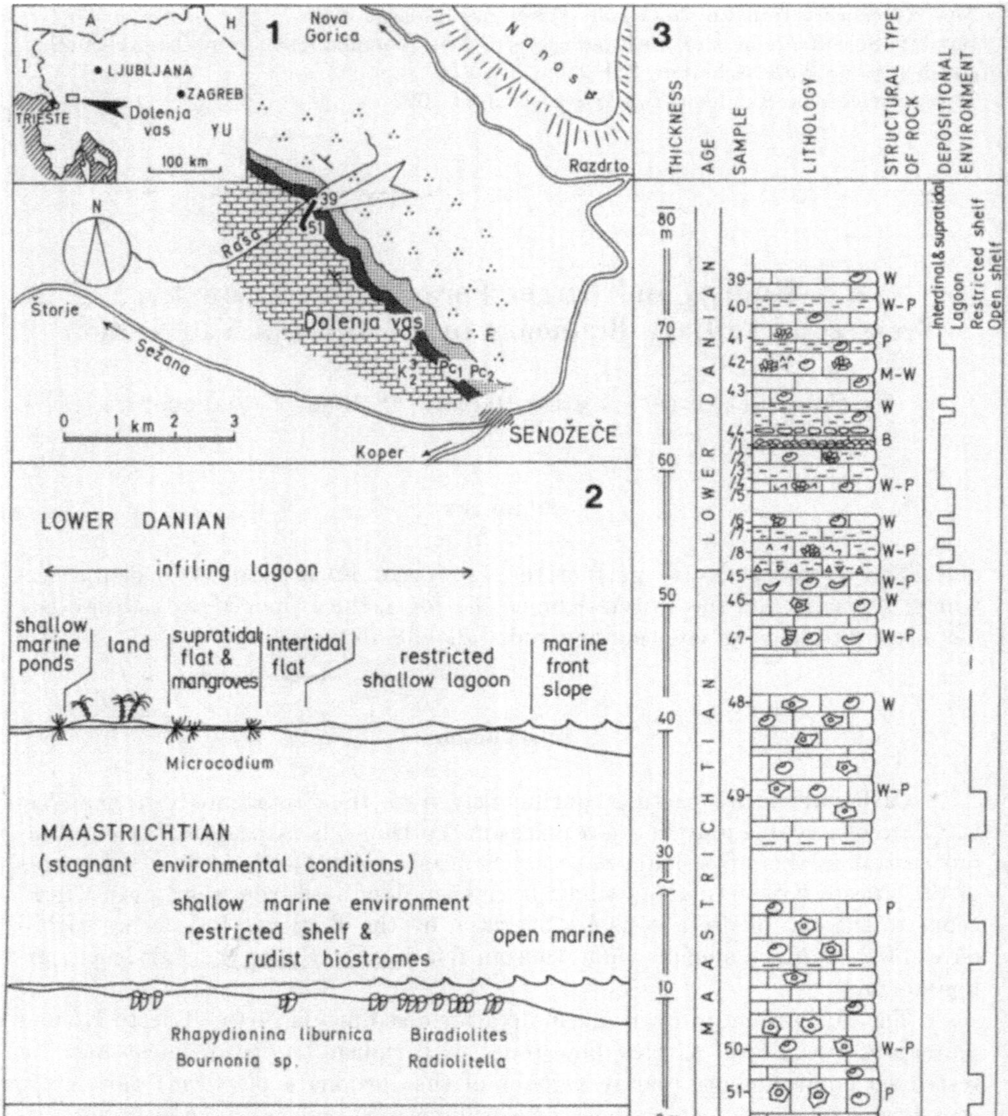

DOLENJA VAS SECTION

SLOVENIA

C-T BOUNDARY

Fig.1. Location sketch map of Dolenja vas section

Fig.2. Depositional environment during Maastrichtian & Lower Danian in the Dolenja vas area

Fig.3. Columnar lithostratigraphic section

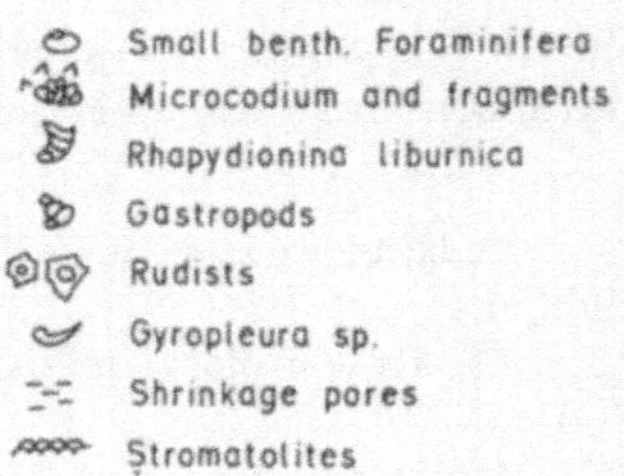

(1889). R. PAVLOVEC (1963) has named the lower part of the Liburnian Formation, for which Maastrichtian age is now proved, Vreme beds (Fig. 1).

At Dolenja vas a sequence of the Liburnian Formation in which Vreme beds pass upwards into the Paleocene has been investigated (Fig. 3). While the rocks between field numbers 51 and 45 have been dated as Maastrichtian, those of point 44 belong already to the Paleocene. The Cretaceous carbonate series is developed in lagoonal facies. It has been deposited in the back of a biostrome reef and consists of limestones which are partly rhytmic and bioturbated. Of the rudists recorded in the limestones, *Biradiolites* and *Radiolitella* have probably lived on the reef under high energy conditions and have been re-deposited in the lagoon. *Bournonia* was supposedly living on the bottom of the lagoon, partly buried in mud.

Description of the section and rudistid fauna

At point 51, in the oldest (lowermost) part of the studied section, rudistid species characteristic for the Upper Senonian, partly for Maastrichtian, were found in grey micritic limestone. The autochthonous species are *Bournonia adriatica* PEJOVIĆ, *B. judaica* BLANCKENHORN, *B. quadripinnae* PEJOVIĆ. Rudistid specimens of reef origin buried here including *Kuehnia* aff. *serbica* MILOVANOVIĆ, *Radiolitella* aff. *maestrichtiana* PEJOVIĆ, *Fossulites undesaltus* ASTRE and *Pseudopolyconites* sp.

Higher up, at point 49, remains of the rudists *Sauvagesia* cf. *slovenica* PLENIČAR, *Durania martellii* PARONA and *Pseudopolyconites* sp. were found in a similar limestone. At various levels between points 51 and 49, fragments of valves belonging to the genus *Biradiolites* were found. Point 48 has yielded valves of the genera *Bournonia, Apricardia* and *Gyropleura*. A similar fauna was recorded at point 47; here, the rudistid fauna was accompanied by foraminifera of the species *Rhapydionina liburnica* (STACHE). At point 46, the remains of rudistid valves of the genera *Gyropleura* and *Apricardia* were found in several layers. They were partly crushed by wave action and transported into a less agitated depositional environment. Therefore, they are found accumulated in lenses and beds of grey micritic limestone. These beds also contain numerous foraminifera of the species *Rhapydionina liburnica* (STACHE) as well as miolids.

The latest Cretaceous beds (point 45) are synsedimentary breccias formed in a lagoonal environment as well as dark grey to dark brown, almost black micritic limestone. These beds have the character of rhythmites and bioturbidites. They have yielded numerous rudistids of the genera *Bournonia, Biradiolites* and *Radiolitella*. The autochthonous species of the genus *Bournonia* found in the cement of breccias were identified as *Bournonia adriatica* PEJOVIĆ, *B. parva* PEJOVIĆ, *B. problematica* PLENIČAR & ZUCCHI-STOLFA, *B.* aff. *retrolata* (ASTRE), *B.* aff. *wiontzeki* PEJOVIĆ and *B. triangulata* PLENIČAR & ZUCCHI-STOLFA (Pl. 1, fig. 1–5). Fragments of *Biradiolites baylei* TOUCAS and *Radiolitella* aff. *maestrichtiana* PEJOVIĆ were also found in this breccia. Several species are characteristic for the Maastrichtian, e. g., *Bournonia adriatica, B. wiontzeki, Biradiolites baylei* and *Radiolitella maestrichtiana,* while others are typical for the Senonian in general (PLENIČAR & ZUCCHI-STOLFA, in DROBNE et al., 1988). At point 44, no more traces of rudistids could be found (Fig. 2).

Foraminifera

In the Maastrichtian, especially in its upper part, Larger foraminifera are very rare on the shallow carbonate platform. In the studied section, foraminifera were studied from sample 47 upwards. The boundary between the Cretaceous and the Tertiary lies between samples 45 and 44/8. According to determinations of algae, foraminifera and corals, the beds above the boundary are of Danian age (Fig. 3, Pl. 2, fig. 1–3. DROBNE at al., 1988, 1989).

Between beds 47 and 45 *Rhapydionina liburnica* (Stache) is most characteristic. The frequency of tests is extremely variable in particular beds. They are quite numerous in micritic limestones and rare in beds with rudistids and *Gyropleura.* A peculiarity in the *Rh. liburnica* population is the occurrence of solely conical tests belonging to the megalospheric generation. The number of chambers and their diameters vary greatly (Pl. 2, fig. 2). Fan-like forms of the microspheric generation or fragments of their tests were not found. This might be an indicatior of rapid changes of depositional conditions in this area, which prevented the development of the entire reproductive cycle. Tests of both generations occur in a western extension of these beds at Vrabče (DROBNE, 1981). *Rh. liburnica* was also associated with *Dicyclina schlumbergeri* MUNIER-CHALMAS and *Moncharmontia* sp. 1.

Among the porcelaneous foraminifera, large miliolids with irregularily arranged chambers are worth mentioning. They have been preliminarily indentified as *"Dargenioella"* (DE CASTRO, 1988).

The appearance of *Rhapydionina liburnica* indicates an Upper Maastrichtian age of the beds. This is supported by the biostratigraphic range of *Bournonia adriatica* and *B. wiontzeki.* Thus, two different fossil groups known as endemic can be used for interpreting the age and for the correlation with similar beds of the Adriatic carbonate platform. On the island of Brač, beds with similar fossil associations were lately attributed to the Maastrichtian (PEJOVIĆ & RADOIČIĆ, 1985/86; GUŠIĆ et al., 1988). Based on ecological adaptation and geographic extension of large foraminifera of the groups Meandropsinidae and Rhapydioninae, E. CAUS and L. HOTTINGER (1986) deduced several bioprovinces in the otherwise uniform western Tethys during the Upper Cretaceous. Differentiation was most probably due rather to oceanic current directions than to barriers in the marine regions. According to these authors, Meandropsinidae are characteristic for the Pyrenean bioprovince, Rhapydioninae for the Mediterranean bioprovince in the central part of the western Tethys. A similar delimitation of bioprovinces on the Adriatic-Apulian platform was proposed by G. BIGNOT (1987). However, paleobioecological data from the Dolenja vas profile show, that this region belonged during the Upper Cretaceous, i. e. at the Cretaceous-Tertiary transition, to the area with Rhapydioninae on the NW part of the Adriatic carbonate platform.

Facies

The Maastrichtian limestone was deposited on a broad and shallow shelf area (Fig. 2). The more open parts were occupied by rudistid biostromes with numerous rudistids (biomicritic and biointramicritic rudistid limestones, "packstone" after DUNHAM's classification). In wave-protected, more restricted parts of the shelf, numerous foraminifera, especially miliolids, as well as tiny snails (the "wackestone" type is prevailing) occur together with rudists and gyropleuras (Pl. 2, fig. 2, 3). The rudistid valves show marks of intensive endolithization (Pl. 2, fig. 3).

At the end of the Cretaceous period, at the boundary to the Lower Danian, the water depth above the carbonate shelf decreased. Local emersion phases and intertidal belts occurred between prevailing lagoons. The supratidal and intertidal depositional environments are characterized by numerous shrinkage pores, thin stromatolithic laminae, nests of fine emersion breccia and numerous Microcodium structures which were formed along contours of roots. Bioturbation structures are also frequent. Limestone, biomicrites ("wackestone" and "packstone"), indicate an environment of very low energy. Pyritic pigment and organic matter which give the rocks a dark to black coloration, indicate local reducing conditions within the sediment (Pl. 2, fig. 1). Characteristic for the passage between the Cretaceous and Tertiary are thin accumulations of the phosphate mineral collophane. These beds show structures of an intra- and supratidal sedimentation.

Conclusions

The Vreme beds of the Liburnian Formation which have been studied in a section at Dolenja vas have been deposited in a highly saline, partly brackish environment. The water was warm and relatively calm. The chemical and biochemical deposition was intensive. At the time of deposition the carbonate platform of the northwest Dinarides was uplifted and desintegrated. The sea retreated temporarily from small resticted areas. As in Vremski Britof, intercalations of fresh water sediments characterized by snails and coal lenses have been recorded in a small number of other localities.

Rudists of the genus *Bournonia* show long, narrow cylindrical lower valves up to 1 cm in diameter. They are reinforced by strong longitudinal ribs (pl. 1, fig. 1–5, pl. 2, fig. 3). The texture of the walls is lamellar-prismatic; transversal barriers are developed between the lamellae. Therefore, the texture of the wall is very dense.

Among the Larger Foraminifera occurring together with the rudists, *Rhapydionia liburnica* is abundant. Like the species of *Bournonia* it proves latest Maastrichtian for the sediments. The strata of the Dolenje vas section have been deposited exclusively on a shallow water carbonate platform. There are no severe changes in sedimentation at the Cretaceous-Tertiary boundary.

References

BIGNOT, G. (1987): Evolution comparée de deux bassins épicontinentaux dans le nord de la plaque Adriatique au Crétacé supérieur. – Mém. Géol. Univ. Dijon, **11:** 183–193. Dijon.

CAUS, E., & HOTTINGER, L. (1986): Particularides de la fauna (macroforaminiferos) del Cretacio superior pirenaico. – Paleontologia i evolucio, **20:** 115–123. Barcelona.

DE CASTRO, P. (1989): On some Foraminifera and Algae in Apennine Upper Cretaceous and Paleocene. – Mem. Soc. Geol. Ital., **40:** 109–124. Roma.

DROBNE, K. (1981): Značilne foraminifere in njih združbe v podlagi danijskih plasti. (Characteristical foraminifers and their associations in the base of Danian beds.) – Simpozij o problemih danija v Jugoslaviji, Zbornik referatov 85–103. Ljubljana.

DROBNE, K., OGORELEC, B., PAVLOVEC, R., PAVŠIČ, J. (1987): Section Golez. In: Danian-Cuisian Sympos. "Evolution of the karstic carbonate platform, Excursion guide III: 17–22. Trieste.

DROBNE, K., OGORELEC, B., PLENIČAR, M., † ZUCCHI-STOLFA, M. L., TURNŠEK, D. (1988): Maastrichtian, Danian and Thanetian beds in Dolenja vas (NW Dinarides, Yugoslavia). Microfacies, Foraminifers, Rudists and Corals. – Razprave SAZU 4. razr., 29, 147–224, Pl. 1–35, Ljubljana.

DROBNE, K., OGORELEC, B., PLENIČAR, M., BARATTOLO, F., TURNŠEK, D., † ZUCCHI-STOLFA, M. L. (1989): The Dolenja vas section, a transition from Cretaceous to Paleocene. NW Dinarides, Yugoslavia. – Mem. Soc. Geol. Ital., **40:** 73–83. Roma.

DUNHAM, R. J. (1962): Classification of carbonate rocks according to depositional texture. – In: W. E. HAM (ed), Classification of Carbonate Rocks. – Am. Assoc. Petrol. Geol., Mem., **1:** 108–121. Tulsa.

KLAPPA, C. F. (1980): Rhizoliths in terrestrial carbonates; classification, recognition, genesis and significance. – Sedimentology, **27:** 613–629. London.

KOCH, R. (1978/79): Fazies und Diagenese eines Biostrom-Komplexes (Oberkreide, Jugoslawien). – Zeitschrift Erdöl – Kohle, Erdgas, Petrochemie. Compendium, 78/79: 41–64. Leinfelden-Echterdingen.

PEJOVIĆ, D. (1988): New Bournonias from the Senonian of Brač. – Bull. Acad. Serbe des Sci, Class sc. nat., **28:** 41–45. Beograd.

PEJOVIĆ, D., & RADOIČIĆ, R. (1985/86): Prilog stratigrafiji gornje krede ostrva Brača. – Jadranska karbonatna platforma. (Contribution to the study of Upper Cretaceous stratigraphy of Brač.) – Geologija **28/2:** 121–150. Ljubljana.

PLENIČAR, M. (1970): Tolmač za list Postojna, osnovna geološka karta SFRJ. Zvezni geoloski zavod, 1–62. Beograd.

REIJERS, T. J. A., & HSÜ, K. J. (1986): Manual of carbonate sedimentology: A lexicographical approach. Academic Press, 302 p. London.

SCHNEIDER, J. F. (1975): Recent tidal deposits of Abu Dhabi, UAE, Arabian Gulf. In: Ginsburg, R. N. (ed.), Tidal Deposits. Springer Verlag, p. 209–214. Berlin.

Plate 1

Fig. 1: *Bournonia triangulata* Pleničar & Zucchi-Stolfa 1988;
transversal section through the lower valve; thin section; Dolenja vas near Senožeče; Maastrichtian.

Fig. 2: *Bournonia* cf. *wiontzeki* Pejović 1968;
transversal section through the lower valve; thin section; Dolenja vas near Senožeče; Maastrichtian.

Fig. 3: *Bournonia problematica* Pleničar & Zucchi-Stolfa 1988;
transversal section through the lower valve; thin section; Dolenja vas near Senožeče; Maastrichtian.

Fig. 4: *Bournonia parva* Pejović 1988;
transversal section through the lower valve; thin section; Dolenja vas near Senožeče; Maastrichtian.

Fig. 5: *Bournonia* aff. *retrolata* (Astre) 1929;
transversal section through the lower valve; thin section; Dolenja vas near Senožeče; Maastrichtian.

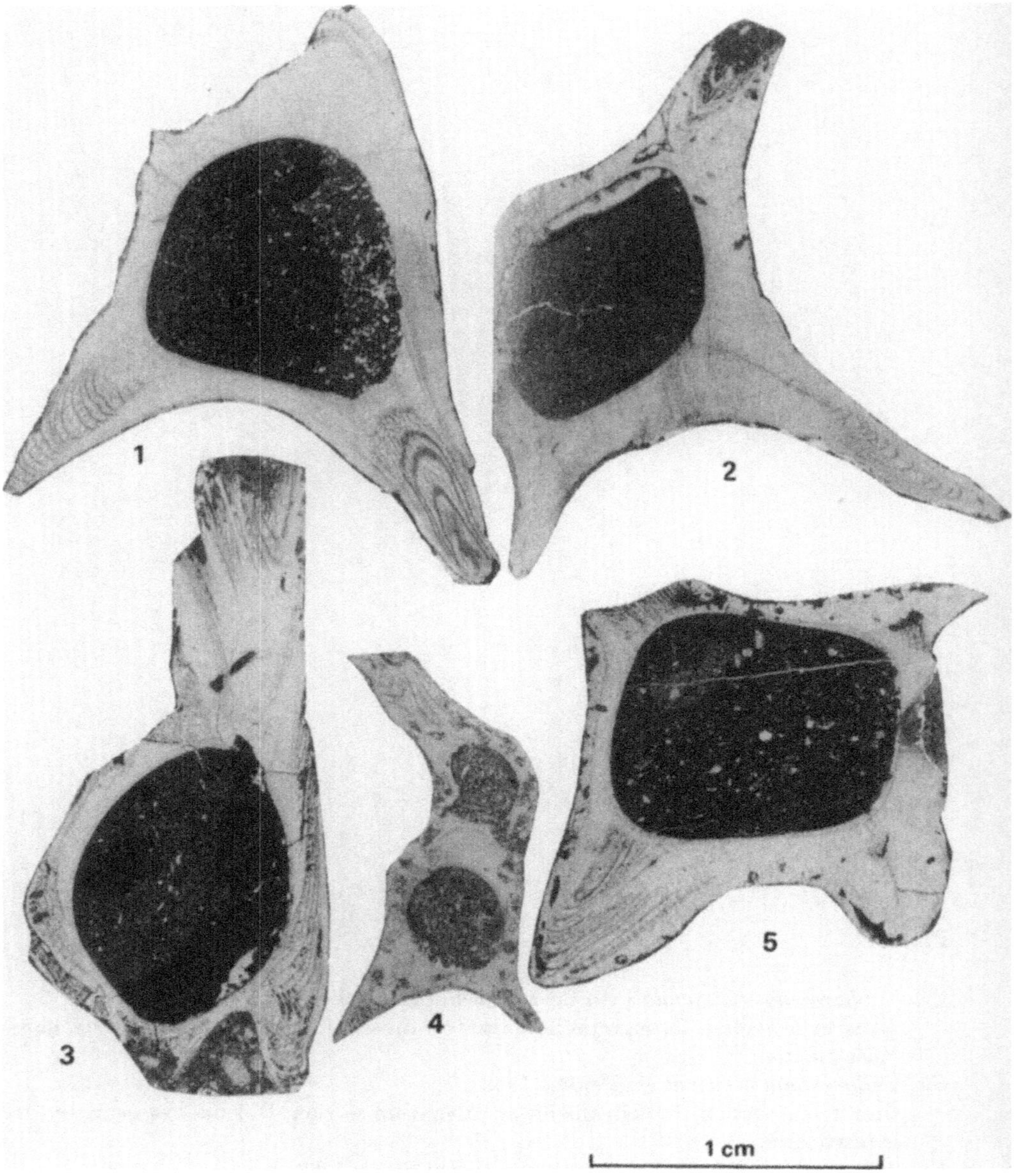
1
2
3
4
5
1 cm

Plate 2

Fig. 1a: Micritic limestone with shrinkage pores integrated with synsedimentary intramicritic breccia. Intertidal sedimentary environment. Evaporite minerals (arrow) are replaced by sparry calcite. Dv 44/8 – 6556.

Fig. 1b: Microcodium structure from the same specimen. Dv 44/8 – 6556.

Fig. 2: Biomicritic limestone with numerous porcelaneous miliolids, *Rhapydionina liburnica* (arrow), Gyropleura fragments, snail with internal sediment (geopetal texture, arrow). Wackestone to packstone type of rock. Dv 1/-5 – 4576.

Fig. 3: Organogenic biomicritic limestone with rare miliolids, *Bournonia* sp. and Gyropleura fragments. Bioturbation of sediment is evident. Packstone. Dv 46/ – 5077.

All enlarged 10 x in incident light.

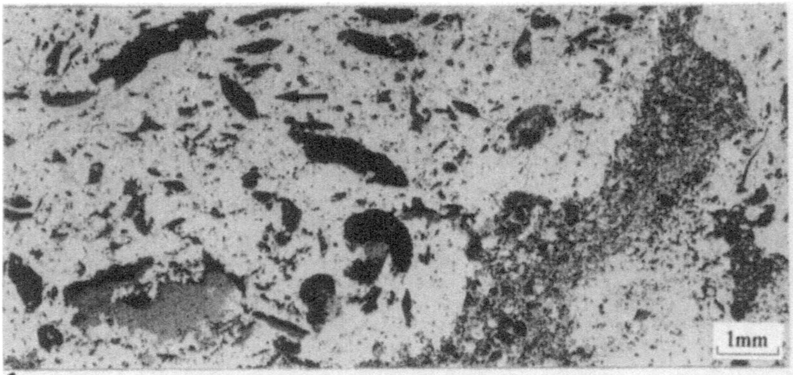

1a

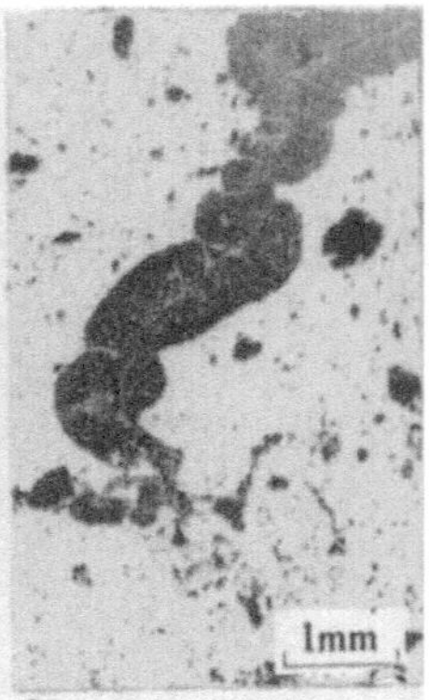

1b

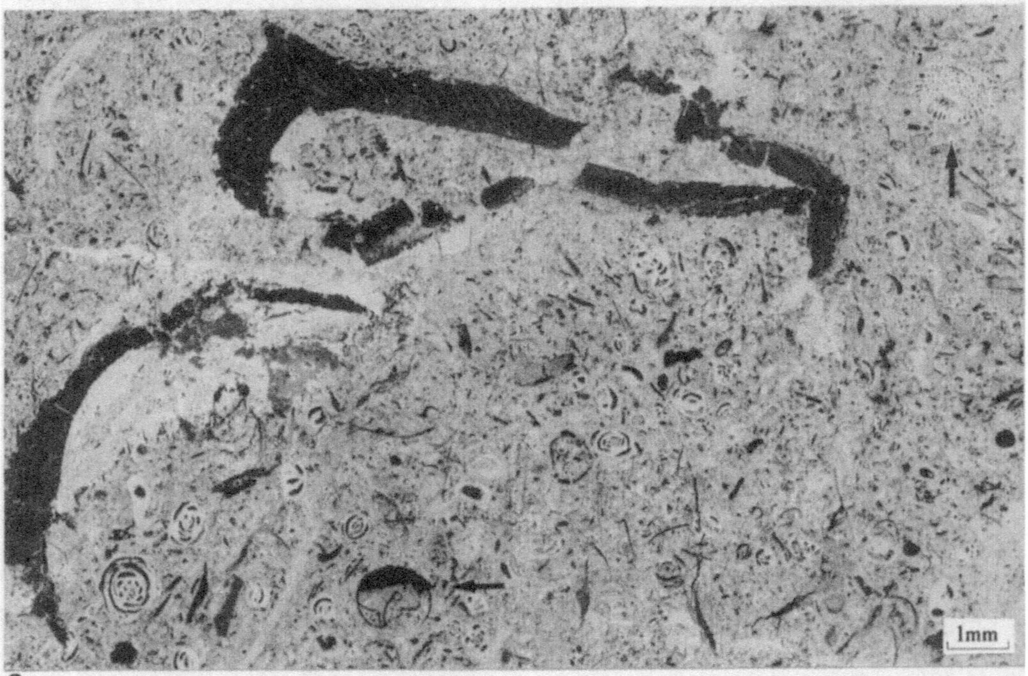

2

3

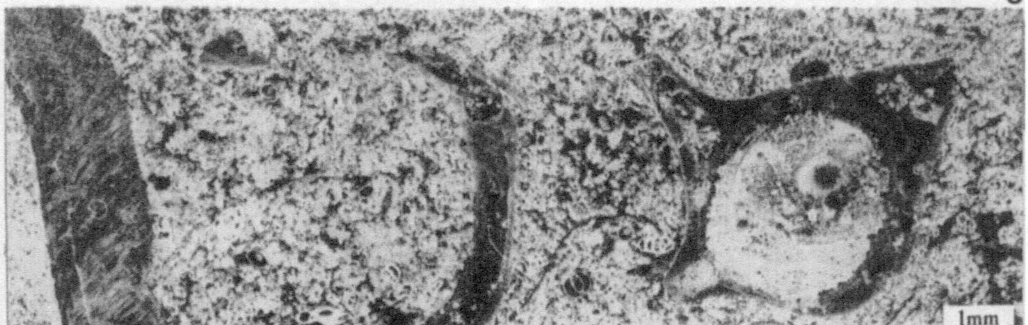

GPSR Compliance

The European Union's (EU) General Product Safety Regulation (GPSR) is a set of rules that requires consumer products to be safe and our obligations to ensure this.

If you have any concerns about our products, you can contact us on ProductSafety@springernature.com

In case Publisher is established outside the EU, the EU authorized representative is:

Springer Nature Customer Service Center GmbH
Europaplatz 3
69115 Heidelberg, Germany

Batch number: 10726298

Printed by Printforce, the Netherlands